BIOLOGICAL AND CHEMICAL HAZARDS IN FOOD AND FOOD PRODUCTS

Prevention, Practices, and Management

Innovations in Agricultural & Biological Engineering

BIOLOGICAL AND CHEMICAL HAZARDS IN FOOD AND FOOD PRODUCTS

Prevention, Practices, and Management

Edited by

Santosh K. Mishra, PhD

Megh R. Goyal, PhD

Manju Gaare, PhD

AAP | APPLE ACADEMIC PRESS

First edition published 2022

Apple Academic Press Inc.
1265 Goldenrod Circle, NE,
Palm Bay, FL 32905 USA

4164 Lakeshore Road, Burlington,
ON, L7L 1A4 Canada

CRC Press
6000 Broken Sound Parkway NW,
Suite 300, Boca Raton, FL 33487-2742 USA

2 Park Square, Milton Park,
Abingdon, Oxon, OX14 4RN UK

Library and Archives Canada Cataloguing in Publication

Title: Biological and chemical hazards in food and food products : prevention, practices, and management / edited by Santosh K. Mishra, PhD, Megh R. Goyal, PhD, Manju Gaare, PhD.
Names: Mishra, Santosh K., editor. | Goyal, Megh R., editor. | Gaare, Manju, editor.
Series: Innovations in agricultural and biological engineering.
Description: First edition. | Series statement: Innovations in agricultural and biological engineering | Includes bibliographical references and index.
Identifiers: Canadiana (print) 20210381612 | Canadiana (ebook) 20210381671 | ISBN 9781774637135 (hardcover) | ISBN 9781774637982 (softcover) | ISBN 9781003189183 (ebook)
Subjects: LCSH: Food—Toxicology. | LCSH: Food—Safety measures. | LCSH: Food industry and trade—Safety measures.
Classification: LCC RA1258 .B56 2022 | DDC 615.9/54—dc23

Library of Congress Cataloging-in-Publication Data

Names: Mishra, Santosh K., editor. | Goyal, Megh R., editor. | Gaare, Manju, editor.
Title: Biological and chemical hazards in food and food products : prevention, practices, and management / Santosh K. Mishra, Megh R. Goyal, Manju Gaare.
Description: First edition. | Palm Bay : Apple Academic Press, [2022] | Includes bibliographical references and index. | Summary: "This volume takes an in-depth look at various biological and chemical hazards in food and food products that pose health threats. It also outlines methods and practices for the diagnosis, prevention, and management of these hazards in food production processes. The new scientific research and case studies presented in the volume cover mycotoxins, foodborne pathogens, antibiotic residues from dairy animals, pesticide residues, the presence of heavy metals in food, and more. Chapters also address food allergy management and offer lessons and practices in food recall situations. The authors discuss the various food toxins, their sources, as well as management, mitigation, and prevention strategies. Also addressed are the specific adverse effects on people with health problems such as diabetes, hypertension, cancer, neurodegenerative diseases, and more. This book is organized in such a way that each chapter treats one major food safety hazard and offers novel control methods for health, food safety, and quality enhancement through various means. Biological and Chemical Hazards in Food and Food Products: Prevention, Practices, and Management will aid researchers and policymakers as it illustrates the various aspects of food safety hazards and how to analyze and control these potential health threats"-- Provided by publisher.
Identifiers: LCCN 2021057356 (print) | LCCN 2021057357 (ebook) | ISBN 9781774637135 (hardback) | ISBN 9781774637982 (paperback) | ISBN 9781003189183 (ebook)
Subjects: LCSH: Food industry and trade--Sanitation. | Food--Safety measures. | Food contamination.
Classification: LCC TX531 .B555 2022 (print) | LCC TX531 (ebook) | DDC 338.4/7664--dc23/eng/20211214
LC record available at https://lccn.loc.gov/2021057356
LC ebook record available at https://lccn.loc.gov/2021057357

ISBN: 978-1-77463-713-5 (hbk)
ISBN: 978-1-77463-798-2 (pbk)
ISBN: 978-1-00318-918-3 (ebk)

ABOUT THE BOOK SERIES: INNOVATIONS IN AGRICULTURAL AND BIOLOGICAL ENGINEERING

Under this book series, Apple Academic Press Inc. is publishing book volumes over a span of 8–10 years in the specialty areas defined by the American Society of Agricultural and Biological Engineers (www.asabe.org). Apple Academic Press Inc. aims to be a principal source of books in agricultural and biological engineering. We welcome book proposals from readers in areas of their expertise.

The mission of this series is to provide knowledge and techniques for agricultural and biological engineers (ABEs). The book series offers high-quality reference and academic content on agricultural and biological engineering (ABE) that is accessible to academicians, researchers, scientists, university faculty and university-level students, and professionals around the world.

Agricultural and biological engineers ensure that the world has the necessities of life, including safe and plentiful food, clean air and water, renewable fuel and energy, safe working conditions, and a healthy environment by employing knowledge and expertise of the sciences, both pure and applied, and engineering principles. Biological engineering applies engineering practices to problems and opportunities presented by living things and the natural environment in agriculture.

ABE embraces a variety of the following specialty areas (www.asabe.org): aquaculture engineering, biological engineering, energy, farm machinery and power engineering, food, and process engineering, forest engineering, information, and electrical technologies, soil, and water conservation engineering, natural resources engineering, nursery, and greenhouse engineering, safety, and health, and structures and environment.

For this book series, we welcome chapters on the following specialty areas (but not limited to):

1. Academia to industry to end-user loop in agricultural engineering.
2. Agricultural mechanization.
3. Aquaculture engineering.
4. Biological engineering in agriculture.
5. Biotechnology applications in agricultural engineering.
6. Energy source engineering.

7. Farm to fork technologies in agriculture.
8. Food and bioprocess engineering.
9. Forest engineering.
10. GPS and remote sensing potential in agricultural engineering.
11. Hill land agriculture.
12. Human factors in engineering.
13. Impact of global warming and climatic change on agriculture economy.
14. Information and electrical technologies.
15. Irrigation and drainage engineering.
16. Micro-irrigation engineering.
17. Milk Engineering.
18. Nanotechnology applications in agricultural engineering.
19. Natural resources engineering.
20. Nursery and greenhouse engineering.
21. Potential of phytochemicals from agricultural and wild plants for human health.
22. Power systems and machinery design.
23. Robot engineering and drones in agriculture.
24. Rural electrification.
25. Sanitary engineering.
26. Simulation and computer modeling.
27. Smart engineering applications in agriculture.
28. Soil and water engineering.
29. Structures and environment engineering.
30. Waste management and recycling.
31. Any other focus areas.

Books in the Innovations in Agricultural & Biological Engineering Series

- Biological and Chemical Hazards in Food and Food Products: Prevention, Practices, and Management
- Bioremediation and Phytoremediation Technologies in Sustainable Soil Management, 4-volume set:
  - Volume 1: Fundamental Aspects and Contaminated Sites
  - Volume 2: Microbial Approaches and Recent Trends
  - Volume 3: Inventive Techniques, Research Methods, and Case Studies
  - Volume 4: Degradation of Pesticides and Polychlorinated Biphenyls
- Dairy Engineering: Advanced Technologies and Their Applications

- Developing Technologies in Food Science: Status, Applications, and Challenges
- Emerging Technologies in Agricultural Engineering
- Engineering Interventions in Agricultural Processing
- Engineering Interventions in Foods and Plants
- Engineering Practices for Agricultural Production and Water Conservation: An Interdisciplinary Approach
- Engineering Practices for Management of Soil Salinity: Agricultural, Physiological, and Adaptive Approaches
- Engineering Practices for Milk Products: Dairyceuticals, Novel Technologies, and Quality
- Field Practices for Wastewater Use in Agriculture: Future Trends and Use of Biological Systems
- Flood Assessment: Modeling and Parameterization
- Food Engineering: Emerging Issues, Modeling, and Applications
- Food Process Engineering: Emerging Trends in Research and Their Applications
- Food Processing and Preservation Technology: Advances, Methods, and Applications
- Food Technology: Applied Research and Production Techniques
- Functional Dairy Ingredients and Nutraceuticals: Physicochemical, Technological, and Therapeutic Aspects
- Handbook of Research on Food Processing and Preservation Technologies, 5-volume set:
    - Volume 1: Nonthermal and Innovative Food Processing Methods
    - Volume 2: Nonthermal Food Preservation and Novel Processing Strategies
    - Volume 3: Computer-Aided Food Processing and Quality Evaluation Techniques
    - Volume 4: Design and Development of Specific Foods, Packaging Systems, and Food Safety
    - Volume 5: Emerging Techniques for Food Processing, Quality, and Safety Assurance
- Modeling Methods and Practices in Soil and Water Engineering
- Nanotechnology and Nanomaterial Applications in Food, Health, and Biomedical Sciences
- Nanotechnology Applications in Agricultural and Bioprocess Engineering: Farm to Table
- Nanotechnology Applications in Dairy Science: Packaging, Processing, and Preservation

- Nanotechnology Horizons in Food Process Engineering, 3-volume set:
  - o Volume 1: Food Preservation, Food Packaging and Sustainable Agriculture
  - o Volume 2: Scope, Biomaterials, and Human Health
  - o Volume 3: Trends, Nanomaterials, and Food Delivery
- Novel Dairy Processing Technologies: Techniques, Management, and Energy Conservation
- Novel Processing Methods for Plant-Based Health Foods: Extraction, Encapsulation and Health Benefits of Bioactive Compounds
- Novel Strategies to Improve Shelf-Life and Quality of Foods: Quality, Safety, and Health Aspects
- Processing of Fruits and Vegetables: From Farm to Fork
- Processing Technologies for Milk and Milk Products: Methods, Applications, and Energy Usage
- Quality Control in Fruit and Vegetable Processing: Methods and Strategies
- Scientific and Technical Terms in Bioengineering and Biological Engineering
- Soil and Water Engineering: Principles and Applications of Modeling
- Soil Salinity Management in Agriculture: Technological Advances and Applications
- State-of-the-Art Technologies in Food Science: Human Health, Emerging Issues and Specialty Topics
- Sustainable Biological Systems for Agriculture: Emerging Issues in Nanotechnology, Biofertilizers, Wastewater, and Farm Machines
- Sustainable Nanomaterials for Biosystem Engineering: Impacts, Challenges, and Future Prospects
- Technological Interventions in Dairy Science: Innovative Approaches in Processing, Preservation, and Analysis of Milk Products
- Technological Interventions in Management of Irrigated Agriculture
- Technological Interventions in the Processing of Fruits and Vegetables
- Technological Processes for Marine Foods, from Water to Fork: Bioactive Compounds, Industrial Applications, and Genomics

OTHER BOOKS ON AGRICULTURAL AND BIOLOGICAL ENGINEERING FROM APPLE ACADEMIC PRESS, INC.

Management of Drip/Trickle or Micro Irrigation
Megh R. Goyal, PhD, PE, Senior Editor-in-Chief

Evapotranspiration: Principles and Applications for Water Management
Megh R. Goyal, PhD, PE, and Eric W. Harmsen, Editors

Book Series: Research Advances in Sustainable Micro Irrigation
Senior Editor-in-Chief: Megh R. Goyal, PhD, PE

Volume 1: Sustainable Micro Irrigation: Principles and Practices
Volume 2: Sustainable Practices in Surface and Subsurface Micro Irrigation
Volume 3: Sustainable Micro Irrigation Management for Trees and Vines
Volume 4: Management, Performance, and Applications of Micro Irrigation Systems
Volume 5: Applications of Furrow and Micro Irrigation in Arid and Semi-Arid Regions
Volume 6: Best Management Practices for Drip Irrigated Crops
Volume 7: Closed Circuit Micro Irrigation Design: Theory and Applications
Volume 8: Wastewater Management for Irrigation: Principles and Practices
Volume 9: Water and Fertigation Management in Micro Irrigation
Volume 10: Innovation in Micro Irrigation Technology

Book Series: Innovations and Challenges in Micro Irrigation
Senior Editor-in-Chief: Megh R. Goyal, PhD, PE

Volume 1: Sustainable Micro Irrigation Design Systems for Agricultural Crops
Volume 2: Principles and Management of Clogging in Micro Irrigation
Volume 3: Performance Evaluation of Micro Irrigation Management

Volume 4: Potential Use of Solar Energy and Emerging Technologies in Micro Irrigation
Volume 5: Micro Irrigation Management: Technological Advances and Their Applications
Volume 6: Micro Irrigation Engineering for Horticultural Crops
Volume 7: Micro Irrigation Scheduling and Practices
Volume 8: Engineering Interventions in Sustainable Trickle Irrigation
Volume 9: Management Strategies for Water Use Efficiency and Micro Irrigated Crops
Volume 10: Fertigation Technologies In Micro Irrigation

ABOUT THE LEAD EDITOR

Santosh Kumar Mishra, PhD, is working as an Assistant Professor in the Department of Dairy Microbiology, College of Dairy Science and Technology, Guru Angad Dev Veterinary and Animal Sciences University, Ludhiana, Punjab, India.

He received his BTech degree in Dairy Technology from Maharashtra Animal and Fisheries Sciences University, Nagpur, India; and his MSc and PhD degrees from the National Dairy Research Institute, Karnal, Haryana. Now he is working in the area of functional foods and dairy products incorporating live probiotics and technology of functional lactic cultures for fermented and non-fermented dairy products. He has also served the dairy industry as Quality Assurance Executive at Mother Dairy, New Delhi.

In India, he has handled externally funded projects granted by DST, MoFPI, and UGC as PI or Co-PI. He received several awards for best papers and posters/presentations. He is the recipient of junior and senior research fellowships during his master's and doctoral programs at the National Dairy Research Institute, Karnal, Haryana. He was a university gold medalist during his graduation program and also earned the best thesis award by his university after his PhD research. Recently he received an Award of Honor at an international conference sponsored by Partap College of Education, Ludhiana, in association with the International Professionals Development Association, UK.

He is a member of various scientific societies: life member of SASNET-Fermented Foods, Anand; member of Indian Dairy Associations, New Delhi. He has published several research, review, and popular articles in national and international journals. He has also published three international edited books, several book chapters, and teaching reviews in various training programs. He has recently completed a young scientist project by DST, SEED Department, Govt. of India, New Delhi, on isolation, and characterization of novel oxalate degrading lactic acid bacteria for potential probiotic management of kidney stone.

Readers may contact him at: *skmishra84@gmail.com*

ABOUT THE SENIOR EDITOR-IN-CHIEF

Megh R. Goyal, PhD, PE, is a Retired Professor in Agricultural and Biomedical Engineering from the General Engineering Department in the College of Engineering at the University of Puerto Rico–Mayaguez Campus; and Senior Acquisitions Editor and Senior Technical Editor-in-Chief in Agriculture and Biomedical Engineering for Apple Academic Press, Inc. He has worked as a Soil Conservation Inspector and as a Research Assistant at Haryana Agricultural University and Ohio State University.

During his professional career of 52 years, Dr. Goyal has received many prestigious awards and honors. He was the first agricultural engineer to receive the professional license in Agricultural Engineering in 1986 from the College of Engineers and Surveyors of Puerto Rico. In 2005, he was proclaimed as "Father of Irrigation Engineering in Puerto Rico for the Twentieth Century" by the American Society of Agricultural and Biological Engineers (ASABE), Puerto Rico Section, for his pioneering work on micro irrigation, evapotranspiration, agroclimatology, and soil and water engineering. The Water Technology Centre of Tamil Nadu Agricultural University in Coimbatore, India, recognized Dr. Goyal as one of the experts "who rendered meritorious service for the development of micro irrigation sector in India" by bestowing the Award of Outstanding Contribution in Micro Irrigation. This award was presented to Dr. Goyal during the inaugural session of the National Congress on "New Challenges and Advances in Sustainable Micro Irrigation" held at Tamil Nadu Agricultural University.

Dr. Goyal received the Netafim Award for Advancements in Microirrigation: 2018 from the American Society of Agricultural Engineers at the ASABE International Meeting in August 2018. VDGOOD Professional Association of India awarded Lifetime Achievement Award at 12th Annual Meeting on Engineering, Science and Medicine that was held on 20–21 of November of 2020 in Visakhapatnam, India. A prolific author and editor, he has written more than 200 journal articles and textbooks and has edited over 95 books. He is the editor of three book series published by Apple Academic Press: Innovations in Agricultural & Biological Engineering, Innovations and Challenges in Micro Irrigation, and Research Advances in Sustainable Micro Irrigation. He is also instrumental in the development of the new book series Innovations in Plant Science for Better Health: From Soil to Fork.

Dr. Goyal received his BSc degree in engineering from Punjab Agricultural University, Ludhiana, India; his MSc and PhD degrees from Ohio State University, Columbus; and his Master of Divinity degree from Puerto Rico Evangelical Seminary, Hato Rey, Puerto Rico, USA.

ABOUT THE EDITOR

Manju Gaare, PhD, is an Assistant Professor in the Department of Dairy Microbiology at GN Patel College of Dairy Technology, Sardarkrushinagar Dantiwada Agricultural University, Dantiwada, Gujarat, India. He received a BTech (Dairy Technology) from the Dairy Science College, KVAFSU, Bengaluru, India; and an MSc (Dairy Microbiology) and PhD (Dairy Microbiology) from the National Dairy Research Institute, Karnal, India. The same institute granted him junior and senior research fellowships during his master and doctoral programs. He has acquired proficiency in the field of food safety and quality assurance through his industrial, research, and teaching experience. He is a life member of several scientific societies, such as the Association of Food Scientists and Technologists India and the Probiotic Association of India. He has received best poster presenter awards at national and international conferences. He has published book chapters and several research, review, and technical articles in journals, and he holds one patent. At present he is working on novel antimicrobials for food preservation, minimal food processing, and food safety.

CONTENTS

CONTRIBUTORS

J. Adeppa
Assistant Professor, Department of Veterinary Parasitology, Veterinary College, Bidar, Karnataka, India, Mobile: +91-9741888087, E-mail: dr.adeppaivri@gmail.com

Dipesh Aggarwal
Assistant Professor, Department of Food Technology, Lady Irwin College, Delhi University, New Delhi, India, Mobile: +91-9050674994, E-mail: dipeshaggarwal87@gmail.com

Santosh Anand
Assistant Professor, Dairy Microbiology Department, Phulo Jhano Murmu College of Dairy Technology, Dumka, Jharkhand, India, Mobile: +91-9896088157, E-mail: santosh.anand87@gmail.com

Nancy Awasti
Quality Assurance Representative, Gate Gourmet, 44845 Package Court, Dulles–20166, Virginia, USA, Mobile: +1-6056913263, E-mail: nancyawasthi10@gmail.com

Manorama Chauhan
Professor, College of Dairy Science and Food Technology, Chhattisgarh Kamdhenu Vishwavidyalaya, Raipur–492012, Chhattisgarh, India, Mobile: +91-9826142168, E-mail: manoramachauhan2@gmail.com

Gaurav K. Deshwal
Scientist, Dairy Technology Division, ICAR-National Dairy Research Institute, Karnal–132001, Haryana, India, Mobile: +91-9034733330, E-mail: ndri.gkd@gmail.com

V. Dharini
PhD Research Scholar, Department of Food Process Engineering, School of Bioengineering, SRM Institute of Science and Technology, Kattankulathur–603203, Chennai, Tamil Nadu, India, Mobile: +91-9488803999, E-mail: dharu.venkat@gmail.com.

Lina Dhote
Assistant Professor, Department of Veterinary Public Health and Epidemiology, Veterinary College, Bidar, Karnataka, India, Mobile: +91-7972413533, E-mail: dhotelina@gmail.com

Ravindra Dombar
Assistant Professor, Department of Animal Nutrition, Veterinary College, Bidar, Karnataka, India, Mobile: +91-9880147177, E-mail: drravivet609@gmail.com

Manju Gaare
PhD, Assistant Professor, Department of Dairy Microbiology, GN Patel College of Dairy Technology, Sardar Dantiwada Agriculture University (SDAU), Village Dantiwada, Sardar Krushinagar–385506, Gujarat, India, Mobile: +91-8971769001, E-mail: manjugdsc@gmail.com

Khokan C. Gayen
Assistant Professor, Department of Clinical Nutrition and Dietetics, Vidyasagar University, Rangamati–721102, Midnapore, West Bengal, India, Mobile: +91-8145103269, E-mail: chandragayen113@gmail.com

Apurba Giri
Assistant Professor and Head, Department of Nutrition, Coordinator Department of Food Processing, Mugberia Gangadhar Mahavidyalaya, Bhupatinagar–721425, Purba Midnapore, West Bengal, India, Mobile: +91-9564289290, E-mail: apurbandri@gmail.com

Megh R. Goyal
Retired Faculty in Agricultural and Biomedical Engineering from College of Engineering at University of Puerto Rico-Mayaguez Campus; and Senior Technical Editor-in-Chief in Agricultural and Biomedical Engineering for Apple Academic Press Inc., USA, E-mail: goyalmegh@gmail.com

Neha Halder
PhD Research Scholar, Department of Biotechnology, School of Bioengineering, SRM Institute of Science and Technology, Kattankulathur – 603203, Chennai, Tamil Nadu, India, Mobile: +91-9791915794, E-mail: nehalder93@gmail.com.

Ahmed R. A. Hammam
PhD Research Scholar, Dairy and Food Science Department, South Dakota State University, Brookings, SD–57007, USA, Mobile: +1-6055926864, E-mail: ahmed.hammam@sdstate.edu

Tanmay Hazra
Assistant Professor, Post Graduate Institute of Dairy Education and Research, Kamdhenu University, Gujarat–365601, India, Mobile: +91-9541053968, E-mail: tanmayhazra08@gmail.com

Saurabh Kadyan
Scientist, Dairy Microbiology Division, ICAR-National Dairy Research Institute, Karnal–132001, Haryana, India, Mobile: 91-9466564543, E-mail: kadyan.saurabh3@gmail.com

Arun Kharate
Assistant Professor, Department of Veterinary Public Health and Epidemiology, Veterinary College, Bidar, Karnataka, India, Mobile: +91979679897, E-mail: kharatearunvet@gmail.com

Pragya Kulkarni
Associate Professor and Head, Department of Microbiology, Government Vishwanath Yadav Tamaskar Post-Graduate Autonomous College, Durg–490001, Chhattisgarh, India, Mobile: +91-9826142086, E-mail: pragya.mic@gmail.com

M. Mahesh Kumar
Assistant Professor, Department of Food Process Engineering, School of Bioengineering, SRM Institute of Science and Technology, Kattankulathur – 603203, Chennai, Tamil Nadu, India, Mobile: +91-9884278027, E-mail: maheshkumar.m@ktr.srmist.ac.in.

Samlesh Kumari
Scientist, Dairy and Food Microbiology, Center of Excellence on Soybean Processing and Utilization, ICAR-Central Institute of Agricultural Engineering, Bhopa–462038, Madhya Pradesh, India, Mobile: +91-9425948072, E-mail: samleshndri.kumari@gmail.com

Rashmi H. Mallappa
Scientist, Dairy Microbiology Division, ICAR-National Dairy Research Institute, Karnal–132001, Haryana, India, Mobile: 91-7082033557, E-mail: rashmi.ndri@gmail.com

G. Mikunthan
Professor, Department of Agricultural Biology, Faculty of Agriculture, University of Jaffna, Ariviyal Nagar, Kilinochchi, Sri Lanka, Mobile: +94-718475539, E-mail: mikunthan@univ.jfn.ac.lk

Diwakar Mishra
Assistant Professor, Dairy Technology Department, Phulo-Jhano Murmu College of Dairy Technology, Dumka, Jharkhand, India, Mobile: +91-7019679209, E-mail: diwakaraaidu@gmail.com

Santosh K. Mishra
Assistant Professor, Department of Dairy Microbiology, College of Dairy Science and Technology, Guru Angad Dev Veterinary and Animal Sciences University (GADVASU), Ludhiana–141004, Punjab, India, Mobile: +91-9464995049, E-mail: skmishra84@gmail.com

Sushil K. Mohapatra
Assistant Professor, Department of Animal Biotechnology, College of Veterinary Science and A.H., Sardar Dantiwada Agriculture University (SDAU), Village Dantiwada, Sardarkrushinagar–385506, Gujarat, India, Mobile: +91-8763086951, E-mail: drsushilmohapatra@gmail.com

Shruti Nayyar
MTech Research Scholar, Dairy Microbiology Division, ICAR-National Dairy Research Institute, Karnal–132001, Haryana, India, E-mail: shrutinayyar2903@gmail.com

Raghav Pandey
Teacher, College of Dairy Science and Food Technology, Chhattisgarh Kamdhenu Vishwavidyalaya, Raipur–492012, Chhattisgarh, India, Mobile: +91-7224866892,
E-mail: raghav.pandey0506@gmail.com

Heena Parveen
Senior Research Fellow (SRF), Dairy Microbiology Division, ICAR-National Dairy Research Institute, Karnal–132001, Haryana, India, E-mail: heenahussain0000@gmail.com

Tapas K. Patbandha
Assistant Professor, College of Veterinary Science and A.H., Junagadh Agricultural University, Junagadh, Gujarat, India, Mobile: +91-9671727812, E-mail: patbandhavet@gmail.com

Sarang D. Pophaly
Assistant Professor, College of Dairy Science and Food Technology, Chhattisgarh Kamdhenu Vishwavidyalaya, Raipur–492012, Chhattisgarh, India, Mobile: +91-9993734325,
E-mail: sarang01@gmail.com

Diwas Pradhan
Scientist, Dairy Microbiology Division, ICAR-National Dairy Research Institute, Karnal–132001, Haryana, India, Mobile: 91-8053319936, E-mail: zawidprd@gmail.com

Vimal M. Ramani
Principal, Post Graduate Institute of Dairy Education and Research, Kamdhenu University, Gujarat–365601, India, Mobile: +91-9426422883, E-mail: vimalramani@gmail.com

Monalisa Roy
Assistant Professor, Department of Food Processing, Mugberia Gangadhar Mahavidyalaya, Village and P.O. Bhupatinagar–721425, Purba Midnapore, West Bengal, India,
Mobile: +91-8116303175, E mail: monalisaroy1997@gmail.com

Sucheta Sahoo
Assistant Professor, Department of Food Processing, Mugberia Gangadhar Mahavidyalaya, P.O. Bhupatinagar–721425, Purba Midnapore, West Bengal, India, Mobile: +91-8293033976,
E-mail: suchetasahoo12@gmail.com

Kushal K. Sandey
Assistant Professor, College of Dairy Science and Food Technology, Chhattisgarh Kamdhenu Vishwavidyalaya, Raipur – 492012, Chhattisgarh, India, Mobile: +91-8602006313,
E-mail: kksandey@gmail.com

S. Periyar Selvam
Associate Professor, Department of Food Process Engineering, School of Bioengineering, SRM Institute of Science and Technology, Kattankulathur – 603203, Chennai, Tamil Nadu, India, Mobile: +91-7904250136, E-mail: periyar.india@gmail.com

Rohit G. Sindhav
Assistant Professor, Post Graduate Institute of Dairy Education and Research, Kamdhenu University, Gujarat–365601, India, Mobile: +91-9016767674, E-mail: rohitsindhav7@gmail.com

Tejinder P. Singh
Assistant Professor, College of Dairy Science and Technology, LUVAS, Hisar, Haryana, India, Mobile: +91-8929970050, E-mail: 88.tejindersingh@gmail.com

Akash Solanki
Assistant Professor, Post Graduate Institute of Dairy Education and Research, Kamdhenu University, Kamdhenu–365601, Gujarat, India, Mobile: +91-9265219592, E-mail: akashkumarks@gmail.com

CH. V. K. Sudheendra
Assistant Professor, Post Graduate Institute of Dairy Education and Research, Kamdhenu University, Gujarat–365601, India, Mobile: +91-8238162268, E-mail: sudheendracvk@yahoo.co.in

Venkateswarlu Sunkesula
Associate Director of Research and Technical Services, Idaho Milk Products, Jerome, ID–83338, USA, Mobile: +1-2087490923, E-mail: vsunkesula@idahomilk.us

Ankit J. Thesiya
Assistant Professor, Post Graduate Institute of Dairy Education and Research, Kamdhenu University, Kamdhenu–365601, Gujarat, India, Mobile: +91-9558247047, E-mail: ankittehsiya@gmail.com

Awadhesh K. Tripathi
Professor, College of Dairy Science and Food Technology, Chhattisgarh Kamdhenu Vishwavidyalaya, Raipur–492012, Chhattisgarh, India, Mobile: +91-7000883138, E-mail: drakt11@gmail.com

ABBREVIATIONS

μECD	micro electron capture detector
μL	microliter
μPAD	microfluidic paper-based analytical devices
AAS	atomic absorption spectroscopy
ABC	ATP-binding cassette
ACBK	acid chrome blue K
AF	aflatoxin
AFB1	aflatoxin B1
AMAs	antimicrobial agents
AMR	antimicrobial resistance
AMV-RT	avian myeloblastosis reverse transcriptase
AOAC	Association of Analytical Communities
AP	alkaline phosphatase
APCI	atomic pressure chemical ionization
AR	antibiotic resistance
ASE	accelerated solvent extraction
ASESPE	automated size-exclusion solid-phase extraction
ASSURED	affordable, sensitive, specific, user-friendly, rapid and robust, equipment-free, and deliverable
ATA	alimentary toxic aleukia
AuNP	gold nanoparticles
BRICS	Brazil, Russia, India, China, and South Africa
CAC	Codex Alimentarius Commission
CAMERA	community cyberinfrastructure for advanced marine microbial ecology research and analysis
CBs	carbamates
CCPs	critical control points
CE	capillary electrophoresis
CFIA	Canadian Food Inspection Agency
CFU	colony forming units
CGMPs	current good manufacturing practices
CL	chemiluminescence
COVID-19	coronavirus disease

CPE	cloud point extraction
CRISPR	clustered regularly interspaced short palindromic repeats
CSR	corporate social responsibility
CZE	capillary zone electrophoresis
DAD	diode array detector
DALYs	disability-adjusted life years
DHFR	dihydro-folate reductase
DLLME	dispersive liquid-liquid micro-extraction
DNA	deoxyribonucleic acid
dNTP	deoxyribonucleotide triphosphate
DON	deoxynivalenol
DPASV	differential pulse anodic stripping voltammetric technique
ds	double-stranded
D-SPE	dispersive solid-phase extraction
dsRNA	double-stranded ribonucleic acid
EBI	European Bioinformatics Institute
EBT	eriochrome black T
ECD	electron capture detector
ED	endocrine disruptors
EEC	European Economic Community
ELISA	enzyme-linked immune-sorbent assay
EMA	economical motivated adulteration
EO	essential oil
ESBL	extended spectrum beta-lactam
ESI	electrospray ionization
ET	enzyme thermistor
ETASS	electrothermal atomic absorption spectroscopy
EU	European Union
FAAS	flame atomic absorption spectrophotometer
FALCPA	Food Allergen Labeling and Consumer Protection Act
FAO	Food and Agricultural Organization
FBP	foodborne pathogens
FDA	Food and Drug Administration
FIA	flow injection analysis
FITC	fluorescein isothiocyanate
FPD	flame photometric detector
FSANZ	Food Standards Australia and New Zealand
FSMS	food safety management systems

FSO	Food Safety Organization
FSSAI	Food Safety and Standards Authority of India
FTA	Flinders Technology Associates
FT-IR	Fourier-transform infrared
GAP	global action plan
GC	gas chromatography
GC-MS	gas chromatography-mass spectrometry
GHP	good hygienic practice
GI	gastrointestinal
GLC	gas-liquid chromatography
GMP	good manufacturing practice
GOI	Government of India
GPC	gel permeation chromatography
GRAS	generally regarded as safe
HACCP	hazard analysis critical control point
HAD	helicase dependent amplification
HAU	hemagglutinating units
HAV	hepatitis A
HCC	hepatocellular carcinoma
HCl	hydrochloric acid
HDA	helicase-dependent amplification
HGT	horizontal gene transfer
HIV	human immunodeficiency virus
HLS	high-level structure
HNB	hydroxy naphthol blue
HNoV	human norovirus
HPLC	high performance liquid chromatography
HPLC-UV	HPLC with ultraviolet detection
HRP	horseradish peroxidase
HRV	heart rate variability
IAC	immune-affinity column
IBS	irritable bowel syndrome
ICAR	Indian Council for Agricultural Research
ICP-AES	inductively coupled plasma atomic emission spectroscopy
ICP-MS	inductively coupled plasma mass spectrometry
IgE	immunoglobulin E
IgG	immunoglobulin G
IPM	integrated pest management

ISO	International Organization for Standardization
JECFA	Joint Expert Committee on Food Additives
kDa	kilo Dalton
kGy	kilo gray
KJ	kilojoules
LAB	lactic acid bacteria
LAMP	loop-mediated isothermal amplification
LC	liquid chromatography
LC-MS	liquid chromatography-mass spectrometry
LFA	lateral flow immunoassay
LLE	liquid-liquid extraction
LOD	limit of detection
MAE	microwave-assisted extraction
MATE	multidrug and toxic compound extrusion
MBLs	metallo-beta-lactamases
MFS	major facilitator superfamily
MGE	mobile genetic elements
MG-RAST	metagenomic rapid annotations using subsystem technology
MIP	molecularly imprinted polymers
MRL	maximum residual limit
MRSA	methicillin-resistant Staphylococcus aureus
MS	mass spectrometry
MSPD	matrix solid-phase dispersion
MSPE	magnetic solid-phase extraction
MT	million tons
MWCNTs	multi-walled carbon nanotubes
NABT	nucleic acid-based technique
NaCl	sodium chloride
NAM	N-acetyl muramic acid
NAP	national action plan
NASBA	nucleic acid sequence-based amplification
NDM	New Delhi-mettalo-beta-lactamase
NDRI	National Dairy Research Institute
ng	nanogram
NIV	nivalenol
NPD	nitrogen and phosphorus
OCC	organochlorine compounds
OCPs	organochlorine pesticides

OIE	Office International des Epizooties
OPPs	organophosphorus pesticides
OPRPs	operational prerequisite programs
OTA	ochratoxin A
OTB	ochratoxin B
OTC	ochratoxin C
PAGE	polyacrylamide gel electrophoresis
PAT	patulin
PBP	penicillin-binding proteins
PCBs	polychlorinated biphenyls
PCR	polymerase chain reaction
PDCA	plan-do-check-act
PEI	polyethylenimine
PEP	post-exposure prophylactic
PHA	phytohemagglutinin
PHF	potentially hazardous food
PLE	pressurized liquid extraction
PPi	pyrophosphate
PRPs	prerequisite programs
PSA	potentiometric and chronopotentiometry stripping analysis
PYRs	pyrethroids
QA	quality assurance
qPCR	quantitative PCR
QTOF	quad time of flight
QuEChERS	quick, easy, cheap, effective, rugged, and safe
RASFF	rapid alert system for food and feed
RCA	rolling circle amplification
RCA	root cause analysis
RFID	radio frequency identification
RNA	ribonucleic acid
RND	resistance-nodulation-division
RP	reversed-phase
RPA	recombinase polymerase amplification
RP-HPLC	reversed-phase high-performance liquid chromatography
RPP	ribosomal protection proteins
RSM	response surface methodology
RT-qPCR	real-time or quantitative polymerase chain reaction
RT-PCR	real-time polymerase chain reaction

SCF	Scientific Committee for Food
SDA	strand displacement amplification
SIM	selected ion monitoring
SMR	small multidrug resistance
SPE	solid-phase extraction
SPME	solid-phase microextraction
SPR	surface plasmon resonance
SPT	skin prick test
ss	single-stranded
SSB	single-stranded DNA binding
SSOPs	sanitation standard operation procedures
ssRNA	single-stranded ribonucleic acid
STEC	Shiga toxin-producing
TA	tenuazonic acid
TBPS	tert-butylbicyclophosphorothionate
TCTC	trichothecenes
TLC	thin layer chromatography
tRNA	transfer ribonucleic acid
U.S. FDA	United States Food and Drug Administration
UN	United Nations
UPLC	ultra-performance liquid chromatography
US	United States
UV	ultraviolet
WGA	wheat germ agglutinin
WHO	World Health Organization
ZEA	zearalenone
μg	microgram

PREFACE

In today's competitive global food business, the greatest challenge is to supply safe food to consumers. Any substance that could pose a threat to the safety to those who consume the food product is a hazard. The concept that food will not cause any harm to the health of consumers when consumed is known as food safety. Food safety during food processing and distribution has become achievable by adopting HACCP (hazard analysis critical control point), a systematic preventive approach towards food safety. For a successful HACCP system, it is essential to understand different hazards that can be controlled in any given food.

Professionals in food processing need to have adequate knowledge and expertise in food chemistry, food processing, preservation, and food analysis and quality control. It is also desirable to be well versed in food microbiology, food laws and sensory evaluation. Professionals may be employed with regulatory and public health agencies as food legislators, food safety officers (inspectors), and food analysts/public analysts. This book will enable readers develop the skills necessary to understand and manage food safety hazards.

This book mainly covers the current scenario of the current research and case studies and contains scientific evidence on the basic concepts and terms associated with food safety and quality. It then provides the fundamental hazard types and a preliminary introduction to risk assessment, main methods of food safety management, and why it is important to make safe food. The findings reported in this book can be useful in food safety and health policy decisions. Moreover, these food safety issues are drawing the attention of researchers/policymakers because of their demonstrated adverse effects on health with high global burdens such as diabetes, hypertension, cancer, and neurodegenerative diseases.

On searching the literature, one can find enormous specialized articles or books on controlling food hazards and ensuring food safety during food processing. While they can be useful, a frequent limitation is that they tend to be imprecise and also often fail to identify and elaborate what are most likely to occur in food products. Unfortunately, most of these publications have dealt with theoretical aspects of these strategies and technologies with little emphasis on real application for consumer and food products safety.

This book volume attempts to illustrate various aspects of food safety hazards and how to analyze and control these potential hazards.

This book has several potential users. It can be a reference book for those students who are taking a college- or university-level pharma, food safety, or quality assurance (QA) course for the first time. Our objectives were to compile information that home science, pharma, dairy, and food science students are expected to be familiar with as part of their college or university programs, before they seek career positions. This book will be further useful to pharma and food industry quality practitioners or employees, who need to become familiar with updated information pertaining to their routine work. This book is organized in such a way that each chapter treats one major food safety hazard and novel control methods for health, food safety, and quality enhancement through various means.

This book aims at illustrating food hazards in the field of food safety by including three main parts, such as: (i) chemical hazards in milk and milk products; (ii) biological hazards in food and food products; and (iii) food safety hazards-practices and management.

We introduce this book volume under book series Innovations in Agricultural and Biological Engineering. This book volume is a treasure house of information and an excellent reference for researchers, scientists, students, growers, traders, processors, industries, and others for quality control and safety of food products during production, processing, and transportation in any food industry. It will help to boost confidence in the area of safety and quality aspects of food products.

This book has surpassed our vision and expectations due to the contributions by all cooperating authors to this book volume, who have been most valuable in this compilation. Their names are mentioned in each chapter and in the list of contributors. We are grateful to all of them for their expertise, commitment, and dedication. We hope that this book will be a useful source for analysis and control of various food safety hazards with a wide range of food and consumer product applications.

We would like to thank Apple Academic Press, Inc. for their valuable help and advice throughout this project, production of this book. We request that readers offer their constructive suggestions that may help to improve the next edition.

Also, we would like to thank our families, who have taught us the importance of working hard, having clear goals, and standing for what we believe is right. It is a lesson that guides us in everything we do.

—*Editors*

PART I

Biological Hazards in Foods and Food Products

CHAPTER 1

POTENTIAL THREAT OF MYCOTOXINS TO FOOD SAFETY AND SECURITY

V. DHARINI, NEHA HALDER, S. PERIYAR SELVAM,
G. MIKUNTHAN, and M. MAHESH KUMAR

ABSTRACT

Providing safe and nutritious food for human consumption is the ultimate aim of all food manufacturing and processing sectors. The occurrence of mycotoxins in food products continues to be the major hazard producing factor in developing and also in developed countries. Satisfying the hunger needs of the expanded population with safe and healthy foods is still a dream for many countries. Mycotoxins, secondary metabolites of fungi have deleterious effects in human metabolism leading to cancer and other genetic disorders. At such a scenario prevalence of mycotoxins in a wide range of food products adds an additional criterion in hazard analysis and determining its critical control point. Effect of climate on agricultural produce plays a significant role in food production, which directly or indirectly incubates the causative agents of mycotoxins, especially fungi. Admitting this complexity, numerous researchers have developed different mitigation technique which helps in the production of safe, sustainable, wholesome foods and agricultural products.

1.1 INTRODUCTION

Mycotoxins are secondary metabolites, having a low molecular weight, produced by a fungus that causes a toxic response upon ingestion. Synthesized by diverse fungal species, such as *Aspergillus*, *Fusarium*,

Penicillium, and *Alternaria*, its harmful effect on human and animal health varies depending upon the varying molecular structures causing significant differences upon entrance into production and food chain [20]. In 1960 the term mycotoxin was used to construe the toxin contaminating peanuts in feed of animals resulting in loss of turkey in England (Turkey-X disease) termed aflatoxin B1 (AFB1). It was defined as "natural products produced by fungi that evoke a toxic response when introduced in low concentrations to higher vertebrates and other animals by a natural route" [12].

Thought to be "chemical defense system," mycotoxins were synthesized to safeguard the mold from insects, microorganisms, nematodes, etc. Till attainment of favorable conditions, the spores produced by fungi can be dormant sustaining adverse conditions. During favorable growth conditions, spores attain vegetative state [7]. Traditionally, toxic fungi contaminating agricultural produce were grouped as 'Field' fungi (*Fusarium*, *Cladosporium*, and *Alternaria* spp.) that affected the crop during development of plant and 'Storage' fungi (*Penicillium* and *Aspergillus* spp.) proliferating during storage. However, it is currently divided into four types [12]:

- Plant pathogen: *Fusarium graminearum* and *Alternaria alternate*;
- Fungi affecting stressed or senescent plants: *Aspergillus flavus*;
- Fungus that colonizes the plant increasing susceptibility after harvest: *A. flavus*; and
- Fungi found on soil or in decomposing plant part on kernel later to proliferate during storage: *Penicillium verrucosum* and *Aspergillus ochraceus*.

Fungi play an important part as pathogens for plants and insects. They cause mycosis and mycotoxicoses. Exposure through diet, inhalation, dermal or other factors leads to mycotoxicoses whereas growth on host leading to disease is called mycoses. Some infections remain localized, while others tend to spread as systemic infections. However, both generally fall under non-communicable diseases. It is very important for the fungal pathogen to be able to grow at human body temperature of 37°C for systemic infection to occur. However, the optimum temperature falls under the mesophilic range of 20–30°C for biosynthesis of mycotoxins. In plants, various secondary metabolites produced by bacteria and fungi cause plant diseases by acting as potential virulence factors [4].

1.1.1 MYCOSES

Generally, for mycoses the pulmonary tract acts as an entry portal. Mycoses causing fungi can be categorized into two namely primary and opportunistic pathogens:

- Primary pathogen affects a person with healthy immune system and examples include *Coccidiodes immitis* and *Histoplasma capsulatum*; and
- Opportunistic pathogen causes illness in immune-compromised individuals and examples include *Aspergillus fumigatus* and *Candida albicans.*

1.2 MYCOTOXICOSES

Mycotoxicoses occur upon poisoning by natural means such as eating contaminated food, dermal contact, and inhalation of infected substrate. The symptoms depend on various factors, such as:

- Amount of toxin;
- Type of mycotoxin;
- Duration of exposure;
- Age and sex of the individual;
- Other synergistic factors like genetics, diet, etc. [4].

1.3 MYCOTOXINS

In 1967, three children died and 27 people were affected by food poisoning, which was due to consumption of rice containing elevated levels of aflatoxin (AF). In 2004, an outbreak of mycotoxin poisoning was reported in Kenya that resulted in 125 deaths out of 317 cases. All mycotoxins are produced by fungus, but all toxins produced by fungus are not labeled as mycotoxins. Toxins produced against bacteria, like penicillin, are called antibiotics. Toxins produced against plants are called phytotoxins. Mushrooms are also fungi that produce metabolites but those are not categorized as mycotoxins. The difference between poisoning from mushrooms and mycotoxin is not only based on the size of fungus but on the intention of humans because mycotoxin ingestion is mostly accidental [4].

1.4 CLASSIFICATION OF MYCOTOXINS

Mycotoxins are difficult to define as well as categorize due to their diversity. Due to this, different groups classify it differently. Clinically depending upon the affected organ, it may be classified as:

- Neurotoxins;
- Hepatotoxins;
- Nephrotoxins;
- Immunotoxins, etc.

As per biologists, they have groups like carcinogenic, mutagenic, teratogenic, and allergic. Chemists classify them based on chemical structures into groups like polyketides, lactones, coumarins, etc. Mycologists differentiate them by producing organisms like Aspergillus toxin, Penicillium toxin, etc. This differentiation is not accurate as it lands some of the toxins into multiple categories like AF is a carcinogenic, mutagenic, hepatotoxin polyketide-derived Aspergillus toxin [4].

1.5 CAUSE OF PRODUCTION

At any stage, during production and processing or transport and storage, mycotoxins can occur in food. There are many factors that contribute to favorable conditions for the growth of molds and mycotoxin production. The main factors are temperature, fungicides, relative humidity, and fertilizers. However, it also depends on factors like interaction between the toxigenic fungal species, geographical location, nutritional factors, genetic requirements, and infestation by insects [7]. Cereals, grains, nuts, oilseeds, vegetables, cocoa, dry fruits, herbs, fruits, spices are the major source for mycotoxicity as it is consumed by a considerable number of people and animals. It possesses a huge risk on the health of humans, as well as causes huge economic losses due to reduced yield of crops and illness in livestock upon consumption. This leads to losses in trade revenues in both international and domestic markets [5].

This chapter will briefly describe the causes, effect, detection methods along with control strategies of mycotoxins. Overview on model-based approaches in finding out the predictive results for the mycotoxin production will also be given importance along with the deep insights of the practices to be followed in producing safe and secure foods.

1.6 TYPES OF TOXINS AND HARMFUL EFFECTS

Around 25% of the estimated world crops get influenced by mold or growth of fungi, causing serious economic consequences. These crops may in turn be contaminated with mycotoxin which poses a risk on human and animal health. It also may cause secondary contamination into meat, milk, and eggs [5]. Contamination of human food may occur at different stages of the food chain. Three significant genera of mycotoxin producing fungi are *Fusarium, Aspergillus, and Penicillium*. Prominent mycotoxins produced by them are: AFs (*Aspergillus*), trichothecenes (TCTC), fumonisins (Fusarium), and ochratoxins (*Aspergillus* and *Penicillium*).

Mycotoxin exposure results in a disease known as mycotoxicosis. Generally, *Aspergillus*, and *Penicillium* grow on food and feed during storage and *Fusarium* propagates in plants infecting the growing crops such as wheat, maize, and barley. Maize is a crop that is largely susceptible to mycotoxin. They are thermally stable and can withstand various food processes like cooking, boiling, frying, roasting, baking, and pasteurization. When an animal consumes contaminated feed, the mycotoxins are liable to pass into the human plate via meat, eggs, and milk [2].

Contamination by mycotoxins may occur at both preharvest and postharvest periods. AFs are found in maize, peanuts, tree nuts and dried fruits. Ochratoxin is found predominantly in cereals but significant levels occur in wine, coffee, spices, and dried fruits. Fumonisins are found predominantly in maize and its products, whereas TCTC and zearalenone (ZEA) are found in grains [1].

1.6.1 AFLATOXINS (AFS)

The exploration of AF began after an outbreak of Turkey-X disease in 1960 in conjunction with the death of 10,000 turkey chicks in poultry farms. It spread to young ducklings, pheasants, and to pigs and calves. Upon tracing, the toxicity was found to be in the feed of peanut meal. A test was developed to monitor the toxicity. Upon illumination under UV light, the toxic extracts transmitted a bright blue fluorescence. It was later determined to be caused by a toxin produced by a mold *Aspergillus flavus*. AFs, difuranocoumarin derivatives, are produced via the polyketide pathway.

There are almost 18 different AFs produced [12] where the most important are the four toxins, namely B1, B2, G1, and G2. Among these, AFB1 is extremely toxic. It is a potent carcinogen and mutagen causing liver cancer in numerous animal species. AFs are present in the tropical and subtropical regions in commodities like maize, cotton, spices, peanuts, and pistachios [1]. AF shows inhibitory effect on synthesis of chlorophyll and carotenoid resulting in virescence or albinism in affected plants. Studies reveal AF suppresses biosynthesis of chlorophyll a, b, and proto-chlorophyllide in seeds of mung beans (*Vigna radiata*). AF inhibits DNA-dependent polymerase activity due to binding during replication. It was also reported to arrest cell cycle inducing apoptosis [12].

1.6.2 CITRININ

A benzopyran compound, citrinin was isolated by Hetheringtonand Raistrick from *Penicillium citrinum.* Numerous species of *Penicillium verrucosum*, *Penicillium expansum*, *Aspergillus terreus* and *Monascus ruber* produce citrinin. It is found to contaminate wheat, maize, rye, oats, barley, and rice. Citrinin caused yellow rice disease in Japan and was found to be nephrotoxic for all the tested species of animals [1, 12]. It mostly affects the kidney and also affects other organs like liver and bone marrow. The toxicity studies of citrinin have mostly been carried out *in vitro*. Phytotoxicity of citrinin is because it binds with cellular components affecting translocation and osmotic balance. It also affects respiration of germinating seedlings and seeds. Citrinin acts synergistically with Ochratoxin A (OTA) depressing RNA synthesis in murine kidneys [1, 12].

1.6.3 OCHRATOXIN

OTA secreted from *Aspergillus ochraceus* and occurs in three forms A, B, and C. Ochratoxin B (OTB) is a non-chlorinated form of OTA and Ochratoxin C (OTC) is an ethyl ester form of OTA. It is produced by *A. carbonarius*, *A. niger* and *P. verrucosum.* The *P. verrucosum* has been isolated from cereal crops, meat, and cheese variants but is a carcinogen as identified by the International Agency for Research on Cancer. It has been suggested in recent studies that OTA causes cellular death in plants. It inhibits protein synthesis, interferes with metabolism of phenylalanine, disruption of calcium homeostasis and DNA damage. *Aspergillus ochraceus* contaminates a wide range of commodities like wine and beer.

Aspergillus carbonarius is the prominent species found on vine fruit. It releases toxin during the juice making process. OTA is nephrotoxic and is linked to cause tumors in the human urinary tract [1, 12].

1.6.4 PATULIN (PAT)

Patulin (PAT) was named as claviform and produced by *Penicillium claviforme*. Ensuring reclusion from the Penicillium *patulum* it was labeled as PAT. PAT has many names such as clavicin, expansin, clavitin, gigantic acid, leucopin, mucoin, tercinin, and penicidin. During early findings, PAT was used to treat common cold, whereas later was found to be too toxic. It caused various health disorders such as convulsions, congestion in lungs, ulcerations, degeneration of epithelial cells, carcinogenic, genotoxic, teratogenic, immunotoxin, and immunosuppressive. The *P. expansum* is found to grow on fruits and vegetables like apricots, peaches, grapes, particularly in apples and figs. It is found in rotten fruit parts after it gets infested with PAT-producing fungus. It infects during storage at 0°C and grows rapidly upon transfer to room temperature. Though PAT is non-carcinogenic, it has been noted to cause harm to the immune system of animals. Phytotoxic PAT inhibits protein, RNA, and DNA synthesis [1, 12].

1.6.5 FUSARIN

Fusarins are polyketides and produced in over 50 species of *Fusarium* spp. Fusarin A, fusarin B, fusarin C and fusarin D were first found in California by Wiebe and Bjeldanes in Berkeley in 1981. Alternative types of *Fusarium* toxins include beauvercin, equisetin, butenolide, and enniatins. The nervous system of horses gets affected by mycotoxin fumonisin. It is also found to cause cancer in rodents. Phytotoxicity of fusarins is not explored whereas *F. moniliforme* produces type C on maize, wheat, soybean, rye, and barley [1, 12].

1.6.6 ZEARALENONE (ZEA)

Zearalenone (ZEA) also known as F-2 toxin is an estrogenic mycotoxin. It is a potent toxin resembling 17β-estradiol. Estradiol is a principal hormone produced by the ovary in humans to bind to estrogen receptors. It is categorized as nonsteroidal estrogen or mycoestrogen. It is synthesized by *Fusarium graminearum*, *F. culmorum*, *F. equiseti* and *F. crookwellense* found contaminating cereals [4]. It evokes hyper-estrogenism in pigs. It is found to contaminate barley, wheat, oats, sorghum, corn, millet, and rice. It is also found in flour, soybeans, malt, and beer. The production of toxin occurs before harvesting but might also happen after post-harvest storage. ZEA is found to be less toxic [12].

1.6.7 TRICHOTHECENES (TCTC)

Abbreviated as TCTC, it was found to cause alimentary toxic aleukia (ATA) toxicosis in the year of 1932. It is mainly produced by *Fusarium* but is also produced by *Myrothecium*, *Trichoderma*, *Trichothecium*, and *Stachybotrys*. It causes Fusarium head blight in plants causing yield reductions in huge amounts of bananas, rye, maize, oats, wheat, rice, soyabean, potatoes, sunflower seeds, peanuts, and barley [2].

1.6.8 DEOXYNIVALENOL (DON) AND NIVALENOL

Deoxynivalenol (DON) is produced by *F. graminearum* and *F. culmorum*. Nivalenol (NIV) is produced *by F. graminearum*, *F. sporotrichioides* and *F. crookwellense*. They impart an important role in the development of disease. DON toxin causes damage to the membrane for release of K and Na ions into solutions. Inhibitory effect on mitosis was observed [12]. DON is often encountered in the field [12].

1.6.9 ERGOT ALKALOIDS

Ergot is a toxic alkaloid mixture formed in the sclerotia. It was first isolated in 1934 having a lysergic acid structure that is common to all ergot alkaloids. It is produced by *Claviceps* commonly occurring in grasses. Ergotism, basically, affects cattle, sheep, pig, and chicken with systems like abortion, gangrene, convulsions, lactation problems and hypersensitivity [4]. Ergotism may occur upon consumption of infected cereals as bread. It has two forms-Gangrenous that impinges blood supply to extremities, and Convulsive altering the central nervous system. These days with the implementation of precautionary measures occurrence of ergotism has reduced in humans, but it still persists as an important problem in veterinary studies [1].

1.7 DETECTION OF MYCOTOXINS

The most common species of concern to food safety among the 400 identified mycotoxins are AF, ZEA, DON, PAT, OTA, tenuazonic acid (TA), etc. Mycotoxin contamination creates a concealed danger in human health and international trade disputes are increasing in recent years due to fungal contamination [29]. Collection of accurate data is necessary to identify the risk constituted on humans and to establish its regulatory limits. In order to attain accuracy in data collection, evaluation of reliable analytical techniques is essential. Determination of masked mycotoxin and low-level mycotoxin concentration in food matrices are critical challenges in maintaining the accuracy in analytical techniques.

Pretreatment of sample (complex food matrices) aids in separation process, indispensable step in most of the analytical procedures [27]. Isolation and enrichment are the two main processes involved in pretreatment procedure. Sample preparation is the crucial step which is done accurately by extracting the mycotoxin using organic solvents like acetonitrile. Instrument automated solvent extraction techniques are used recently in the analysis of mycotoxins. Methods like supercritical fluid extraction, microwave-assisted extraction (MAE) and accelerated solvent extraction (ASE) are emerging techniques that have several advantages [19].

Extraction is followed by clean-up steps like filtration and centrifugation to remove interfering particles. Different sample clean-up techniques are generally followed. Techniques like solid-phase extraction (SPE) [8], dispersive solid-phase extraction (D-SPE) [22], QuEChERS (quick, easy, cheap, effective, rugged, and safe) [21], solid-phase micro-extraction (SPME) [11], magnetic solid-phase extraction (MSPE) [14] are used. Other techniques like immune-affinity column (IAC) [15] and dispersive liquid-liquid micro-extraction (DLLME) [6] are also used for filtration and centrifugation [29]. Figure 1.1 depicts the several techniques involved in sample preparation.

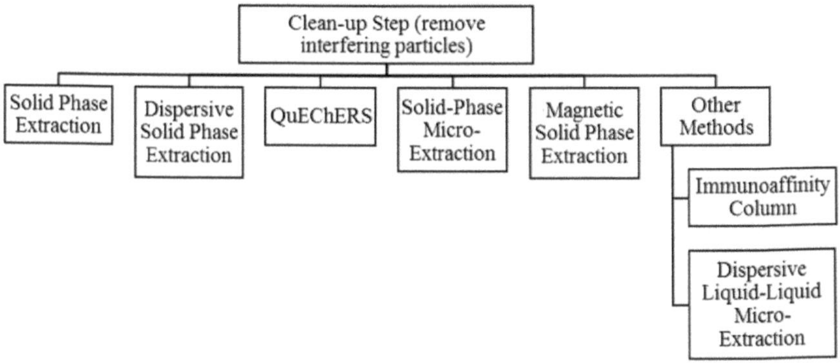

FIGURE 1.1 Methods in sample preparation.

Association of Analytical Communities (AOAC) have developed standard methods based on HPLC (high performance liquid chromatography), which is widely accepted against all other methods in the analysis of mycotoxins. Liquid chromatography (LC) can concurrently determine a number of mycotoxins with the application of several detectors. Less frequently, gas chromatography (GC) is also used to measure mycotoxins. HPLC-MS/MS technology is the fundamental and foremost option in the determination of mycotoxin due to its high sensitivity and accuracy. Sensor technology is the upcoming method that involves the use of certain nanoparticles and electrochemical detectors for rapid and accurate estimation of mycotoxins [29]. Figure 1.2 depicts the common technologies used in the determination of mycotoxins.

1.7.1 COMMON TECHNIQUES IN SAMPLE EXTRACTION

SPE was introduced in the year 1970, normally used for mycotoxin extraction. It is the most common and universally followed technique for sample preparation. The target compounds are adsorbed on a solid sorbent through enrichment of analytes. The basic concept behind SPE is the analyte partitioning between solid sorbent and liquid phase. In order to improve the selectivity, efficiency, and sorption capacity of the SPE technique, appropriate choice of sorbent selection is essential. Carbon-based materials, polymer-based sorbents, and chemically bonded silica is ideally used as sorbents in the past for the SPE technique. At present new solid sorbents are gaining importance to avoid certain physical and chemical limitations in the extraction procedure [10]. Though SPE is a familiar method used in the extraction of target samples but it has several disadvantages like low selectivity, high cost, complicated, and time-consuming operations.

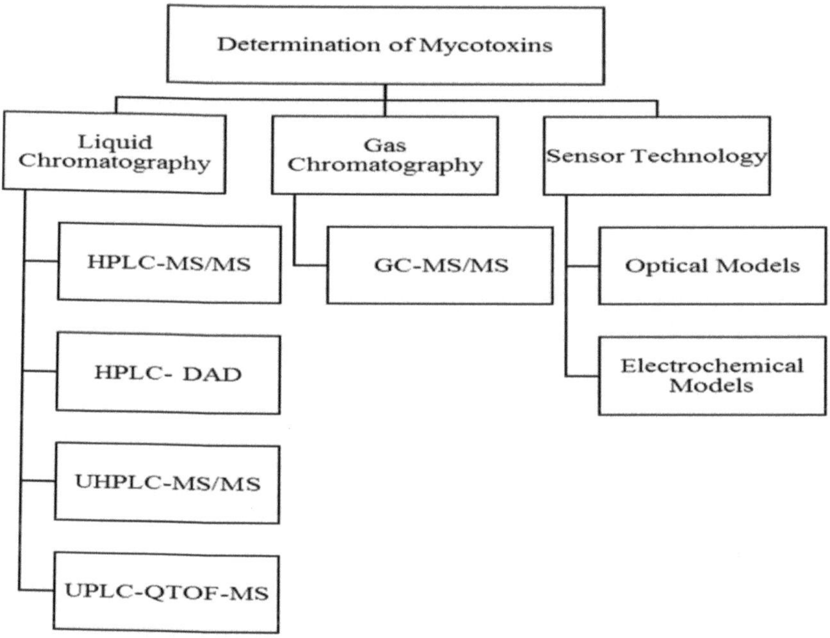

FIGURE 1.2 Technologies used in the separation and determination of mycotoxins.

In the regard of high specificity and efficiency, IAC is a technique frequently used in the determination of mycotoxins. IAC can be effectively used in the determination of AFs when coupled with HPLC along with various detectors. Though IACs offer high selectivity and reproducibility, they are comparatively expensive and could not be reused. On the other hand, molecularly-imprinted polymers along with SPE can be reused and the recovery rates are higher compared with IAC. D-SPE is the pre-treatment technique used for in detection of mycotoxin wherein the sorbents are directly added to the analytical solution creating a strong interaction between the sorbents and the analytes. Low time consumption, feasible operating conditions including low cost and less utilization of organic solvent are the few advantages of D-SPE over traditional SPE technique [29].

SPME is an approved and commonly used green method for the pretreatment and enrichment of a wide variety of analytes. This method is well-established and more readily used for complicated sample analysis. Efficiency of SPME depends on the fiber coating materials which are used for extracting the target compounds from the sample solution. Hence more research is being focused on the area of providing simple, rapid, sensitive, and environmental friendly sampling techniques to improve the efficiency of SMPE. It is essential to extract and enrich the target compounds prior to instrumental detection in order to avoid the presence of impurities and interfering compounds [24]. Synthetic sorbents like molecularly-imprinted polymers have solved many issues relating to mechanical stability, reusability, and thermal strength. Its use is discovered to have wide applications as an efficient sorbent in various matrices of environment and food for the detection of trace levels of mycotoxins [28].

Magnetic nanoparticles are used in MSPE technique which has a larger specific area and shorter diffusion distance of analytes. This new technique MSPE can be called magnetically modified or simply magnetic SPE which enables rapid separation of magnetic materials and solution. This rapid separation is through the application of magnetic fields which aids in reducing the time for sample processing. Magnetite, the most frequently used magnetic particle which is coated with various functional composites like alumina, silica, chitosan, and surfactants to improve the sorption properties by functionalizing with coating materials. MSPE has become the most popular method due to its features like reusability, high recovery and is considered an environmentally-friendly approach developed in recent years [10].

QuEChERS method of reduced procedure in sample preparation technique is essential in order to minimize the random errors associated with the analytical step of the extraction process. This method is known for its simplicity and applicability in mycotoxin analysis. QuEChERS is proposed to be a reliable analytical tool for routine analytical control by allowing a faster extraction process [13]. This technique involves a simple two-step process of extraction (using solvent and salts) and D-SPE clean up. QuEChERS can be coupled with UHPLC-MS/MS for the simultaneous detection of multiple mycotoxins. Similarly, when QuEChERS coupled with LC-MS/MS can be used in the detection of AFB1, B2 in cereal-derived products and Fusarium species in beers and beans. QuEChERS when combined with HPLC-DAD can be used to determine PAT in strawberries. Incorporation of UPLC-MS/MS is effective in the determination of mycotoxins, especially AFB1 in eggs [29]. Hence, analytical techniques used after the extraction of target samples using QuEChERS are also important in the determination of mycotoxins.

1.7.2 COMMON TECHNIQUES IN SAMPLE DETERMINATION

LC is considered as the backbone of mycotoxin analysis and is often used with different detectors like MS, DAD, etc. AOAC has developed certain standard methods based on HPLC, which is widely accepted for the determination of mycotoxins. LC can determine different mycotoxins irrespective of their chemical structures or biological activity. Column and mobile phase are used in the separation of analytes and matrix components. Effective use of MS detection in LC systems is successful in the determination of multi-residues mycotoxins and accepted as the most favorable method. Effectiveness and high sensitivity in the analysis of mycotoxins can be attained with the application of HPLC-MS/MS and is considered as the first choice of mycotoxin determination. HPLC-DAD can also be used to increase the effectiveness and accuracy during the mycotoxin determination. The composition of the mobile phase is also considered as one of the important parameters that influences the separation efficiency [29].

Application GC technique coupled with flame ionization detection and the use GC-MS/MS is employed in the screening of mycotoxins. Derivatization reactions like silylation and acylation reactions are essential in the analysis by GC due to the high polarity and low volatility of mycotoxins. Though GC has proven to have many advantages, this method is not widely used because of the essential derivatization step which is prone to degradation issues. Techniques involving the combination of UPLC-MS/MS and UPLC-QTOF-MS can be used in the determination and screening of multiple mycotoxin metabolites. Here, UPLC-MS/MS is used for the routine determination of mycotoxin whereas UPLC-QTOF-MS is used to identify the non-target mycotoxin metabolites. Combined methods showed great potential in the determination of trace amounts of mycotoxins [23].

1.8 CONTROL STRATEGIES

Mycotoxins have always impaired the quality of agricultural products. Based on storage conditions and growth, commodities like corn, rice, wheat, soy, and peanuts get contaminated with mycotoxins. Consumption of mycotoxin contaminated products may cause acute as well as chronic toxic health disorder on humans and animals. There are copious legislations for mycotoxins in place protecting consumer health and animals worldwide. Mycotoxins are naturally occurring contaminants in food. It is spread throughout the product at very low levels. Before harvest, production gets affected by weather changes and agricultural practices. Contamination occurs on inappropriate post-harvest storage conditions. This knowledge of controlling points during and after harvest, drying, and storage plays essential roles to prevent mycotoxin contamination [17].

Control of mycotoxin is important for securing public health. It involves, generally, three steps:

1. **Primary Prevention:** It should be initially carried out before contamination. This can be achieved by establishing fungal resistant variants in plants, reducing moisture content in seeds during the storage period, using fungicides and preservatives, insect infestation control by using suitable insecticide use, etc.
2. **Secondary Prevention:** When invasion occurs at early phase, certain preventive measures can be taken to check further deterioration with the need to eliminate the existing toxic fungi. Further, some measures such as re-drying of produce, removing contaminated seeds, detoxification of contaminated products and protection from favorable conditions needed for fungal growth.
3. **Tertiary Prevention:** Upon heavy infestation of molds, the entire produce has to be destroyed and detoxification has to be carried out.

Contaminated mycotoxins should be detoxified by physical, biological, and chemical methods. However, the treatment process is limited by the fact that treated products should be safe to consume. Physical method is labor-intensive and time consuming. It could be done manually by picking or photoelectric detecting machines. Radiations such as ɣ-rays check the growth of spoilage microbes. Studies reported that 5–10 M rad reduced incidence of AF. But the treated product should be assessed for mutagenicity. Chemical treatment is the most effective means of arresting mycotoxin contamination in commodities. The common chemicals used are acetic acid, ammonia gas, hydrogen peroxide, sodium bicarbonate, etc.

[25]. In a study, it was found that control of mycotoxins in crops in Africa involved a few steps:

- Preventing fungal growth in crops and other feedstuffs;
- Decontamination of mycotoxin contaminated feed and food;
- Continuous surveillance of crops, feed of animals and food for humans.

This can be enacted by following a precautionary measure during harvest, storing, and processing. Appropriate drying and storage of crops prove to be an effective tool to reduce mold growth followed by production of mycotoxin. This includes various physical, biological, and chemical approaches. Physical approach includes: sorting, washing, crushing dehulled maize grains which prove to be an effective tool to eliminate AF and fumonisin. Chemical approach incorporates the application of fungicides such as prochloraz, cyproconazole, propiconazole, tebuconazole, epoxiconazole, chlorophyllin, azoxystrobin, and oltipraz to reduce incidence of fumonisin and AF. Development of toxigenic fungi, which then competes with toxigenic fungi is one biological approach [26].

1.9 MODEL-BASED APPROACH IN THE DETERMINATION OF MYCOTOXINS

1.9.1 OVERVIEW OF THE APPROACH

In order to improve the safety and the quality of the food that is produced, various technologies provide tools for predicting the fungal growth and mycotoxin production. These tools play a vital role in the assessment of risk and implementing the mitigation strategies. Decision making can be accurate based on the results obtained and predictive tools are found to be promising in the field of food safety. The core aim of modeling is to simulate the conditions to which the food is exposed so as to forecast its microbial behavior and then optimizing those conditions in order to suggest detailed predictive studies [3]. However, modeling of mycotoxin could be impractical while perceiving food safety as there is high variability in the mycotoxin-producing species [18]. Table 1.1 refers to the factors that are primarily responsible for fungal and mycotoxin production.

TABLE 1.1 Factors Responsible for Mycotoxin Production

Conditions	Intrinsic Factors	Implicit Factors
Pre-harvest	Water activity of the crop	Climatic conditions
	Plant variety	Agronomic practices
	Nutrient composition	Presence of insects, fungi, spore load and plant diseases
Harvest/ Drying	Rewetting properties	Spore load
	Final moisture content	Mechanical damage
Storage	Addition of preservatives	Nature of substrates
	Temperature	Composition of substrates

1.9.2 AFLATOXIN (AF) MODELS: CASE STUDY

Common problem encountered during harvest and storage of food commodities is the production of AFs. Under specific conditions, *Aspergillus flavus* and *Aspergillus parasiticus* contaminate the majority of the food crops like maize and peanuts. Several available models relate the effect of water activity (a_w) and temperature in the mycotoxins production on synthetic and real food matrices. Baranyi and Luedeking-Piret models predict AF production over time and are considered as the primary models. AF formation can be estimated through the radius of the colony formed, surface characteristics of the colony and dry weight of the biomass [9]. Polynomial approach can be used to determine the secondary modeling of AFs. Due to high variability in the toxin concentration among strains, probability models are rarely used in the determination of AFs.

1.9.3 CONSIDERATIONS FOR CONSTRUCTION OF A MODEL

- Deep understanding in the biosynthesis of mycotoxin is necessary to overcome the intraspecific variability in the mycotoxin production;
- Though most of the models deal with interaction between aw and temperature as the critical factors, inclusion of pH as one of the variables could have an impact with the microbial interactions—accurate model designing;
- Validation of models in real substrates—mycotoxin production is different when compared with agar system and real food and feed

system. Validation of already generated and developed models is essential.

1.10 MYCOTOXIN AND ITS IMPLICATIONS ON HUMAN HEALTH

1.10.1 ALLERGENS

Molds have been associated with allergies for a long time. They reside in air conditioning vents and filters and in indoor environments where they grow due to water damage. The commonly found indoor genera are *Alternaria*, *Cladosporium*, *Aspergillus*, and *Penicillium*. However, in modern times with advancement in building techniques, the 'compact' building systems have aggravated the symptoms which get better when outdoors. This is known as Sick-building syndrome. The symptoms include respiratory tract and eye irritation, headache, skin irritation, fatigue, non-specific hypersensitivity, peculiar odor, and taste sensations [4].

It has no specific definition based on identified etiological factors. The etiological agents are poor ventilation, chemicals from various sources such as cleaning supplies, aerosols, and different microbial contaminations. However, exposure to *Stachybotrys* spp. is generally associated with sick building symptoms. It was found to have caused idiopathic pulmonary hemorrhage in infants in Cleveland, Ohio in the year of 1994. It is found to flourish on wet surfaces with high cellulose content such as ceiling tiles, wooden boards, dust-lined AC ducts [2].

Early studies on mycotoxin exposure from indoor air were done by Hendry and Cole. It was reported that air-borne dust carries spores that cause ochratoxin exposure. Similarly, aerosols have been found having TCTC.

1.10.2 IMPACT ON GUT MICROBIOTA

Ingestion of mycotoxin causes perturbations in the intestinal epithelial tissues. When introduced via food, they first interact with the gastrointestinal (GI) tract, where the gut microbes reside. Generally, GI tract functions as the filter to mycotoxins. However, it can interrupt with normal functioning of the intestine altering barrier functions and affecting nutrient absorptions.

The gut microbes are responsible for a healthy digestion and immune system. Recent studies have reported that there exists a direct relation between gut microbiome and mycotoxin. A person with a healthy microbiota has an overall better health. Gut microbiota have the potential to remove mycotoxin from the host naturally. It has also been studied that gut microbiota and mycotoxin play an important role in the development of hepatocellular carcinoma (HCC) [2].

1.10.3 TOXICOLOGICAL STUDIES OF MYCOTOXINS

Upon DON toxication, decreased absorption of glucose was observed in the GI tract. It was also found to induce diarrhea due to reduced number of SGLT1 transporters that help with water reabsorption. During animal studies, TCTC was found to cause necrotic lesions. It shortened the villi height leading to mal-absorption of nutrients. Studies of ZEA showed its effect on reproductive health. It caused reduced fertility, enlarged uterus with abnormal progesterone and estradiol levels. Ingestion during pregnancy decreased the rate of survival of the fetus [2].

Studies on fumonisin toxicity revealed that it caused esophageal cancer upon ingestion of contaminated corn in the populations of South Africa, Italy, and China. It also caused a neural tube defect which is a defect in the spinal cord, brain, and spine. Studies on OTA revealed that it caused kidney disease. It is also associated with chronic tubulointerstitial disease leading to renal failure. The mode of toxic action of OTA is by its effect on protein synthesis. AF B1 is the most hazardous hepatocarcinogen in mammals. It mainly affects the liver. Acute aflatoxicosis caused abdominal pain, edema, vomiting, and death. Cytotoxic effect of AFB1 is by generating reactive oxygen that damages cell membrane and DNA [2].

1.10.4 PROBIOTICS AND GUT HEALTH

It is known that probiotics improve gut health and restores the microflora in the intestines. These probiotics, in turn, reduce the level of mycotoxins via binding. A study reported that chronic mycotoxicosis such as HCC is caused by *Helicobacter hepaticus* that alters DNA by generating DNA adducts, regulating DNA methylation and alters gene expression.

Treatment with probiotics helped relieve hepatic inflammation decreasing tumor growth in the liver.

1.11 FUTURE PROSPECTS OF MYCOTOXIN

By 2030, the world population will reach 8.2 billion and with increasing population, there will be an increased demand for food. This will provide a challenge, and with the presence of hazardous substances such as mycotoxins, there will be limitations to the marketability of products. The earth is experiencing climate change at an alarming level due to global warming and changing trends. This in turn affects the production and geographic distribution of food as well as the phyllosphere microflora of crops. For example, *Aspergillus flavus* is showing resilient growth under extremely hot and dry conditions. In Serbia, in 2012, 69% of maize was found to be contaminated with *A. flavus* despite drought-like conditions. This was also the case found in Hungary. Study of mycotoxins and their effect on consumption paves a way to a better understanding of toxicosis to find novel ways of prevention and therapy upon exposure [2].

1.12 PRODUCTION OF SAFE FOODS: GENERAL PRACTICES

Control of food contamination is the ultimate aim of any food producing Nation. The extensive occurrence of mycotoxins in many countries created an initiative for the development of appropriate regulations to mitigate its high toxicity nature. The scientific advisory board of WHO and FAO (Food and Agricultural Organization) is known as Joint Expert Committee on Food Additives (JECFA). This committee is responsible for the evaluation of mycotoxin and its related risks. Scientific Committee for Food (SCF) substantiated the maximum level of AF in foods. Mycotoxins is positioned second as per the number of hazard notifications by the rapid alert system for food and feed (RASFF), established by the European Union (EU) to check and notify the prevalence of mycotoxins in food and animal feed [16]. Food security is of higher importance and its key factors are availability of sufficient, safe, and nutritious food for all people at all times. Tables 1.2–1.4 display the reduction strategies of mycotoxin in food and feed.

TABLE 1.2 Prevention of Mycotoxin Formation

Factors	**Description**
Environmental factors	Appropriate control in temperature, moisture content and humidity are necessary
Good agricultural and manufacturing practice (GMP)	HACCP plays a crucial role in prevention and management of mycotoxin. Farmers should be aware of the antiquated agricultural techniques.
Storage management	Maintaining appropriate levels in moisture and humidity of warehouses decreases the mold growth.
	Low O_2 concentration and relatively high CO_2 concentration can mitigate fungal growth and mycotoxin production.

TABLE 1.3 Detoxification of Mycotoxin

Traditional Physical Method	**Description**
Filtering and adsorption	Various adsorbents like activated charcoal have the capability to adsorb mycotoxins and is widely explored
Immersing and washing	Washing is done to discard the floating fragments, which clears away some amounts of aflatoxins
Irradiation	Radiation is found to inhibit of fungal growth and is effective in decontamination of aflatoxins
Sorting and separation	Elimination of decayed/rotten fruits or cutting off spoiled segments especially in fruits
Chemical Method	**Description**
Bases (ammonia)	Ammonization reduces several mycotoxins and inhibits fungal growth
Oxidizing agents	Effective oxidizing agents like ozone help in detoxification process and its usage-approved as safe for human consumption
Biological Method	**Description**
Bacteria, yeast, and mold	Bacteria, yeast, and mold have the ability to biodegrade mycotoxins. Wide variety of various species are identified as biocontrol agents
Detoxification	Description
Detoxification by adsorption	Lactic acid bacteria (LAB) can adsorb AF, OTA, and DON along with other vast range to mycotoxins

TABLE 1.4 Innovative Reduction Strategies of Mycotoxin

Strategies to Reduce Mycotoxin	Description
Magnetic materials and nanoparticles	• Use of magnetic materials that could remove mycotoxins are proven to be effective for use as adsorbents in food industries.
Natural essential oils (EO)	• Natural EO advantages: High efficiency, ecofriendly, and low-drug-resistance tool.
	• The antifungal activity is due to the disruption of the fungal cell membrane.
Polyphenols and flavonoid inhibitors	• Gallic acids: protects from the occurrence of AFB1.
	• Quercetin and umbelliferone: Prevents the accumulation of PAT effectively even at the transcript level.

1.13 SUMMARY

Prevalence of mycotoxin in various foods is a major concern in the production of safe and wholesome food for human consumption. Mycotoxin poses an unavoidable threat in the food processing sector and has to be scrutinized thoroughly to prevent the interruption in the food chain. Various environmental factors that influence the production of fungal and mycotoxin growth in food commodities have to be supervised separately. Temperature and humidity of the agricultural produce should be maintained at appropriate levels from the harvest till the storage and transportation to terminate the causative factors of mycotoxin production. Pre-harvest and post-harvest contamination of mycotoxins can be controlled by effective implementation of Hazard Analysis and Critical Control Point throughout the entire process of food production. Most of the contributing factors of mycotoxin production can be nullified by good manufacturing practices (GMPs). Educating framers on the production of safe crops from cultivation to harvest can inculcate Good Agricultural Practices at the farm level. Accurate detection of mycotoxin is complicated due to its variant toxicity nature. Highly sensitive and accurate technologies are developed to determine the mycotoxin content in food and feed. Validation of such technologies is based on its repeatability and reproducibility. Predictive models are gaining importance in the current scenario and have a wide research area to explore. Preventing the occurrence of mycotoxin is the best control technique that is ever encountered. Together with the implication of all possible means of mycotoxin mitigation strategies, production of safe and secure food is achievable.

KEYWORDS

- **aflatoxin**
- **control strategies**
- **food chain**
- **food safety**
- **food security**
- **hazard**
- **mycotoxicity**
- **mycotoxin**

REFERENCES

1. Adeyeye, S. A., (2016). Fungal mycotoxins in foods: A review. *Cogent Food and Agriculture, 2*(1), 1–11.
2. Ahmad, A., & Yu, J. H., (2017). Occurrence, toxicity and analysis of major mycotoxins in food. *International Journal of Environmental Research and Public Health, 14*(8), 632–645.
3. Aldars-García, L., Ramos, A. J., Sanchis, V., & Marín, S., (2016). Modeling postharvest mycotoxins in foods: Recent research. *Current Opinion in Food Science, 11*, 46–50.
4. Bennett, J. W., & Kilch, M., (2003). Mycotoxins. *Clinical Microbiology Reviews, 16*(3), 497–516.
5. Bryden, W. L., (2007). Mycotoxins in the food chain: Human health implications. *Asian Pacific Journal of Clinical Nutrition, 16*(1), 95–101.
6. Chen, F., Luan, C., Wang, L., Wang, S., & Shao, L., (2017). Simultaneous determination of six mycotoxins in peanut by high performance liquid chromatography with a fluorescence detector. *Journal of the Science of Food and Agriculture, 97*(6), 1805–1810.
7. Cinar, A., & Onbasi, E., (2019). Mycotoxins: *The Hidden Danger in Foods* (p. 22). IntechOpen.
8. Dong, H., Xian, Y., Xiao, K., Wu, Y., Zhu, L., & He, J., (2019). Development and comparison of single-step solid-phase extraction and QuEChERS clean-up for the analysis of 7 mycotoxins in fruits and vegetables during storage by UHPLC-MS/MS. *Food Chemistry, 274*, 471–479.

9. Garcia, D., Ramos, A. J., Sanchis, V., & Marín, S., (2013). Modeling kinetics of aflatoxin production by *Aspergillus flavus* in maize-based medium and maize grain. *International Journal of Food Microbiology, 162*(2), 182–189.
10. Hashemi, B., Zohrabi, P., & Shamsipur, M., (2018). Recent developments and applications of different sorbents for SPE and SPME from biological samples. *Talanta, 187*, 337–347.
11. Hu, Q., Liu, S., Chen, X., Xu, J., Zhu, F., & Ouyang, G., (2019). Enhancing enrichment ability of a nanoporous carbon based solid-phase microextraction device by a morphological modulation strategy. *Analytica Chimica Acta, 1047*, 1–8.
12. Ismaiel, A. A., & Papenbrock, J., (2015). Mycotoxins: Producing fungi and mechanisms of phytotoxicity. *Agriculture, 5*(3), 492–537.
13. Juan, C., Mañes, J., Font, G., & Juan-García, A., (2017). Determination of mycotoxins in fruit berry by-products using QuEChERS extraction method. *LWT- Food Science and Technology, 86*, 344–351.
14. Li, N., Wu, D., Li, X., Zhou, X., Fan, G., Li, G., & Wu, Y., (2020). Effective enrichment and detection of plant growth regulators in fruits and vegetables using a novel magnetic covalent organic framework material as the adsorbents. *Food Chemistry, 306*. Article ID: 125455.
15. Liu, H., Luan, Y., Lu, A., Li, B., Yang, M., & Wang, J., (2018). Oligosorbent-based aptamer affinity column for selective extraction of aflatoxin B2 prior to HPLC with fluorometric detection. *Microchimica Acta, 185*(71), 1–10.
16. Luo, Y., Liu, X., & Li, J., (2018). Updating techniques on controlling mycotoxins: A review. *Food Control, 89*, 123–132.
17. Magan, N., & Aldred, D., (2007). Postharvest control strategies: Minimizing mycotoxins in the food chain. *International Journal of Food Microbiology, 119*(1, 2), 131–139.
18. Magan, N., Sanchis, V., & Aldred, D., (2004). Role of spoilage fungi in seed deterioration. In: Arora, D. K., (eds.), *Fungal Biotechnology in Agricultural, Food and Environmental Applications* (Vol. 21, pp. 311–323) Taylor & Francis Inc.
19. Pereira, V. L., Fernandes, J. O., & Cunha, S. C., (2014). Mycotoxins in cereals and related foodstuffs: A review on occurrence and recent methods of analysis. *Trends in Food Science and Technology, 36*(2), 96–136.
20. Pitt, J. I., (2013). Mycotoxins. Chapter 10. In: Morris, J. G., & Potter, M. E., (eds.), *Foodborne Infections and Intoxications* (4th edn., pp. 409–418). Academic Press.
21. Qian, M., Yang, H., Li, Z., Liu, Y., Wang, J., Wu, H., & Xu, J., (2018). Detection of 13 mycotoxins in feed using modified QuEChERS with dispersive magnetic materials and UHPLC–MS/MS. *Journal of Separation Science, 41*(3), 756–764.
22. Ran, C., Chen, D., Ma, H., & Jiang, Y., (2017). Graphene oxide adsorbent based dispersive solid-phase extraction coupled with multi-pretreatment clean-up for analysis of trace aflatoxins in traditional proprietary Chinese medicines. *Journal of Chromatography B, 1044*, 120–126.
23. Romera, D., Mateo, E. M., & Mateo-Castro, R., (2018). Determination of multiple mycotoxins in feedstuffs by combined use of UPLC–MS/MS and UPLC–QTOF–MS. *Food Chemistry, 267*, 140–148.
24. Spietelun, A., Marcinkowski, Ł., De La Guardia, M., & Namieśnik, J., (2013). Recent developments and future trends in solid-phase microextraction techniques towards green analytical chemistry. *Journal of Chromatography A, 1321*, 1–13.

25. Suttajit, M., (1991). *Prevention and Control of Mycotoxins*. http://www.fao.org/3/x5036e/x5036E0q.htm (accessed on 21 September 2021).
26. Tola, M., & Kebede, B., (2016). Occurrence, importance and control of mycotoxins: A review. *Cogent Food and Agriculture, 2*, 1–12.
27. Tulipani, S., Llorach, R., Urpi-Sarda, M., & Andres-Lacueva, C., (2012). Comparative analysis of sample preparation methods to handle the complexity of the blood fluid metabolome: When less is more. *Analytical Chemistry, 85*(1), 341–348.
28. Vasapollo, G., Sole, R. D., Mergola, L., Lazzoi, M. R., Scardino, A., Scorrano, S., & Mele, G., (2011). Molecularly imprinted polymers: Present and future prospective. *International Journal of Molecular Sciences, 12*(9), 5908–5945.
29. Yang, Y., Li, G., Wu, D., Liu, J., Li, X., Luo, P., & Wu, Y., (2020). Recent advances on toxicity and determination methods of mycotoxins in foodstuffs. *Trends in Food Science and Technology, 96*, 233–252.

CHAPTER 2

POTENTIAL CONCERN OF FOODBORNE PATHOGENS IN THE FOOD INDUSTRY

SAMLESH KUMARI and PRAGYA KULKARNI

ABSTRACT

Renaissance and emergence of foodborne pathogens (FBP) is a serious matter that needs to be addressed by the food industry, consumer, and public health organizations. Pathogens reported in food typically belong to bacteria, viruses, and various parasitic microbes. Parameters, which are responsible for the emergence of the pathogens in food are adaptive or versatile nature of microbes and consumer handling, post-harvest processing measures, increase in urbanization and traveling, distribution of food in global and local markets, and climate changes. Since the decade's food safety has become an integral part of food processing and preservation to eliminate the load of FBP from food. However, the capability to survive in harsh environments of food processing and human digestive tract is the big cause of increasing incidences of foodborne illnesses. The reasons responsible for their survival in these sub-lethal conditions (processing treatment) are the alter gene expression or adaptation mechanism, resulting in resistance to the bactericidal insults. Furthermore, advances in detection techniques for FBP also faster the process of new or emerging pathogen recognition and help in reducing the risk of occurrence of FBP in food. To handle the issue of the pathogens emergence or re-emergence required a worldwide collaboration among the various food processing industries, food safety assurance bodies, etc. In this present chapter, pathogens related to the food industry, their survival during processing and innovative methods for their detection, advanced food preservation methods are discussed.

2.1 RENAISSANCE AND IMPACT OF FOODBORNE PATHOGENS (FBP)

Over time change in the behavior of bacterial pathogens has been noticed and these pathogens are "typically" known as emerging or re-emerging pathogens, it is not an exception in case of foodborne pathogens. Study of microbiology since the last two centuries, it has been observed that microbiological contamination in the food is a source of foodborne pathogens (FBP) for human beings. These observations helped us to understand the evolution and spread of the pathogens in foods and also facilitated to recognize the new emerging pathogens in the food products. With globalization, the lifestyle is getting changed over time and this change leading to the change in food habits and these changing foods habits demanding more variety of processed food products.

Increase in the population with advancement in technology is also a major cause for the demand of processed and preserved food to fulfill the growing need of human hunger. The quality and safety of food depend upon the specified technologies adopted during processing. Range of temperature is a major factor affecting the presence and growth of the pathogenic microbes in food and alteration in the temperature of storage (refrigeration) of raw or finished food products is the most frequent method for preservation of food. The availability of suitable means of transport of food material is another factor; decides the presence and growth of the FBP in food products. The change in the processing technologies, storage, and transportation system are the important factors in the food chain responsible for the evolution of the pathogens and gradually giving them genetic diversity and taking advantage of conquering new colonization niches. Furthermore, change in the human behavior and human expansion into new land, opening new doors for FBP.

Distribution and supply of food at global levels has formed parameters for the appearance, existence, re-emergence, and distribution of foodborne pathogens around the globe. This has given a new question to be solved, noticing, and careful handling of the health threats generated from these uninvited concerns. In the present chapter, the points of discussion are basic details of FBP, which have been found as emerging, the significance of these foodborne pathogens in food processing industry, existing, and advanced detection methods, possible strategies for food preservation to eliminate the risk of FBP.

2.2 PATHOGENS THAT CAUSE FOODBORNE ILLNESSES

2.2.1 GRAM-POSITIVE SPORE FORMING BACTERIAL PATHOGENS (Table 2.1)

TABLE 2.1 Systematic P\osition of Some Selected Gram-positive Spore Forming Bacterial Pathogens

Pathogenic Bacteria	Phylum	Class	Order	Family	Genus	Species
Bacillus	Fermicutes	Bacilli	Bacillales	Bacillaceae	*Bacillus*	*cereus*
Clostridium	Firmicutes	Clostridia	Clostridiales	Clostridiaceae	*Clostridium*	*botulinum, perfringens, tetani, sordellii*

2.2.1.1 BACILLUS spp.

Most *Bacillus spp.* are found throughout the environment, including soils, fresh, and marine water environments.

Gram-Positive, form spores, rode with motility. Spores possess appendages and pili on their surfaces. The spores are more hydrophobic than any other *Bacillus* spores [10]. These properties enable the spores to adhere to many different types of surfaces and to resist removal during cleaning and sanitation [52]. Vegetative cells optimum temperature range for growth is 5–40°C. The organism grows at pH 4.9–9.3, but the toxic effect of pH is reduced in foods at pH 4.35 [7].

Foods that are regularly contaminated with *B. cereus* resulted in diarrheic food poisoning. Such foods include meat products, soups, vegetables, puddings, sauces, milk, and milk products [52].

The pathogens commonly present in food manufacture environments due to the formation of hydrophobic adhesive endospores. The toxic factors produced are emetic (vomiting) and diarrheal toxin that may cause gastrointestinal (GI) tract diseases in humans. These results with two types of illness, the first one is caused by toxin releases at the log phase of bacteria in the contaminated food. The other type of illness is caused by the toxins released in the log phase of pathogenic cells in the human gut. Symptoms are characterized after an incubation period of approximately 8–16 hrs. Diarrheal syndrome symptoms generally persist no longer than

12–24 hrs. Emetic syndrome symptoms including primarily nausea and vomiting appear after 1–5 h incubation period, and persist for 6–24 h [34]. The rapid onset of the emetic type is characterized by nausea and vomiting while the late onset of the diarrheal type is characterized by diarrhea and abdominal discomforts [7]. Both syndromes are a result of endospores surviving the cooking process, after which germination and subsequent proliferation of vegetative cells occur at some point during storage.

2.2.1.2 CLOSTRIDIUM spp.

Clostridium species dwell in soils and also exist in the abdominal tract of animals, including humans. The thermotolerant endospores are frequently observed in soil to endure in the extreme environment [55].

It is a Gram-positive, form spores, rods in shape, obligate anaerobic (do not grow in the presence of oxygen) bacteria. It grows in anaerobic conditions, forms bottle-shaped endospores during growth [55]. *Clostidium botulinum* is a diverse group of disease-causing, able to release botulinum toxins, classified in four different categories groups, I to IV. Some other species of *Clostridium* are also responsible for producing botulinum toxin.

It is frequently observed in canned foods causing swelling in cans because of gas produced by the bacteria [35].

Different species of *Clostridium* may cause diverge indications after their infection. *C. botulinum* can cause botulism due to release of botulinum toxin in food or wounds. The symptoms of *C. perfringens* are like gas-gangrene and food-poisoning, and tetanus is caused by *C. tetani*, however the infection caused by *C. sordellii* is fatal but less common [55]. The symptoms of foodborne botulism originate from 18 to 36 hrs. after the toxins enter into the body, and persist depending on the amount of toxin ingested, from a few hours to several days. The sign and symptoms include twofold vision, hazy vision, baggy eyelid, vomiting, and intestinal cramps, indistinct speech, anxiety, strain in swallowing, dry mouth, fatigue, and constipation. Botulinum disease caused in infants is mainly due to honey or honey-like foods given to the babies. The indications of disease can be observed within 18–36 hrs. after the invasion of toxins inside the body. The symptoms start with constipation, trouble in governing the head, irritability, baggy eyelids, tiredness, difficulty of feeding and also leading to paralysis. The botulinum toxin is known as Botox and is practiced in numerous therapeutic applications including esthetic health centers, to reduce the signs of aging.

2.2.2 *GRAM-POSITIVE NON-SPORE FORMING BACTERIAL PATHOGENS* (Table 2.2)

TABLE 2.2 Systematic Position of Some Selected Gram-Positive Non-Spore Forming Bacterial Pathogens

Pathogenic Bacteria	Phylum	Class	Order	Family	Genus	Species
Staphylo-coccus	Fermicutes	Bacilli	Bacillales	Staphylococ-caceae	Staphylo-coccus	*aureus*
Listeria	Fermicutes	Bacilli	Bacillales	Listeriaceae	*Listeria*	*monocytogenes*

2.2.2.1 *STAPHYLOCOCCUS AUREUS*

Staphylococcus spp. is widespread throughout nature, commonly found on the upper layer or skin of our body, also inside the skin of other mammals and birds. Its habitat is in harmony with other microflora of the body and is often observed in the upper part of the respiratory tract [7].

Staphylococcus aureus is non-motile, gram-positive, coccus, which appears as single or in pairs or tetrads and sometimes form short chains also. It also comes into view as staphylococci (grape-like clusters) and the colonies appearing on the blood agar are round golden yellow in color [57]. It is habitually positive for catalase enzyme production and able to reduce nitrate. *Staphylococcus aureus* is a facultative anaerobe that can grows at temperatures ranging from 7 to 48°C. Some of the strains also produces enterotoxins at the temperature range 10 to 46°C but the optimum temperature for growth is 40–45°C. The bacterium grows between pH 4.5 and 9.3, with an ideal between 7.0 and 7.5, and is very tolerant to more than 10% of NaCl [7].

Staphylococcus aureus is found accompanying food that gets contaminated during preparations due to inadequate refrigeration, insufficient cooking, or non-proper heating. This organism is identified in different varieties of foodstuff like raw milk products, high salt foods, salads, bakery foods, sandwiches, sausage of pork, ground beef, salmon fish dishes, shrimps, etc. [7].

Humans are the major reservoir for *S. aureus*. Being a common cause of skin infections including boils, cellulite, and postoperative wound infections and respiratory infections, it is also responsible for food poisoning and linked with other serious infections meningitis, osteomyelitis, and boils or cyst inside genital tracts, in CVS [57]. Food can be contained by this organism via contact of skin or through sneezing or coughing [7]. Pathogenic strains often encourage infections by producing powerful protein toxins during foodborne infections. The incubation period during infection persists one to six hours, with the disease itself lasting from 30 minutes to 3 days. Food storage below 5 or above 63°C is best for reducing the ill effects.

2.2.2.2 LISTERIA MONOCYTOGENES

Listeria spp. are found in soil, leading to contamination in vegetables. Animals can be carriers for further spread. *L. monocytogenes* is a common habitat of soil, river, and sewage water silage, plants, and food materials.

Genus *Listeria* causing listeriosis, are Gram-positive, facultatively anaerobic, catalase-positive rods. They do not produce endospores. The species emerge as small rods, sometimes arranged in short chains. It is a typical psychrophilic organism able to grow at 0°C temperature and it can multiply in 0–8°C and grow in human foodstuffs. The cells are motile through flagella and usually lose motility at 37°C. They may look coccoid in direct smear and can be mistaken for streptococci. Longer cells may resemble *Corynebacteria. Listeria monocytogenes* can be distinguished on the basis of hemolytic activity on blood agar from others. It can be easily identified on the media containing sheep blood on which a small zone showing hemolysis around the colonies. For better isolation of the organism, the sample should be pre-incubated under refrigeration temperature. *Listeria* utilizes a variety of carbohydrates and produces acid but no gas has been reported [12]. *L. monocytogenes* can be growing up at very lower temperatures during storage in the refrigerator [22].

The food sources of *Listeria* may be raw meats, fresh uncooked vegetables, fruits, pasteurized or unpasteurized milk, milk products, cheeses, particularly soft-ripened cheeses, and managed foods. At the temperature of pasteurization and cooking the bacterial cells can be destroyed but the post-processing contamination due to handling and transportation of food may again contaminate the food. If the organisms enter into the food before packing, it can easily survive and grow due to its ability to sustain wide temperature variation without forming the spores.

The bacteria move around inside the host cell using host cellular machinery to survive. This organism uses a 'zipper like' mechanism to enter inside the host cells most commonly non-phagocytic cells and the main site for infection is epithelial cell lining. The infection is most prevalent in the old persons, pregnant ladies, and patients with HIV (human immunodeficiency virus) infections. This pathogen is a highly lethal pathogen among the foodborne, it is reported that around 30% of listeriosis infection resulted in death of the individual [53].

2.2.3 *GRAM-NEGATIVE BACTERIAL FOODBORNE PATHOGENS (FBP)* (Table 2.3)

2.2.3.1 CAMPYLOBACTER JEJUNI

Campylobacter genes are a common habitat of intestinal microflora of healthy hosts and other domestic and wild animals. These organisms are frequently observed in the water of troughs and streams [10]. *C. jejuni* is commonly found in animal excreta and environmental samples from groundwater and other sources commonly contain this pathogen [58].

It is a Gram-negative, does not form spores, non-fermenting, micro-aerophilic rods, motile in nature have a single polar flagellum. The cells are oxidase-positive and grow optimally at 37 to 42°C [57]. The bacteria are small ranging from 0.2–0.9 μm wide and 0.2–5.0 μm long, with 18 species, six subspecies and two biovars. They are essentially microaerophilic, growing best in an atmosphere containing approximately 10% CO_2 and approximately 5% O_2 [10].

The infection is observed with consumption of contained poultry and meat due to cross-contamination of other foods. Unpasteurized dairy products and untreated water may also be possible sources of infection. The number of contaminants may be reduced by freezing the raw food including meat. Proper heating of foods may also reduce the incidence. The symptoms of Campylobacteriosis appear more often in the summers, particularly in infants and young children.

TABLE 2.3 Systematic Position of Some Selected Gram-Negative Bacterial Foodborne Pathogens

Pathogenic Bacteria	Phylum	Class	Order	Family	Genus	Species
Campylobacter	Proteo-bacteria	Epsilonproteobacteria	Campylo-bacterales	Campylo-bacteraceae	*Campylobacter*	*jejuni*
Shigella	Proteo-bacteria	Gammaproteobacteria	Enterobaceterales	Enterobacteriaceae	*Shigella*	*boydii, dysenteriae, flexneri, and sonnei*
Cronobacter	Proteo-bacteria	Gammaproteobacteria	Enterobaceterales	Enterobacteriaceae	*Cronobacter*	*sakazakii, malonaticus, turicensis, muytjensii, dublinensis*
Escherichia	Proteo-bacteria	Gammaproteobacteria	Enterobaceterales	Enterobacteriaceae	*Escherichia*	*coli*
Salmonella	Proteo-bacteria	Gammaproteobacteria	Enterobaceterales	Enterobacteriaceae	*Salmonella*	*bongori, enterica, enterica*
Vibrio	Proteo-bacteria	Gammaproteobacteria	Vibrionales	Vibrionaceae	*Vibrio*	*cholerae, adaptatus, aerogenes*
Yersinia	Proteo-bacteria	Gammaproteobacteria	Enterobaceterales	Yersiniaceae	*Yersinia*	*enterocolitica*

Campylobacter is a common cause of diarrhea. The bacteria colonize the small and large intestines, causing inflammatory diarrhea with fever. The infection initiates with minimum 800 organisms, in the form of infective dose in the food products. The organism gets penetrated the GI mucus, using its high motility. The bacteria then adhere to the gut enterocytes and induce diarrhea by toxin release. It releases enterotoxin and cytotoxins. Symptoms develop within two to five days of infection and sickness typically lasts in seven days. The transmission to the human host takes place through infected food items and the infection commonly observed in the intestine. The transmission to humans occurs via contaminated food products. Diarrhea results from damage to the epithelial cells [10]. Increase in the concentration of immunoglobulins occurs during infection. Stools contain leukocytes and blood. The antigens may also be responsible for triggering the Guillain-Barré syndrome.

2.2.3.2 SHIGELLA spp.

Shigella spp. are most commonly observed under poor hygienic and ill-treated sanitized conditions ill sanitation and poor hygiene. Primarily it gets transferred via personal contacts causing Shigellosis, which can occur after the consumption of infected food items [7].

The organism is Gram-negative, do not form spores, cells are rod shapes and non-motile in nature and, facultative anaerobe. The growth temperature ranges between 5 to 48°C and the optimum growth occurs at 37°C. *S. sonnei* is able to tolerate lower temperatures better than the other serogroups [10]. Best growth occurs between pH 6.0 and 8.0. Some species can grow between the pH 4.8 and 9.3.

Shigella spp. most commonly found in the food items like salads, chicken, shellfish, milk, and other fresh products served with restaurants, homes, schools, sorority houses, commercial airlines, cruise ships, and military mess halls.

All *Shigella* serogroups may cause GI infections after an incubation period of 12–50 hrs. with watery diarrhea in combination with fever, fatigue, malaise, and abdominal cramps [10]. *Shigella spp.* invades the gut epithelia of the small intestine, choosing the basolateral side. This results in tissue damage, large, and painful bowel movements which are characteristic pathology of shigellosis. The species possesses four serogroups:

serogroup A: *Shigella dysenteriae*, serogroup B: *Shigella flexneri*, serogroup C: *Shigella boydii* and serogroup D: *Shigella sonnei*.

2.2.3.3 CRONOBACTER spp.

Cronobacter spp. are usually found associated with infant milk formula. The presence observed in kitchen items like spices, herbs, and other food items mainly infant formula, and dust.

The organism gives negative Gram staining, facultative anaerobe, rods are motile. They can reduce nitrate, positive for L-ornithine decarboxylation and produce acid during carbohydrate fermentation. *Cronobacter spp.* are catalase and VP positive and oxidase and MR negative [32].

Cronobacter turicensis is usually a foodborne bacterium. *C. dublinensis* is originally reported from powder milk plants. The disease may appear in both adults and infants and it is usually associated with powdered formula for infants. The bacteria may survive in the infant foods for 1–3 years [8]. Some species have also been isolated from t, soured tea, lettuce, and vegetables. The bacterial cells are resistant to osmotic pressure and can grow under low water activity and may tolerate slightly higher temperature due to encapsulation.

The infectious cells were first observed in the case of severe neonatal meningitis. The species can cause meningitis and necrotizing enterocolitis through the consumption of contaminated foods. The symptoms develop in infants, catheter-associated elders, and immune-compromised people [68].

2.2.3.4 ESCHERICHIA COLI

The common habitat of *E. coli* is the intestine of warm-blooded animals. It grows enormously in fresh fecal matter under the conditions of availability of oxygen. It persists in natural environments as; it is able to grow and survive as a biofilm in association with other microbial species with an example of cooperative existence [65].

The organism belongs to a coliform group, Gram-negative rods which are facultative anaerobe, non-spore-forming. It is capable of transferring to fermentation or anaerobic respiration in the absence of oxygen. Some strains are motile and possess peritrichous flagella. The bacteria grow optimally at the temperature range between 37 °C to 48 °C. Its competence to grow under an oxygen-depleted environment makes it capable of surviving in environments with pervasiveness of water.

Some serotypes of *E. coli* may cause fatal illness due to consumption of contaminated food [65]. Shiga toxin-producing (STEC) strain may result in fatal illness arising from this foodborne pathogen which occurs due to the consumption of raw milk, meat, and vegetables. Enterotoxigenic species most commonly cause traveler's diarrhea in developing countries due to contaminated food and drinking water [20].

The illness is self-limiting in most common cases, however can result in a dangerous effect like *Hemolytic Uremic* Syndrome in children and old persons. The young children are more vulnerable to severe symptoms however; healthy people may also be infected and affected and leads to fatal illness [20].

2.2.3.5 *SALMONELLA spp.*

These are mainly host-associated bacteria and found to persist in a bathroom setting, most commonly observed in samples from water sources and pools as cells can easily transfer to different hosts through personal contacts [77]. There are chances of transmission from the carrier animal to human host and the bacterial cells easily pass from contaminated host to food commodities.

Salmonella spp. are gram-negative non-spore former, motile rods with cell dimensions between 0.7–1 μm to 5 × 2–5 μm. The bacterium is a facultative anaerobe, capable of growing with oxygen or without oxygen [23].

The bacterial frequency can be associated with numerous foods such as processed foods, vegetables, fruits, and dry fruits, chicken, etc. Incidences of *Salmonella* spp. have been reported in smoked finfish and smoked shellfish [45].

Salmonella species are facultative intracellular pathogens and inhabit in different cells (epithelial cells, macrophages, and dendritic cells) of their hosts. *Salmonella* genes have two types of serotypes: typhoid (*S. typhi*) causing and non-typhoidal (*Sal. paratyphi)* which affects humans only, however the most frequent serotype which does not cause typhoid and this serotype causes self-limiting infection. Toxins released by *Salmonella* strains are endotoxins, and they act on vascular and nervous apparatus, which leads to an increase in permeability, disturbances in body temperature, vomiting, and diarrhea.

2.2.3.6 *VIBRIO CHOLERAE*

The bacteria are usually found in salty water (brackish). The cells of the bacterium easily inhabit shells of crabs, shellfish, and shrimps.

It is a gram-negative, comma-shaped, highly motile, halophilic bacterium with a flagellum at one cell pole and pili. It is a facultative anaerobe.

Cholera is usually spread through contaminated water. The most common sources of cholera infection are standing water and food types including seafood, raw fruits and vegetables, and grains are additional sources of infection. Physical conditions like cold storage, high pH, humidity, and if competing flora is not present the chances of *V. cholerae* survival is high in food items [4]. The cells of bacterium may be found in raw marine foods, sweet fruits, meat, and milk products [51]. The food products, salad leaves and water infected with fecal contamination [1].

The disease initiated as a proteinaceous toxin is secreted during infection which results in abundant watery diarrhea. Cholera toxin has two subunits: 5B and 1A, and the functions of 5B are attached in the epithelial lining of the intestine and 1A facilitate the toxic activity. The bacteria present in the intestinal lumen, uses fimbriae to attach to the intestinal mucosa and secretes cholerae toxin causing the symptoms. Indications of symptoms are abrupt onset of watery diarrhea with a gray and cloudy liquid, infrequent vomiting, and abdominal cramps [51]. Common symptoms associated with are dry mucous membrane, reduction in skin turgor, hypotension, hoarse voice, and failure of renal [70]. Sever infections sometimes appear as coma followed by death.

2.2.3.7 *YERSINIA ENTEROCOLITICA*

The bacterium is dispersed widely in the surroundings may be found in soil, raw milk, sewage water, seafood, and inside warm-blooded animals, etc.

These are variable in nature in gram staining, so also known as Gram variable, rods in shape, non-spore former and are usually considered facultative anaerobes. All *Yersinia spp.* possess peritrichous flagella and are motile at 22–30C, the cells lost motility at 37C during growth. *Y. pestis* is the exceptional one having no flagella and motility [7]. The genus includes 10 established species, out of which only 03 are considered pathogenic to humans or animals. Optimal growth temperature of *Y. enterocolitica* ranges between 25–30°C, however can be grown at between 0 and 45°C. Best growth occurs between pH 4.0 and 10.0, pH 7.6 being optimum for survival in alkaline conditions. Acidic environments are not good for its growth. Furthermore, *Y. enterocolitica,* able to tolerate alkaline pH and can grow in NaCl concentrations up to 5% [7].

Being a psychrotroph, infection of *Y. enterocolitica* may be a health threat in contaminated refrigerated foods, although under refrigeration temperatures, the pathogen is usually outgrown by other competing psychrotrophs [42]. Ingestion of contaminated water or food, more specifically raw or undercooked pork, is a source of foodborne infection in humans [10].

The symptoms give the impression after a few days to a week. Diarrhea with fever, vomiting, and abdominal cramps are the common indicators of the infection. *Y. enterocolitica* has surfaced as a cause of foodborne gastroenteritis in humans and other spp. Like *Y. pseudotuberculosis* mainly infect or inhabit animals, however consumption of contaminated food may also infect humans [7, 15]. Immuno-compromised persons and children under the age of 15 are most frequently get ill with extraintestinal infection related to yersiniosis [15]. The effect of disease varies with season, race, and age groups and overall differences in epidemiology by civilization [17]. The disease also shows seasonal variation and annual incidence tendencies by race, age group, and overall.

2.2.4 SOME LESS COMMON BACTERIAL PATHOGENS

2.2.4.1 *BRUCELLA* spp.

These are Gram-negative coccobacilli, which grow intracellularly as facultative pathogens. The cells are small, starting from 0.5 to 0.7 by means of 0.6 to 1.5 μm, without pill or motile appendages. Brucellosis, resulting from *Brucella* spp. are transmitted by way of consuming contaminated ingredients like unpasteurized milk products, direct contact with an inflamed animal, or through aerosols. They are intracellular inside the host organism, and can survive in surroundings outside the host. They also survive in extremes temperature, pH, and humidity, and also in ice-cold materials. The GI tract is affected causing anorexia, stomach pain, vomiting, diarrhea, constipation, hepatomegaly, and splenomegaly.

2.2.4.2 *CORYNEBACTERIUM* spp.

It is a rod-shaped, aerobic, and Gram-positive bacteria. The rods are slightly curved in shape, bacteria do not show any motility, and do not form spores and give catalase test positive. *Corynebacterium ulcerans* reported frequently in soil, contaminated water, plants, and in food commodities *C. diphtheriae* is a most common infectious agent of humans, causing diphtheria.

2.3 SOURCES OF FOODBORNE PATHOGENS (FBP) IN FOOD PROCESSING

The association between the consumption of food and human diseases was recognized very early with reports that there is a strong connection between food consumed and human illness [31]. Sources of pathogens entrance depicted in Figure 2.1. These pathogenic agents may get access into the food processing area or industry via various sources chiefly: (i) dairy animals/poultry farms; (ii) human food handlers; (iii) surrounding environment; (iv) contaminated water sources; (v) soil or dirty in fresh produces; (vi) packaging material.

The prevention of pathogenic agents in food is the most common problem of the food processing industry; it is also a major concern for the food safety departments due to various illnesses caused by the FBP.

Since the last three decades, emerging pathogens are continuously reported these tiny pathogens includes various bacterial species, viruses, parasitic organisms, endotoxins, and prions. Outbreaks through foodborne diseases, defined as the existence of similar kind of two or more illnesses emerging from the consumption of the same or common food items [9]. Different types of pathogenic agents gain access through various sources and contaminate the food, which may cause mainly three types of illness (food intoxication, food infection and toxi-infection). Not much difference observed in the appearance and other sensory parameters of contaminated and safe food (food which is safe for consumption) but contaminated may carry diverse disease-causing agents.

The onset of symptoms and the most common responsible microorganisms for the major foodborne illnesses are studied and revealed that the most severe cases have a tendency to arise in the person with weak immune functionality due to age effects, but in case of healthy individual high dose is required to cause infection [9].

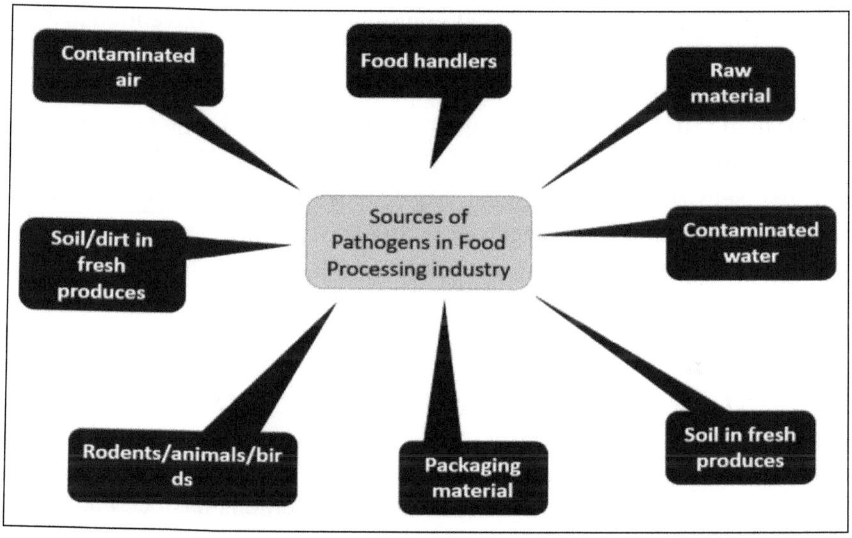

FIGURE 2.1 Various sources of pathogen in the food processing industry.

2.4 FOOD SAFETY PRACTICES BY FOOD PROCESSORS, HANDLERS, AND CONSUMERS

Food safety assurance is not only the responsibility of the food industry, it is the responsibility of links existing in our food chain (farm to dining table). The typical role of the food processing industry is to process, prepare, label, and market the best quality, safe food within the technical, moral, and financial constraints. In the latest legislations, focus is only safety along with quality, which is ensured with the implementation of hazard identification policy and other risk management measures that requires for adoption of hazard analysis critical control point (HACCP) principles. HACCP, is a tool for management and assurance of food safety that can be adopted by even pilot-scale food industries. In addition, the HACCP is implemented with some prerequisite programs (PRPs) like good hygiene and manufacturing practices (GHP and GMP).

2.5 RESEARCH ADVANCES IN FOODBORNE PATHOGEN DETECTION TECHNIQUES

Most of the laboratories of developing countries use conventional type of methods for foodborne pathogen detection. These are mainly bacterial pathogens and these are based on agar plate morphological colony identification then followed by biochemical confirmation [43]. The conventional methods are typically simple and inexpensive but quite cumbersome [82] and less sensitive may give false-negative results, which could increase the risk of transmission. Advanced methods have been developed to save the time as well as to increase the accuracy of the detection process. In this section, some of the advances in methods of detection are discussed.

2.5.1 NUCLEOTIDE SEQUENCE-BASED METHODS

Many FBP are not cultural or their handling is more risk-taking for this type of pathogens nucleotide sequence-based methods has been adopted. For the toxins of some of the toxins producing pathogens such as gram-positive *(Staphylococcus aureus, Clostridium botulinum)* gram-negative *(Escherichia coli* O157, *Vibrio cholerae*) [3, 24, 63, 64] toxin-related gene-based nucleic acid methods has been developed.

These types of methods are depending on the function of the specific gene and the detections of the specified genes are the main target to detect pathogens. Hence, accuracy of the results increases. In the present time, various methods have been adopted on the basis of genetic material like PCR (polymerase chain reaction), RT-PCR (real-time or quantitative polymerase chain reaction), nucleic acids sequence-based methods, etc.

2.5.2 BIOSENSOR BASED METHODS

The basic definition of a biosensor is a simple and easy to handle analytical device which is made up of two basic components; (1) a baroreceptor and (2) a transducer. This device gives results in the form of signal, which are proportional to the concentration of a particular analyte (a pathogen or a toxin produced by it). Biosensors are user friendly equipment, easy to operate, no sample preparation is required as needed in various detection techniques [62]. Electrochemical biosensors, mass-based biosensors, optical biosensors are the most commonly used biosensors in recent times, which are helpful in the identification of various type non-cultural pathogenic microbes [81, 82]. The most commonly used biosensors in the food industry for detection of pathogens are electrochemical type, optical type and mass-based. Some of the examples of biosensors for pathogen detection are given in Table 2.4.

2.5.3 IMMUNOLOGICAL BASED METHOD

The basic principle behind the immunological methods is antibody and antigen interactions, in that a specific antibody is bound to its particular antigen. ELISA (Enzyme-Linked Immunosorbent Assay) and LFA (lateral flow immunoassay) are most commonly used immunological methods for the detection of FBP in recent times. For ELISA various enzymatic treatments enzymes can be used and HRP (horseradish peroxidase), AP (alkaline phosphatase) and beta-galactosidase, some of the most commonly used enzymes [78].

For the identification of food pathogens more advanced ELISA systems have been developed like VIDAS from BioMerieux and assurance-EIA from BioControl) [27]. VIDAS is a kind of (VITEK immunodiagnostic assay) immuno-assay based diagnostic system, which automatically runs the entire procedure of the ELISA and this system takes only 45 m^{-2} h for completing an assay cycle. In this assay, enzyme-linked fluorescent immunoassay is used that is quite similar to the ELISA, however based on reported results, its sensitivity is quite higher as compared to ELISA. Some selected cases of the ELISA and LFA based-methods used for the identification of Gram-negative enteric pathogens (*Salmonella spp.* and *Escherichia spp.*) in different food systems are given in Table 2.5.

2.6 ADVANCED STRATEGIES FOR FOOD PRESERVATION TO ELIMINATE THE RISK OF FOODBORNE ILLNESSES

To ensure the safety of food, to enhance hygienic quality and to extend its shelf-life many protection technologies are adopted by food processing industries. Various chemicals and natural preservatives like bacteriocins, essential oils (EOs) and herbal extracts are used to delay the spoilage and to prevent the growth of pathogenic microflora [44, 48]. There are a

TABLE 2.4 Biosensors Used for the Detection of Food Pathogens

Pathogen	Food	Method for Detection	Detection Time	Limit of Detection	References
Salmonella choleraesuis serotype typhimurium, *Listeria monocytogenes*, *Campylobacter jejuni*, and *Escherichia coli* O157:H7	Artificially contaminated apple juice	Optical biosensors	Not stated	4.4×10^4 cfu//ml for *Salmonella choleraesuis* serotype typhimurium, 3.5×10^3 cfu /ml for *Listeria monocytogenes*, $1.1{\times}10^5$ cfu/mL for *Campylobacter jejuni* and $1.4{\times}10^4$ cfu / ml for *Escherichia coli* O157:H7 in PBS	[69]
Campylobacter jejuni	Artificially contaminated broiler meat		45 mins	103 cfu/mL	[75]
Escherichia coli O157:H7	Artificially contaminated cucumber and ground beef		Not stated	3 × 103 cfu/mL	[74]
Escherichia coli, *Listeria monocytogenes* and *Campylobacter jejuni*	Artificially contaminated milk and chicken extract	Electrochemical biosensors	30 mins	50 cells/mL for *Escherichia coli*, 10 cells/ml for *Listeria monocytogenes* and 50 cells/ml for *Campylobacter jejuni*	[18]
Escherichia coli O157:H7	Artificially contaminated ground beef		15 min without enrichment; 6 h after enrichment	1.6 × 101–7.23 × 107 cells/mL without enrichment and 8.0 × 100–8.0 × 101 cells/mL with enrichment	[73]
Listeria monocytogenes	Liquid milk, ground beef and lettuce		3 hrs.	103 cfu/ ml	[33]
Salmonella typhimurium	Artificially contaminated chicken meat	Mass-based biosensors	Not stated	105–106 cells/mL	[66]
Escherichia coli O157:H723 CFU	Artificially contaminated milk		4 hrs.	23 cfu/ml in phosphate buffer saline and 53 cfu/ml in liquid milk sample	[59]

variety of methods used to lessen the population of pathogenic microbes in processed food. The methods used for the protection of the food have their own benefits and limitations, which depend on the protocols used and other parameters in the process. The most commonly used strategies used for preservation of food without affecting its national value are biocontrol preservation and non-thermal preservation methods.

TABLE 2.5 Selected Applications of ELISA and LFA Based Methods for Pathogen Identification in Food Systems

Pathogen	Food	Methods for Identification	Detection Time	Limit of Detection	References
Escherichia coli O157:H7	Vegetables, milk, and ground beef samples	ELISA	3 h	68 cfu/mL in PBS and 6.8 × 103 cfu /ml in food samples	[60]
Salmonella typhi	Milk, vegetables, chicken, and meat rinses	Lateral flow immunoassay	10 h	104–105 cfu/ mL	[40]
Salmonella typhimurium	Tomato fruit sample		Not stated	30 cells /25 g	[61]

2.7 BIOPRESERVATION METHODS FOR FOODBORNE PATHOGENS (FBP)

The use of bio-preservatives in the food system is an emerging area of research. Biopreservation methods are slowly getting acceptance because of their eco-friendly or chemical-free approach. There are many reports on the use of biocontrol agents to prevent the growth of FBP in food processing.

2.7.1 BACTERIOCINS

Bacteriocins are the antimicrobial peptides (sometimes proteins) synthesized ribosomally by bacterial machinery. In the past, these were described as the colicins produced by *E. coli* and are relatively larger in size as proteins (up to 80 kDa). These bacteriocins were commonly having the potential to destroy the cells of closely related species of bacteria. They bind into the inner layer of the cell wall and other components of the cytoplasm [25].

Against Gram-positive bacterium and these bacteriocins possess a wider spectrum of inhibition [46]. Conventionally lactic acid bacteria (LAB) have GRAS (generally regarded as safe) status for their application in the food industry, hence for bio-preservation the most common bacteriocins are from LAB.

2.7.2 BACTERIOPHAGES

Phages are viruses, which have a specific target, get injected inside the bacterial cell and multiply within the cell. However, bacteriophages do not affect human cells. The commonly found bacteriophages belong to the order *Caudovirales,* they are double composed of double-stranded DNA tail and icosahedral head. They infect bacteria so named as bacteriophages they are most our earth is rich in and widespread on foods of various origins [13].

These bacteriophages are classified on the basis of their tail structure (*Myoviridae*, *Siphoviridae*, and *Podoviridae)* in three different families: Some bacteriophages are classified on the bases of their genetic material [2]. Virulent phages, in common follow the lytic life cycle whereas the temperate phages follow lysogenic life cycle (infecting the bacterial cell chromosome with phage DNA) and these phages burst the bacterial cell of host bacterium and serve as a biocontrol or antibacterial agents [29, 67].

These bacteriophages have been successfully used as anti-pathogenic agents in farms for livestock (to remove the pathogen from milking and slaughter). In chicken and ruminants, Gram-negative bacteria pathogenic *E. coli, Salmonella, and Campylobacter* can be effectively removed by the use of phages [5, 54]. In the food processing industry, phages can be used as bio sanitation agents to remove or reduce biofilms [6] and by reducing or eradication of colonization of *S. aureus* in food handlers. Bacteriophages are a potential bio-control agent in food processing industries (for prevention of biological contamination on food contact surfaces.

2.7.3 ENDOLYSINS

Endolysins are the lysines obtained from phages and also known as potential antimicrobial agents (AMAs); these lysine attacks the peptidoglycan layer of the cell wall and damage the linkage between peptidoglycan in the cell wall of the bacteria. Finally, endolysins facilitate the process of cell bursting at the end of the replication process. Endolysins can easily break the barrier of Gram-positive bacterial cells because of the absence of the outermost membrane of the cell wall, they get penetrated into peptidoglycan [80]. As endolysins are derived from phage host bacterial activity and most commonly limited to this hence, possess narrow-spectrum lytic activity. Nevertheless, there are lysins, which can attack *Streptococcus species like Strep. pyogenes, Staphylococcus aureus* and Enterococcal cell walls and are capable to lyse the bacterial cells [79]. Hence, these endolysins have potential agents for bio-preservation in food effectively.

2.7.4 NON-THERMAL PRESERVATION METHODS

2.7.4.1 HIGH-PRESSURE TECHNIQUE IN FOOD PRESERVATION

High-pressure processing is a non-thermal process for preservation of food and modifies it. In this process food is subjected for a few minutes to high hydrostatic pressure (300–700 MPa). In this process, the vegetative cells of the undesirable microbes in food are inactivated, and also the function of some major microbial enzymes is effected at room temperature. But the other valuable ingredients (vitamins, minerals, flavors, and colors, etc.), weights in the food system are not affected.

Nevertheless, the use of high-pressure processing is gradually intensifying in the area of food processing for the production of fresh and safe food products with increased shelf life as well as functional, nutrient profile. In the present scenario, more than 200 HPP systems are in use around the globe and these systems are producing > 300,000 tons of processed food annually.

Vegetative form of microbial cells can be destroyed by targeting the cell wall of the viable cell or due to food processing techniques enzyme inactivation (alter the metabolic actions) occurs in the microbes which may leads to structural changes in the cell wall of the bacteria, ultimately affect the viability of the vegetative cells of the microorganisms [30]. The viable cells of microorganisms (bacteria, yeast, and mold and virus) significantly destroy at higher pressure more than 300 MPa without alternating the temperature of surroundings.

The impact of applied processes revolves around other parameters also like the surrounding environment (acidity, water activity and other chemicals). Some of the spoilage organisms such as *Lactobacillus spp.* (especially in acidic foods) and foodborne pathogenic microbes like entero-hemorrhagic *Escherichia coli* and *Listeria monocytogenes,* commonly tolerate the high-pressure processing if we compare them with other microbes. Perhaps this is due to their high physical and chemical tolerance stressors (genetically). And also, bacterial cells have a nature of adaptability due to gradual exposure to the stress environment (prior exposure and growth history), for example, *L. monocytogenes* was found resistant at stationary phase cells to high pressure [16] and higher temperatures [14].

2.7.4.2 PULSED ELECTRIC FIELD TECHNIQUE IN FOOD PRESERVATION

Pulsed electric field is a type of non-thermal technique; this technique includes the use of high voltage pulses (for a very short time) into food placed between two electrodes. Generally, the electric field applied is several 1,000 volts per cm (20 to 1,000 ms) and it is required to remove or inactivate the microbial cells effectively. The effectiveness of pulsed electric field on microbial cells depends on the structural composition of cell wall and surface area of microbial cells [72].

Furthermore, other factors in the surrounding environment such as pH, acidity, salt, TSS, and electric potential) in the food system also affect the antimicrobial or decontamination efficacy of the pulsed electric field processing such as penetration of microbial cell membrane and opening forced for the protein network within in the cell wall. Due to application of electric fields, the porosity of microbial cell membranes gets disturb because of change in the structural arrangement of the cell wall phospholipids, ultimate results are destruction of microbial cells.

2.7.4.3 COLD PLASMA TECHNIQUE IN FOOD PRESERVATION

Cool plasma in a non-thermal food safety technology used in food processing industries, it is nothing but an ionized gas state, which is used for the inactivation of the microorganisms in the food system. This is generated with treatment gas or liquid with a high-power source in this case partial ionization of the substance occurs in the temporarily excited state. Cold plasma technology is easily adopted at large scale food processing industry as it does not require additional air pressure management equipment, because of this reason its acceptance is widely increasing. In the food processing area, nitrogen, oxygen, or air is normally utilized and the main application of the cold plasma processing is extension of shelf life of the food products by decontamination food contact surfaces. In addition, it is an effective process for the removal or inactivation of microbial cells, bacterial spores on food contact surfaces like stainless steel, glass, and plastic surfaces [36].

It is not an easy task to apply cold plasma techniques in the food system, food matrices are complex surfaces and it becomes difficult to penetrate cold plasma into the food. In addition, in the case of meat products, destruction of pathogenic microbes has been observed [19, 49]. The impact on sensitivity of cold plasma treatment due to the influence of various factors (microbe's growth, microbes load, biofilm formation and other processing parameters) still need to be studied.

2.7.4.4 ULTRAVIOLET (UV) LIGHT IN FOOD PRESERVATION

Ultraviolet (UV) light is commonly applied in the food-processing sector as it has a number of applications. The UV (200–310 nm) light is mainly used as a disinfection in the food processing industry, most commonly used in the packaging area. UV treatment is similar to cold plasma; a specific target is the genetic material of the cell by interfering in the process of DNA replication and damaging the microbial cell. The effectiveness of the UV treatment varies from microbes to microbes, as some microbial cell has mechanisms to repair UV damage. Compared with other microbes, the most sensitive bacteria to UV treatment are gram-negative ones. Treatment, but in the case of vegetative cells bacteria either *E. coli K-12* and *Listeria innocua*, effect of treatment is not considerable [26]. However, other microbes like algae and protozoa, probably possess well developed DNA damage systems as they are not sensitive to UV treatment [39].

The efficiency of the UV light treatment mainly depends on the composition of the food or food matrix, etc., (of the food organization because of that penetration power of the UV light decided. However, the UV light has some limitations as it can degrade the quality of the food, during UV light processing photochemical reactions occur. In addition, the nutritional value of the light-sensitive compounds or nutrients also affects.

2.8 OTHER FOODBORNE PATHOGENIC AGENTS

2.8.1 FOODBORNE PATHOGENIC PRIONS

The word Prion derives from 'proteinaceous infectious particles, which are misfolded to spread their shape onto normal alternatives of the same protein. Unlike all other known disease-causing agents, Prions do not have nucleic acid, i.e., DNA or RNA. Apart from this, prions are able to cause hereditary infectious and other sporadic forms of diseases. Prion proteins can act as infectious agents, when transmitted to another organism, or they may arise due to inherited mutation.

Prion diseases appear in the population at random without any fundamental molecular process of the origin of the prion protein. Consumption of contaminated meat may cause disease in human beings. The signs of prion diseases include: dementia, difficulty walking, illusions or confusion, muscle stiffness, fatigue, and difficulty in speaking [50, 76]. Various types of diseases caused by prions like Kuru, Fatal insomnia, Creutzfeldt Jakob disease, Mad cow disease (a variant form of CDJ), and Gerstmann-Straussler Scheinker disease.

2.8.2 FOODBORNE VIRUSES

2.8.2.1 HEPATITIS A (HAV)

Hepatovirus A is a naked, picornavirus. It holds a single-stranded RNA in a protein coat [28]. It has only one serotype with multiple genotypes [7]. The virus is stable under a wide variety of environmental conditions such as very low, very high, with strong chemicals and extreme dryness. They are considered to be transmitted by infected water, food items, table surfaces, utensils, and through personal contact with infected people [10]. Humans and vertebrates are the accepted hosts. The virus transmits by the fecal and oral routes under poor sanitation and overcrowding conditions. Improperly cooked Shellfish are a fairly common source of infection [10].

The virus can be commonly transmitted through foodborne outbreaks and incorporation through polluted water. They enter the host cell through receptor-mediated endocytosis and invade the bloodstream [38]. It targets the liver, multiplies within hepatocytes and liver macrophages (Kupffer cells). It follows cytoplasmic replication and exits the host cell by lysis.

Initially, the infection appears with no or few symptoms in young ones, but between two to six weeks, it develops with the symptoms like nausea, vomiting, diarrhea, jaundice, fever, fatigue, appetite loss, dark amber urine, diarrhea, clay-colored feces, and abdominal discomfort [28, 38]. Symptoms may be recurrent to six months after the initial infection. Serious liver failure is more common in the adults [28].

2.8.2.2 ENTEROVIRUS

The *Enteroviruses* are members of the picornavirus family containing +ve-sense ssRNA (single-stranded ribonucleic acid) and it contains around 7,500 bases in its genetic sequence. Because of slow replication and high recombination rate, the virus has been reported as a high mutational rate variant [71]. The genus includes 15 species namely, Enterovirus A to L and Rhinovirus A to C which are associated with several human and mammalian diseases causing intestinal infections (enteric meaning intestinal) [28]. The virion spread through the oral and fecal routes and infection may show a variety of symptoms such as common cold, meningitis, acute conjunctivitis, myocarditis, sepsis, and paralysis [28]. This infection may also cause rashes in small children, encephalitis, blurred vision, and pericarditis in adults [11].

2.8.2.3 NOROVIRUSES (HNoV)

Noroviruses (HNoV) are a single-stranded positive-sense RNA virus with icosahedral geometries. The virion is non-enveloped, belonging to the family Caliciviridae [7, 11]. It is the most genetically diverse group with a wide-ranging size of capsid diameters (23 to 40 nm). The RNA genome is non-segmented, linear, and of approximately 7.5 kbp size. The virus enters the host by attachment to host receptors followed by endocytosis. The virus goes after Positive stranded RNA virus transcription in cytoplasmic places. Sources include Shellfish and salad ingredients, fresh food produce, fruit, berries, green onions, ready-to-eat foods, sandwiches, and cookies handled by pre-infected persons [10].

The virions are highly contagious to initiate infection of fewer than 20 virus particles [11]. HNoV infections arise all year round in people of all ages [71]. The viral infection is characterized by diarrhea, vomiting, with no occurrence of blood, but with complications of dehydration. Fever or headaches and stomach pain, loss of taste may also appear in some cases [37].

2.8.2.4 ROTAVIRUS (HRV)

Rotavirus is an RNA virus of the family Reoviridae (ds RNA virus with icosahedral capsid protein). It is a non-enveloped virus and maximally 76.5 nm in diameter [71]. The genus comprises nine species (referred as A to I) with 11 unique molecules of RNA (dsRNA) genome made up of 18,555 nucleotides in total. In each helix, 1 to 11 numbered genes are arranged by reducing in size [10]. *Rotavirus A* is the most common species among all, causing > 90% of human infections.

Diarrhea is the most common symptom of Rotavirus infection usually appears in infants and young children [28, 37]. The infection damages the cells of small intestine linings and results in gastroenteritis (stomach flu). Some symptoms leading to nausea, vomiting, diarrhea, fever, and dehydration [71]. The symptoms start by four to eight days of infection. Immunity develops with subsequent infections, resulting in less severe symptoms in adults [10].

2.8.3 FOODBORNE PARASITES

2.8.3.1 TOXOPLASMA GONDII

It is a parasite member of protozoan Phylum-Apicomplexa. It is an obligate pathogen growing intracellularly causing toxoplasmosis in humans. The cats are the primary reservoir of pathogens and another warm-blooded animal as an intermediate host [10]. The protozoan may be present as three stages of its life cycle, tachyzoites, bradyzoites, or sporozoites. The motile tachyzoites are responsible for increasing the population and quickly multiply in the host. They flourish in the body tissues and destroy infected host cells. Bradyzoites are the slow dividing phase of the parasite making tissue cysts. Sporozoites are excreted, in excreta within oocysts, sporulate for further 1–5 days and survive for months based on their ability to resist the adverse conditions like disinfectants, low, and high temperatures [10].

After getting the new host, the Sporozoites transform into tachyzoites and begin to multiply rapidly within the cells leading to cell death. The tachyzoites invade adjacent cells and repeat the reproduction process. The tachyzoites may be transformed back into bradyzoites by means of the host immune response, and form cysts in the local tissue, to remain in the host organism [10].

This parasite can enter into host in several possible ways, including contaminated meat, plant originated food or water containing the oocyst, contaminated blood transfusion or organ transplantation, or accidental tachyzoite inoculation. Serious health problems may arise in individuals with high risk for food poisoning [7, 56]. Symptoms of toxoplasmosis are similar to flu, including fever, headache, muscle aches and pain, etc. This infection also affects the eyes, causing reduced or blurred vision, pain, and redness or tearing [10].

2.8.3.2 CRYPTOSPORIDIUM

Cryptosporidium is also a genus of apicomplexan phylum. It is a microscopic parasite that causes a respiratory and GI illness (cryptosporidiosis) with watery diarrhea. The disease also called intestinal cryptosporidiosis, with or without a persistent cough, respiratory cryptosporidiosis in humans with both normal and deficient immune responses [7]. The infection begins with drinking contaminated water and swimming in contaminated water. Consumption of contaminated or uncooked food which may contains Cryptosporidia pathogen and contact of contaminated hand to mouth may also originate disease [56]. Transmission by ingestion or inhalation of coughed maybe the second route of infection [7]. However, Cryptosporidiosis is an acute, short-term infection, but can be recurrent in the same host. Some strains can cause more serious illness.

2.8.3.3 SARCOCYSTIS spp.

Sarcocystis spp. is a parasite of the apicomplexan family with heteroxenous (more than one obligatory host) habit. The mature oocysts of the pathogen released out in the feces, divides asexually for the formation of two sporocysts and again releases the sporocysts after lysis. Sporocysts characteristically contain four sporozoites with a size range of 15–19 by 8–10 μm. Cattle or pigs act as intermediate hosts, to receive sporocysts and the particular stage continues about 15–16 d till conversion of merozoites.

Succeeding generations of merozoites enters the blood flow and reach throughout the body and finally reach to muscles final asexual generation. Merozoites rounded up to form metrocytes and start sarcocyst formation. Sarcocystis grow as unicellular bodies, undergo asexual multiplication, and develop into numerous metrocytes.

The non-infectious metrocytes soon converted to crescent-shaped infectious bodies called bradyzoites get ready for secondary infection. The symptoms appeared 3–6 hours after eating infected beef. The symptoms of the first series of infection include nausea, abdominal pain, vomiting, and tachycardia. However, such symptoms are short-lived. The second series the symptoms appear as abdominal pain, distension, and watery diarrhea which disappear after 3 weeks.

2.8.3.4 CYCLOSPORA CAYETANENSIS

Cyclosporra cayetanensis are protozoan parasites, belonging to the family Eimeriidae, that inhabit the small intestine of humans. They spend the intermediate life cycle stages in the cytoplasm of enterocytes and subsequently produce oocysts containing two sporocysts encapsulating four sporozoites [28]. After subsequent shedding of the oocysts, 7–15 days are required for sporulation to occur. *C. cayetanensis* is capable of causing prolonged illness (6 weeks or longer) in both healthy and having weak immunity individuals [71]. The characteristic symptoms include non-bloody diarrhea, nausea, vomiting, anorexia, bloating, abdominal cramping, malaise, fever, and fatigue [28].

2.8.3.5 GIARDIA

Giardia genus belongs to protozoan flagellated microscopic parasites of the phylum metamonada. It is a diplomonad, containing two nuclei, with four associated flagella. *G. lamblia* are found in animal and human feces [47].

These parasites also survive in contaminated food, water, and soil, and also grow outside a host for long periods of time. Infection commences with accidental consumption of these parasites. The most common way is drinking water that contains infectious cysts. The sources of contamination include animal feces, diapers, and agricultural runoff, which lead to contaminate water in swimming pools, spas, and lakes. Foodborne giardiasis may appear due to the use of poor-quality water used for irrigation and for washing of fruits and vegetables. Food handlers may also transmit infection through prepared foods [21].

2.8.3.6 TRICHINELLA SPIRALIS

Trichinella spiralis is a roundworm, belonging to the phylum Nematoda, it causes trichinosis infection in humans. The average sizes of worms are 1.4 to 1.8 mm and they mate and are found in the host epithelial lining of the small intestine [28].

Adult worms may also live in the duodenal and jejunal mucosal epithelium, for up to 8 weeks. During this period, adult females release large numbers of larvae into the bloodstream, reaching to muscle fiber for the formation of viable cysts, before they are excluded [10]. The larvae develop, mature, and undergo encapsulation in skeletal muscle. Encysted larvae survive and stay alive around 10 years, and are fed by stomach enzymes. Symptoms of the disease appear after 3–14 days of incubation due to the migration and cyst formation of larvae. Nonspecific gastroenteritis, nausea, vomiting, headaches, fever, visual deficiencies, difficulty breathing, chills, night sweating and eosinophilia may also include in common symptoms [28]. The worms can be inactivated by heat treatments of cooking products at high temperatures.

2.8.3.7 ASCARIS LUMBRICOIDES

Ascaris lumbricoides is a common intestinal roundworm commonly found in areas without satisfactory hygienic conditions. Humans are the only known hosts. The infection arises in the very early age of babies within months after birth resulting in stunting growth and diarrheal symptoms. The worms cause irritation in the epithelial lining and also alter the absorption of other components (fats and proteins) after digestion. Many infected adults do not show symptoms [21]. Feces contaminated with eggs may be transferred to another person through drinking water, eating with dirty hands, or through uncooked vegetables. After incorporation, the eggs hatch in the intestine and migrate to other organs of the body. They return to the intestine after attaining maturity and to repeat the cycle [56].

2.8.3.8 *FASCIOLA HEPATICA*

Fasciola hepatica, also known as Liver fluke, is a parasite of particularly domesticated ruminants. Human cases usually occur in parts where poor sanitary conditions prevail and contaminated water is available. Major reports are available from South America, North Africa and the Middle East In some areas of Bolivia [21]. Humans are occasionally infected through contaminated salad greens.

2.9 SUMMARY

With globalization, the lifestyle is getting changed over time and this change leading to the change in food habits and these changing food habits demanding more variety of processed food products. Increase in the population with advancement in technology is also a major cause for the demand for processing of food and its preservation to fulfill the increasing need of human hunger. These processing techniques raise the concern for safe or pathogen-free foods. Along with the growth of the food processing sector, the chances of emergence, re-emergence, and occurrence of FBP also increases. However, technologies have been developed to preserve the food during processing, and for detection of post-processing contamination, many new technologies such as biosensors, nucleic acid-based methods, and advanced immunological methods have been developed. In addition, FBP are always a matter of concern in the area of food safety, because their nature is quite versatile.

KEYWORDS

- **detection methods**
- **essential oils**
- **food processing**
- **foodborne pathogens**
- **hazard analysis critical control point**
- **preservation**
- **ultraviolet**

REFERENCES

1. Abigail, C., Joanna, C. Y., Nicholas, C., & Gad, F., (2012). Infection strategies of enteric pathogenic *Escherichia coli*. *Gut Microbes, 3*(2), 71–87.
2. Ackermann, H., (2007). 5500 phages examined in the electron microscope. *Archives of Virology,* 152, 227–243.
3. Akbulut, D., Grant, K. A., & McLauchlin, J., (2004). Development and application of real-time PCR assays to detect fragments of the *Clostridium botulinum* types A, B, and eneurotoxin genes for investigation of human foodborne and infant botulism. *Foodborne Pathogens and Diseases, 1,* 247–257.
4. Albert, M. J., Neira, M., & Motarjemi, Y., (1997). The role of food in the epidemiology of cholera. *World Health Statistics Quarterly, 50*(1, 2), 111–118.
5. Atterbury, R. J., Dillon, E., Swift, C., & Connerton, P. L., (2005). Correlation of *Campylobacter* bacteriophage with reduced presence of hosts in broiler chicken ceca. *Applied and Environmental Microbiology, 71*, 4885–4887.
6. Azeredo, J., & Sutherland, I., (2008). The use of phages for the removal of infectious biofilms. *Current Opinion in Biotechnology, 9*, 261–266.
7. Bacon, R. T., & Sofos, J. N., (2003). Characteristics of biological hazards in foods. Chapter 1. In: Schmidt, R. H., & Rodrick, G. E., (eds.), *Food Safety Handbook* (pp. 157–195). New Jersey: John Wiley & Sons, Inc.
8. Barron, J. C., & Forsythe, S. J., (2007). Dry stress and survival time of *Enterobacter sakazakii* and other *Enterobacteriaceae* in dehydrated powdered infant formula. *Journal of Food Protection, 70*, 2111–2117.
9. Behravesh, C. B., Williams, I. T., & Tauxe, R. V., (2012). Emerging foodborne pathogens and problems: Expanding prevention efforts before slaughter or harvest. In: *Improving Food Safety Through a One Health Approach - Workshop Summary*. Institute of Medicine (US); Washington (DC): National Academies Press, A14.
10. Bintsis, T., (2017). Foodborne pathogens. *AIMS Microbiology, 3*(3), 529–563.
11. Bosch, A., Pinto, R. M., & Guix, S., (2016). Foodborne viruses. *Current Opinion in Food Science, 8*, 110–119.
12. Brooks, G., Carroll, K. C., Butel, J., & Morse, S., (2007). *Jawetz, Melnick & Adelberg's Medical Microbiology* (24th edn., p. 832). New York - USA: The McGraw-Hill.
13. Brüssow, H., & Kutter, E., (2005). Phage ecology. Chapter 6. In: Kutter, E., & Sulakvelidze, A., (eds.), *Bacteriophages: Biology and Applications* (pp 129–163). Boca Raton - FL: CRC Press.
14. Bull, M. K., (2005). Effect of prior growth temperature, type of enrichment medium, and temperature and time of storage on recovery of *Listeria monocytogenes* following high pressure processing of milk. *International Journal of Food Microbiology, 101*, 53–61.
15. Cary, J. W., Linz, J. E., & Bhatnagar, D., (2000). *Microbial Foodborne Diseases: Mechanisms of Pathogenesis and Toxin Synthesis* (p. 573). Lancaster - PA: Technomic Publishing Co, Inc.
16. Casadei, M. A., (2002). Role of membrane fluidity in pressure resistance of *Escherichia coli* NCTC 8164. *Applied and Environmental Microbiology, 68*, 5965–5972.
17. Chakraborty, A., Komatsu, K., & Roberts, M., (2015). The descriptive epidemiology of yersiniosis: A multistate study 2005–2011. *Public Health Reports, 130*(3), 269–277.

18. Chemburu, S., Wilkins, E., & Abdel-Hamid, I., (2005). Detection of pathogenic bacteria in food samples using highly-dispersed carbon particles. *Biosensors and Bioelectronics, 21*, 491–499.
19. Critzer, F. J., (2007). Atmospheric plasma inactivation of foodborne pathogens on fresh produce surfaces. *Journal of Food Protection, 70*, 2290–2296.
20. Croxen, M. A., Law, R. J., & Scholz, R., (2013). Recent advances in understanding enteric pathogenic *Escherichia coli*. *Clinical Microbiology Reviews, 26*(4), 822–880.
21. Doyle, M. E., (2003). *Foodborne Parasites: Review of the Scientific Literature Review* (p. 32). https://fri.wisc.edu/files/Briefs_File/parasites.pdf (accessed on 21 September 2021).
22. Dykes, G. A., & Dworaczek, M., (2002). Influence of interactions between temperature, ferric ammonium citrate and glycine betaine on the growth of *Listeria monocytogenes* in a defined medium. *Letters in Applied Microbiology, 35*(6), 538–542.
23. Fabrega, A., & Vila, J., (2013). *Salmonella enterica* serovar typhimurium skills to succeed in the host: Virulence and regulation. *Clinical Microbiology Reviews, 26*(2), 308–341.
24. Fusco, V., Quero, G. M., Morea, M., Blaiotta, G., & Visconti, A., (2011). Rapid and reliable identification of *Staphylococcus aureus* harboring the enterotoxin gene cluster (*egc*) and quantitative detection in raw milk by real-time PCR. *International Journal of Food Microbiology, 144*, 528–537.
25. García-Bayona, L., Guo, S. M., & Laub, T. M., (2017). Contact-dependent killing by *Caulobacter crescentus* via cell surface-associated, glycine zipper proteins. *eLife, 6*. Article ID: e24869.
26. Geveke, D. J., (2005). UV inactivation of bacteria in apple cider. *Journal of Food Protection, 68*, 1739–1742.
27. Glynn, B., Lahiff, S., & Wernecke, M., (2006). Current and emerging molecular diagnostic technologies applicable to bacterial food safety. *Internal Journal for Dairy Technology, 59*, 126–139.
28. Gordon, B., (2019). *Most Common Foodborne Pathogens*. https://www.eatright.org/homefoodsafety/safety-tips/food-poisoning/most-common-foodborne-pathogens (accessed on 21 September 2021).
29. Hanlon, G., (2007). Bacteriophages: An appraisal of their role in the treatment of bacterial infections. *International Journal of Antimicrobial Agents, 30*, 118–128.
30. Heinz, V., & Buckow, R., (2010). Food preservation by high pressure. *Journal of Consumer Protection and Food Safety, 5*, 73–81.
31. Hutt, P. B., & Hutt, P. B., (1984). History of government regulation of adulteration and misbranding of food. *Food and Drug Law Journal, 39*, 2–73.
32. Iversen, C., Lehner, A., & Mullane, N., (2007). The taxonomy of *Enterobacter sakazakii*: proposal of a new genus *Cronobacter* gen. nov. and descriptions of *Cronobacter sakazakii* comb. nov. *Cronobacter sakazakii* subsp. *sakazakii*, comb. nov., *Cronobacter sakazakii* subsp. *malonaticus* subsp. nov., *Cronobacter turicensis* sp. nov., *Cronobacter muytjensii* sp. nov., *Cronobacter dublinensis* sp. nov. and *Cronobacter genomospecies*. *BMC Evolutionary Biology, 17*(7), 64–73.
33. Kanayeva, D. A., Wang, R., & Rhoads, D., (2012). Efficient separation and sensitive detection of *Listeria monocytogenes* using an impedance Immuno sensor based on magnetic nanoparticles, microfluidic chip, and an inter-digit at edmicro electrode. *Journal of Food Protection, 75*, 1951–1959.

34. Kathleen, R. T., & Mikolajcik, E. M., (1987). Characteristics of selected strains of *Bacillus cereus*. *Journal of Food Protection, 50*, (3), 199–205.
35. Keith, G. M., & James, G. F., (2019). Bacterial diseases. Chapter 16. In: Marini, R. P., Wachtman, L. M., Tardiff, S. D., Mansfield, K., & Fox, J. G., (eds.), *The Common Marmoset in Captivity and Biomedical Research* (pp. 265–287). New York - USA: Academic Press.
36. Klämpfl, T. G., (2012). Cold atmospheric air plasma sterilization against spores and other microorganisms of clinical interest. *Applied and Environmental Microbiology, 78*, 5077–5082.
37. Koopmans, M., & Duizer, E., (2004). Foodborne viruses: An emerging problem. *International Journal of Food Microbiology, 90*(1), 23–41.
38. Koopmans, M., & Vonbonsdorff, C. H., (2002). Foodborne viruses. *FEMS Microbiology Reviews, 26*(2), 187–205.
39. Koutchma, T., (2009). Advances in ultraviolet light technology for non-thermal processing of liquid foods. *Food Bioprocess Technology, 2*, 138–155.
40. Kumar, S., Balakrishna, K., & Batra, H. V., (2008). Enrichment - ELISA for detection of *Salmonella typhi* from food and water samples. *Biomedical and Environmental Sciences, 21*, 137–143.
41. Lee, H. J., (2011). Inactivation of *Listeria monocytogenes* on agar and processed meat surfaces by atmospheric pressure plasma jets. *Food Microbiology, 28*, 1468–1471.
42. Longenberger, A. H., & Gronostaj, M. P., (2011). *Yersinia enterocolitica* infections associated with improperly pasteurized milk products: Southwest Pennsylvania. *Epidemiology and Infection, 142*(8), 1640–1650.
43. Mandal, P. K., & Biswas, A. K., (2011). Methods for rapid detection of foodborne pathogens: An overview. *American Journal of Food Technology, 6*, 87–102.
44. Martínez, B., García, P., & Rodríguez, A., (2019). Swapping the roles of bacteriocins and bacteriophages in food biotechnology. *Current Opinion in Biotechnology, 56*, 1 6.
45. Maxine, L. H., & Janelle, M. J., (1998). The incidence of *Listeria* spp., *Salmonella* spp., and *Clostridium botulinum* in smoked fish and shellfish. *Journal of Food Protection, 61*(3), 318–323.
46. Mokoena, M. P., (2017). Lactic acid bacteria and their bacteriocins: Classification, biosynthesis and applications against uropathogens: Mini-review. *Molecules, 22*, 1255–1260.
47. Murrell, K. D., (1995). Foodborne parasites. *International Journal of Environmental Health Research, 5*(1), 63–85.
48. Nazari, M., & Ghanbarzadeh, B., (2019). Garlic Essential oil nanophytosomes as a natural food preservatives: Its application in yogurt as food model. *Colloid Interface Science Communications, 30*. Article ID: 100176.
49. Noriega, E., Shama, G., Laca, A., Diaza, M., & Kong, M. G., (2011). Cold atmospheric gas plasma disinfection of chicken meat and chicken skin contaminated with *Listeria innocua*. *Food Microbiology, 28*(7), 1293–1300.
50. Prusiner, S., (1998). Prions. *Proceedings of the National Academy of Sciences of the United States of America, 95*(23), 13363–13383.
51. Rabbani, G. H., & Greenough, W. B., (1999). Food as a vehicle of transmission of cholera. *Journal of Diarrheal Diseases Research, 17*(1), 1–9.

52. Rajkowski, K. T., & Smith, J. L., (2001). Update: Food poisoning and other diseases induced by *Bacillus cereus*. Chapter 4. In: Hui, Y. H., Pierson, M. D., & Gorham, J. R., (eds.), *Foodborne Disease Handbook* (pp. 61–76). New York: Markel Dekker, Inc.
53. Ramaswamy, V., & Cresence, V. M., (2007). Listeria: Review of epidemiology and pathogenesis. *Journal of Microbiology, Immunology and Infection, 40*(1), 4–13.
54. Raya, R. R., Varey, P., & Oot, R., (2006). Isolation and characterization of a new T-even bacteriophage, CEV1 and determination of its potential to reduce *Escherichia coli* O157:H7 levels in sheep. *Applied and Environmental Microbiology, 72*, 6405–6410.
55. Riley, V., (2012). Clostridium: Gas gangrene; tetanus; food poisoning; pseudomembranous colitis. Chapter 22. In: Greenwood, D., Barer, M., Slack, R., & Irving, W., (eds.), *Medical Microbiology* (18th edn., pp. 245–255). Churchill Livingstone: Elsevier.
56. Robertson, L. J., & Van, D. G. J. W. B., (2013). Have foodborne parasites finally become a global concern? *Trends in Parasitology, 29*(3), 101–103.
57. Ryan, K. J., & Ray, C. G., (2004). *Sherris Medical Microbiology: An Introduction to Infectious Diseases* (4th edn., p. 997). New York - USA: The McGraw-Hill Co., Inc.
58. Schaffner, N., Zumstein, J., & Parriaux, A., (2004). Factors influencing the bacteriological water quality in mountainous surface and groundwaters. *Acta Hydrochimica et Hydrobiologica, 32*, 225–234.
59. Shen, Z. Q., Wang, J. F., Qiu, Z. G., Jin, M., Wang, X. W., & Chen, Z. L., (2011). QCM immune sensor detection of *Escherichia coli* O157:H7 based on beacon immune magnetic nanoparticles and catalytic growth of colloidal gold. *Biosensors and Bioelectronics, 26*, 3376–3381.
60. Shen, Z., Hou, N., Jin, M., Qiu, Z., Wang, J., & Zhang, B. A., (2014). novel enzyme-linked immune sorbent assay for detection of *Escherichia coli* O157:H7 using immune magnetic and beacon gold nanoparticles. *Gut Pathogens, 6*, 14–20.
61. Shukla, S., Leem, H., Lee, J. S., & Kim, M., (2014). Immuno Chromatographic strip assay for the rapid and sensitive detection of *Salmonella* typhimurium in artificially contaminated tomato samples. *Canadian Journal of Microbiology, 60*, 399–406.
62. Singh, A., Poshtiban, S., & Evoy, S., (2013). Recent advances in bacteriophage-based biosensors for foodborne pathogen detection. *Sensors, 13*, 1763–1786.
63. Singh, D. V., Matte, M. H., Matte, G. R., & Jiang, S., (2001). Molecular analysis of *Vibrio cholerae* O1, O139, non-O1, and non-O139 strains: Clonal relationships between clinical and environmental isolates. *Applied and Environmental Microbiology, 67*, 910–921.
64. Son, I., Binet, R., Maounounen, L. A., Lin, A., Hammack, T. S., & Kase, J. A., (2016). Detection of five shigatoxin-producing *Escherichia coli* genes with multiplex PCR. *Food Microbiology, 40*, 31–40.
65. Steven, L. P., Percival, S. L., & Chalmers, R., (2014). *Microbiology of Waterborne Diseases* (2nd edn., p. 696). New York: USA: Academic Press.
66. Su, X. L., & Li, Y., (2005). AQCM immune sensor for *Salmonella* detection with simultaneous measurements of resonant frequency and emotional resistance. *Biosensors and Bioelectronics, 21*, 840–848.
67. Sulakvelidze, A., & Kutter, E., (2005). Bacteriophage therapy in humans. Chapter 14. In: Kutter, E., & Sulakvelidze, A., (eds.), *Bacteriophages: Biology and Applications* (pp. 381–436). Boca Raton - FL: CRC Press.

68. Tall, B. D., Grim, C. J., & Franco, A., (2013). *Cronobacter* Species (formerly *Enterobacter sakazakii*). Chapter 16. In: Morris, J. G., & Potter, M., (eds.), *Foodborne Infections and Intoxications* (4th edn., pp. 251–258). New York - USA: Academic Press.
69. Taylor, A. D., Ladd, J., Yu, Q., Chen, S., Homola, J., & Jiang, S., (2006). Quantitative and simultaneous detection of four foodborne bacterial pathogens with a multi-channel SPR sensor. *Biosensors and Bioelectronics, 22*, 752–758.
70. Tirumale, S., & Tessy, A. T., (2018). Pathogenesis of cholera: Recent prospective in rapid detection and prevention of cholera. Chapter 7. In: Kırmusaoğlu, S., (ed.), *Bacterial Pathogenesis and Antibacterial Control* (pp. 129–144). IntechOpen, online.
71. Todd, E. C. D., & Grieg, J. D., (2015). Viruses of foodborne origin: A review. *Virus Adaptation and Treatment, 7*, 25- 45.
72. Toepfl, S., (2006). Applications of pulsed electric fields technology for the food industry. Chapter 7. In: Raso, J., & Heinz, V., (eds.), *Pulsed Electric Fields Technology for the Food Industry- Fundamentals and Applications* (pp. 197–222). New York-USA: Springer.
73. Varshney, M., Yang, L., Su, X. L., & Li, Y., (2005). Magnetic nanoparticle-antibody conjugates for the separation of *Escherichia coli* O157:H7 in ground beef. *Journal of Food Protection, 68*, 1804–1811.
74. Wang, Y., Ye, Z., Si, C., & Ying, Y., (2013). Monitoring of *Escherichia coli* O157:H7 in food samples using lectin based surface plasmon resonance biosensor. *Food Chemistry, 36*, 1303–1308.
75. Wei, D., Oyarzabal, O. A., & Huang, T. S., (2007). Development of a surface plasma on resonance biosensor for the identification of *Campylobacter jejuni. Journal of Microbiological Methods, 69*, 78–85.
76. Weissmann, C., Enari, M., Klöhn, P. C., Rossi, D., & Flechsig, E., (2002). Transmission of prions. *Proceedings of the National Academy of Sciences of the United States of America, 99*(4), 16378–16383.
77. Winfield, M., & Eduardo, G., (2003). Role of nonhost environments in the lifestyles of Salmonella and *Escherichia coli. Applied and Environmental Microbiology, 69*(7), 3687–3694.
78. Yeni, F., Acar, S., Polat, O. G., & Soyer, Y. H., (2014). Rapid and standardized methods for detection of foodborne pathogens and mycotoxins on fresh produce. *Food Control, 40*, 359–367.
79. Yoong, P., Schuch, R., Nelson, D., & Fischetti, V., (2004). Identification of a broadly active phage lytic enzyme with lethal activity against antibiotic-resistant *Enterococcus faecalis* and *Enterococcus faecium. Journal Bacteriology, 186*, 4808–4812.
80. Young, R. Y., Wang, I. N., & Roof, W., (2000). Phages will out: Strategies of host cell lysis. *Trends in Microbiology, 8*, 120–128.
81. Zhang, G., (2013). Foodborne pathogenic bacteria detection: An evaluation of current and developing methods. *Educator, 1*, 15–19.
82. Zhao, X., Lin, C. W., Wang, J., & Oh, D. H., (2014). Advances in rapid detection methods for foodborne pathogens. *Journal of Microbiology Biotechnology, 24*, 297–312.

CHAPTER 3

RAPID ISOTHERMAL NUCLEIC ACID DEPENDENT TECHNIQUES FOR ON-SITE DIAGNOSIS OF FOODBORNE PATHOGENS

SAURABH KADYAN, GAURAV K. DESHWAL, RASHMI H. MALLAPPA, and DIWAS PRADHAN

ABSTRACT

There is a growing demand for quick and precise diagnosis of foodborne pathogens (FBP) in order to control and avert human foodborne illness. Nucleic acid-based technique (NABT) offers sensitive detection by rapidly amplifying the target gene to millions of copies. Though the conventional NABT such as PCR, qPCR for specific detection of FBP are comparatively rapid but have obvious drawbacks of requiring trained personnel and expensive instruments, impeding their application in on-site and poor resource settings. Related NABT based on isothermal amplification such as LAMP (loop-mediated isothermal amplification), NABSA, RCA, etc., coupled with compatible sample preparation to detection steps and fully integrated with a microfluidic platform can offer a swift, low-cost, and portable solution for foodborne pathogen detection with minimal infrastructure and manpower requirements. This chapter discusses isothermal NABT along with simple sample pre-treatment and colorimetric detection steps and the role of microfluidics in designing such sample-to-answer devices for achieving the wider acceptance of rapid and point-of-need testing for FBP.

3.1 INTRODUCTION

Contaminated food and water with disease-causing pathogens, once consumed by humans give rise to foodborne illness [76]. Although, there are 31 identified pathogens capable of delivering foodborne illnesses, yet *Norovirus*, *Staphylococcus aureus*, *Campylobacter*, *Salmonella*, *Listeria monocytogenes*, *Clostridium perfringens*, *Escherichia coli* O157:H7 and *Toxoplasma gondii* are leading cause of vast majority of infections and deaths [86]. According to the World Health Organization (WHO), bacteria, parasite, and virus-associated diarrheal diseases kill nearly 2 billion people globally due to consumption of contaminated food or water [40].

In order to establish preventive and control strategies for food-associated illnesses, efforts are underway by the agriculture and food industry in developing robust and affordable analytical technologies for improving the food safety and quality. The main motto is to generate such technologies which fulfill the ASSURED criteria (affordable, sensitive, specific, user-friendly, rapid, and robust, equipment-free, and deliverable to food producer), processor, and consumer, thereby providing us with future substitutes to the standard benchtop detection technologies [62].

The DNA (deoxyribonucleic acid) or RNA (ribonucleic acid) are ubiquitous and unique in every single eukaryotic and prokaryotic organism, such as, human, animal, plant cells and microorganisms. Their unique fingerprint is universally exploited for biomarker-based detection in various biological studies and diagnosis. NABT involves the operations of nucleic acid extraction, amplification to millions of folds and specific detection to provide swift and precise results than traditional immunoassays and culturing methods [26]. Further, the classical nucleic acid analysis involving the cyclic principle of denaturation, annealing, and extension, require sophisticated and expensive equipment which restrict their use to laboratories involving professional operators. This has created demand for development of alternative nucleic acid detection strategies for low cost, user-friendly, and on-site pathogen detection.

Isothermal NABT is a promising tool for pathogen detection at remote settings, since DNA can be amplified at constant temperature using a simple heating element. Isothermal methods used in past decade include: nucleic acid sequence-based amplification (NASBA), rolling circle amplification (RCA), recombinase polymerase amplification (RPA), loop-mediated isothermal amplification (LAMP), helicase-dependent amplification (HDA) and strand displacement amplification (SDA) [5, 18, 77, 86]. These methodologies, when combined with appropriate sample preparation and detection steps, can be used at the field level or in food industries for rapid and timely monitoring and release of products.

The chapter provides detailed insights on the various isothermal NABT, rapid sample pretreatment and detection methods, and microfluidic devices devised using these techniques. The overview of chapter is represented in Figure 3.1.

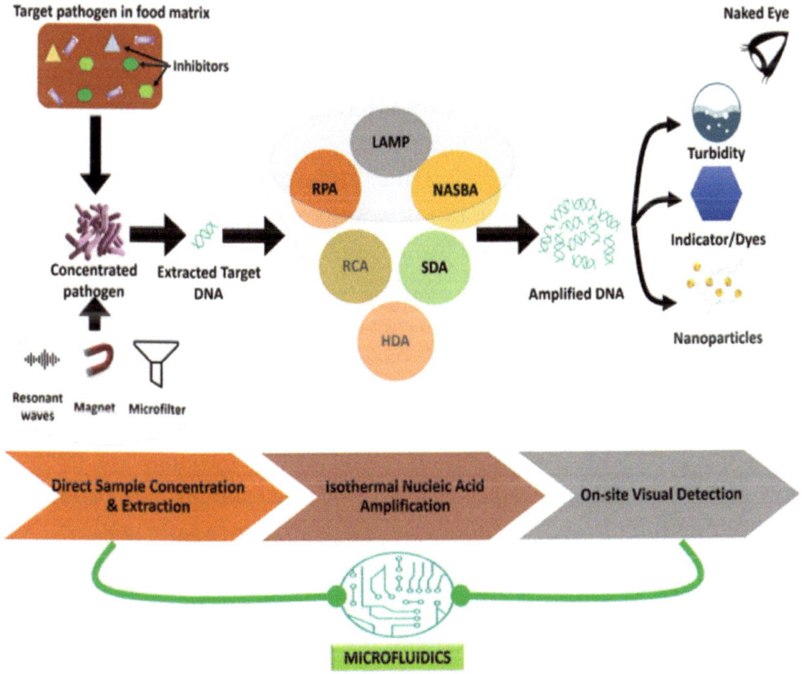

FIGURE 3.1 Overview of rapid sample pretreatment, amplification, and detection techniques and their integration with microfluidics for on-site foodborne pathogen detection.

3.2 LIMITATIONS OF CLASSICAL NUCLEIC ACID-BASED APPROACHES

The rapid and frequent detection of a causative agent responsible for foodborne illness is a crucial step for assuring food safety throughout various production, distribution, and retail points. Leaving aside traditional plating techniques, even the classical nucleic acid amplification approaches are not compatible for in-field identification of these notorious microbes. Although the onset of polymerase chain reaction (PCR) has revolutionized molecular analysis through rapid amplification of specific DNA targets and is still apprised as the gold standard of molecular detection but it lacks the acceptance for field testing in poor resource areas due to thermal cycler requirement [50].

Additionally, PCR-based detection of pathogens usually involves time consuming enrichment steps to enhance sensitivity of qualitative detection followed by gel electrophoresis to present the specific result corresponding to different lengths of amplicon. Thus, the requirement of specialized equipment like thermal cycler, electrophoresis apparatus and UV illuminator for endpoint detection along with skilled staff makes the PCR better suited for central laboratories.

Quantitative PCR (qPCR) is another advanced nucleic acid amplification approach capable of generating results in real-time independent of agarose gel electrophoresis. The formation and accumulation of amplicons is continuously monitored and measured using common DNA binding fluorescent dyes (SYBR Green).

Hydrolysis probes (Molecular beacon and TaqMan) consisting of reporter and quencher molecules are also employed to trace fluorescent signals. The qPCR is also adept in providing multiplex detection of pathogens. Despite its advantages, commercial qPCR systems are normally larger in size, expensive, and sophisticated instruments with significant power requirements making them less applicable in remote settings [82].

The thermal cycling steps significantly impede the potential of these classical PCR techniques for rapid point-of-need testing. Isothermal NABT can serve as a desirable alternative for these technical obstacles as they greatly simplified the amplification process by employing enzymes for performing strand separation at fixed temperature rather than repeated denaturation and annealing steps of thermal cycler [31].

3.3 ISOTHERMAL NUCLEIC ACID AMPLIFICATION TECHNIQUES

Isothermal NABT have the capability to outweigh the technological limitations in Section 13.2 of this chapter as they can be operated without any requirement of thermal cycler for continuous amplification, gel electrophoresis and UV transilluminator for visualization of DNA bands.

These techniques work on the principle of amplification of nucleic acids (DNA or RNA) at fixed temperature using specific enzymes and primers within a considerable short time, free of any instrument and skilled manpower requirements. Further, these tools are ideal for their integration with microfluidic devices, making their applications even broader in healthcare, agri-food, and environmental sectors for unearthing the pathogens. The various rapid isothermal NABT and their applications for identification of foodborne pathogens (FBP) are outlined in Table 3.1.

3.3.1 LOOP-MEDIATED ISOTHERMAL AMPLIFICATION (LAMP) PROCESS

LAMP is a robust and widely researched gene amplification process developed in 2000 by Notomi et al. [49]. This mode of DNA amplification, unlike PCR is free of cyclic processes of heating and cooling, and polymerize DNA with high specificity, sensitivity, and proficiency [49]. Isothermal amplification is based on continuous strand displacement DNA synthesis using *bst* DNA polymerase using two sets of outer and inner primers that bind to six to eight specific sequences on the target DNA. LAMP amplification process covering template synthesis cycle, loop amplification and repeated elongation yields numerous bands of multi-ring and stem-loop DNAs containing inverted repeats of target region within the completion cycle of 30–60 minutes [86].

The higher sensitivity of LAMP enables detection of pathogens directly from the food matrix without sample preparation [75]. LAMP is tenfold timed extra sensitive than PCR-based assays and can better tolerate the biological inhibitors present in food matrices. LAMP is usually employed at a fixed temperature range of nearly 60°C to 65°C, requiring only water bath instead of thermal cycler. LAMP also offers the option of multiplexing, but the need for multiple primers per assay increases the complexity of reaction. The rapid assimilation of sequence-specific DNA concatemers in reaction tubes can be monitored easily with the naked eye using turbidity or colorimetric dyes without further molecular analysis [22].

The endpoint signal for LAMP products can also be detected using calcein based fluorescent dyes or electrochemically [29, 57]. This versatile tool is rapid, cheaper, sensitive, and user-friendly for use in low resource settings. This technique firstly demonstrated success to detect Shiga toxin genes of *E. coli* O157:H7 [41]. Further, LAMP detected enterotoxigenic strains of *E.*

TABLE 3.1 Isothermal Nucleic Acid Amplification Techniques (NABT) for Detection of Foodborne Pathogens.

Technique	Pathogen	Matrix	Sensitivity	Detection Time	References
LAMP	Enterotoxigenic *E. coli*	Raw milk	547 cfu/mL	60 min	[79]
	E. coli (viable but not culturable)	Chicken meat	25 cfu/g	5 h	[9]
	V. parahaemolyticus	Seafood	14 copy/g	30 min	[88]
	L. monocytogenes	Chicken meat	2.82 cfu/g	60 min	[66]
NASBA	*E. coli*	Drinking water	40 cfu/mL	4 h	[43]
	L. monocytogenes	Dairy and egg products	10 cfu/mL	90 min after 48 h enrichment	[7]
	S. enteritidis	Eggs and meat	10 cfu/25 g food sample	90 min after 18 h enrichment	[20]
	S. enteritidis and S. typhimurium	Pure culture	10 cfu/mL	90 min	[44]
RPA	*E. coli* (Shiga toxin-producing)	Pure culture	5–50 cfu/mL	5–10 min	[45]
	Salmonella	Pure culture	10^5 copy/μL plasmid DNA	20 min	[83]
HDA	*Helicobacter pylori*	Pure culture	10 cfu/mL	1 h	[25]
	S. aureus	Pork and dairy samples	5×10^1 and 5×10^0 cfu/mL	2 h	[11]
RCA	*Salmonella and*	Milk	6 cfu/mL	2 h	[89]
	L. monocytogenes	Pure culture	0.0002 ng/μL genomic DNA	4 h	[38]
SDA	*E. coli* O157:H7	Pure culture	10 cfu/mL	45 min	[77]
	S. enteritidis	Pure culture	10 cfu/mL	35 min	[23]

coli from raw milk at limit of detection (LOD) of 547 cfu/mL [79]. Other pathogen detection involves *L. monocytogenes*, *E. coli* and *V. parahaemolyticus* in foods like chicken meat and seafood respectively [9, 67, 88].

3.3.2 NUCLEIC ACID SEQUENCE-BASED AMPLIFICATION (NASBA) TOOL

NASBA is a transcription mediated amplification tool for sensitive detection of target RNA under isothermal conditions independent of thermal cycler [15]. The reaction mixture is composed of three enzymes-RNase H, Avian Myeloblastosis Reverse Transcriptase (AMV-RT) and DNA dependent T7 RNA polymerase; a forward primer (complementary to target RNA region) carrying 5' promoter region that is identified by T7 RNA polymerase; a reverse primer similar to RNA region; dNTPs (deoxyribonucleotide triphosphate) for AMV-RT activity; ribonucleotide triphosphates for T7 RNA polymerase and buffering agents. The whole reaction process takes around one to two hours at 41°C.

Firstly, AMV-RT leads the extension process as forward primer hybridizes to the target region of RNA creating an RNA-cDNA hybrid complex. RNase H degrades the RNA component of the complex, liberating the single-stranded (ss) cDNA. Subsequently, this cDNA anneals with reverse primer, again elongated by AMV-RT to yield double-stranded (ds) copies of cDNA. RNA polymerase delivers its part by binding to the promoter region of ds cDNA and carrying out transcription to generate multiple copies of the target sequence. These ss transcribed copies utilized for hybridization of the next round of amplification leading to exponential generation of RNA amplicons which can be perceived through different means such as hybridization probes, lateral flow assays or by electrophoresis [73]. Unlike RT-PCR, the NASBA rules out the possibility of false-positive amplification due to genomic ds cDNA copies as cDNA remains in its intact form at 41°C and cannot replace the amplification buffer [15].

NASBA is able to provide sensitive detection of up to 1 cfu/mL [17]. Further, this technique can generate millions of copies of the target sequence within 4–5 cycles in contrast to 20 cycles of PCR. This technique is operated at the same pace and accuracy as PCR but provides additional benefit of detection of viable pathogens due to amplification of RNA molecules. Its application can lead to the development of promising portable and handy diagnostic kits even suitable for low resource settings [60].

NASBA demonstrated selective amplification of mRNA sequences for sensitive determination of viable cells of *E. coli* in drinking water [43]. Moreover, NASBA proved successful for recognizing *L. monocytogenes* and *S. enteritidis* in dairy, egg, and meat products [7, 20]. A novel NASBA kit was developed to concurrently detect *S. typhimurium* and *S. enteritidis* at LOD of 10 cfu/mL through single specific antisense primer [44].

3.3.3 RECOMBINASE POLYMERASE AMPLIFICATION (RPA) METHOD

RPA recently appeared as a preferred molecular approach for swift, specific, and cheap identification of pathogens [64]. Patented in 2006 by Piepenburg et al. [51], this process generates multiple copies of DNA to detectable levels at isothermal temperature between 37–42°C within 10–20 min by utilizing three main enzymes-recombinase, polymerase, and single-stranded DNA binding (SSB) protein. The recombinase after forming a complex with oligonucleotide primers facilitates their hybridization with the complementary sequences in the target gene via homologous recombination. SSB attaches to displaced DNA strand followed by DNA synthesis via strand displacing polymerase at site of primer binding to target DNA. ds DNA templates produced in each reaction are used as templates for ensuing reaction creating an explosive amplification cycle [51].

The final product can be visualized in real-time by means of probes consisting of nucleotide analog with endonuclease IV and exonuclease III enzyme for signal detection via lateral flow and fluorescent assays, respectively [51]. This technique due to its minimal sample-preparation requirements and low operating temperatures have the potential for application in field conditions and poor resource facilities, and integration with automated handled microfluidic devices [59]. RPA kits for determination of Shiga toxin-producing *Escherichia* have been developed [45]. An instant RPA test for diagnosis of *Salmonella* at 39C within 20 min was reported [83].

3.3.4 HELICASE DEPENDENT AMPLIFICATION (HDA) TOOL

Helicase dependent amplification (HAD) utilizes accessory protein DNA helicase to unwind dsDNA into ssDNA followed by primer annealing and elongation via DNA polymerase [66]. HDA employs constant temperature and simple reaction mechanisms for continuous amplification of target DNA and can be used as efficient point-of-care devices for online food pathogen diagnosis [34]. HDA was employed for colorimetric sensing of *Helicobacter pylori* [66]. The detection of *S. aureus* in pork and dairy samples at LOD of 5×10^1 and 5×10^0 cfu/mL at 65°C within 90 min was established using HDA [11].

3.3.5 ROLLING CIRCLE AMPLIFICATION (RCA)

Evolved from rolling circle replication of circular microbial DNA, this isothermal method of pathogen detection also called hyperbranched RCA involving a padlock probe, i.e., a linear oligonucleotide probe flanked by two target specific sequences of pathogen of interest which after annealing is circularized or ligated by DNA ligase, forming the base circular template for subsequent polymerization and elongation by DNA or RNA polymerase nuclease enzyme using the pair of primers. After multiple rounds of adding nucleotides and displacing sequences coming in way during polymerization, the reaction ceases when all the dissociated ssDNA is transformed into dsDNA [6, 8].

This single binding event enables replication of template DNA over a 1,000-folds at a single temperature point which increases sensitivity of method significantly. RCA is immune to contaminants and requires minimal assay optimization. The method can be used for in-situ and micro-array hybridization experiments [60]. RCA was successfully exploited for electrochemical sensing of DNA from *Salmonella* in milk samples using DNA-AuNPs probe [89]. A sensitive RCA coupled with magnetic bead guided electro-chemiluminescence was exploited to specifically detect *L. monocytogenes* at a level of 0.0002 ng/ μL of genomic DNA [38].

3.3.6 STRAND DISPLACEMENT AMPLIFICATION (SDA) TOOL

Strand displacement amplification (SDA) is also an isothermal amplification method firstly reported in 1992 [68]. This method produces around 10^9 copies of target DNA in a single step within 2 hours at broad temperature varying from 37°C to 70°C. The method employs two primers harboring restriction site (GTTGAC) for HincII exonuclease at 5' end, two bumper primers, a template DNA and a DNA polymerase-1 devoid of exonuclease activity for elongation of primers.

Firstly, the primers containing restriction sites are hybridized with template DNA followed by their strand displacement by bumper primers by annealing adjacently to previous primers. These bumper primers attached to original template DNA as well as to displaced template containing restriction sites are polymerized by DNA polymerase to form ds DNA templates.

Secondly, ds DNA containing restriction sites at one end are nicked by Hinc II restriction endonuclease providing free 3'end for further polymerization. This operation of nicking, extension, and displacement initiates the amplification cycle. The process has advantages of being specific along with high amplification efficiency [88]. A lateral flow aptasensor based system has been devised for sequence specific recognition of *E. coli* O157:H7 and *S. enteritidis* by isothermal SDA [23, 77]. The advantages and disadvantages associated with respective techniques are summarized in Table 3.2.

3.4 RAPID PRE-CONCENTRATION METHODS

All the analytical and high-tech methods for detecting this food borne pathogens require certain sample preparation or analyte concentration step, which is the major technological limitation of these novel rapid techniques. Moreover, several components available in the food matrix may affect the sensitivity of a particular technique. Hence, studies focused on the separation cum concentration methods of microorganisms directly from the food itself, preceding to their determination by immunological, nucleic acid-based, or biosensor assays. Pre-concentration increases the number of target analytes (majorly microorganisms) per unit volume several fold, thereby enhancing investigation sensitivity.

The preconcentration methods, which bypass the time-consuming enrichment protocols, include: size fractionation, filtration, centrifugation, magnetic bead-based separation, or combinations of these methods. Certain factors or characteristics, which are to be considered or desired for the development of ideal enrichment/pre-concentration/pre-treatment process, include:

- Compatibility of the enriched sample and detection technique;
- Reduction of detection time due to pre-treatment;

TABLE 3.2 Advantages and Disadvantages of Nucleic Acid-Based Detection Techniques

Method	Advantages	Disadvantages	References
HDA	• Sensitive • Tolerance to biological inhibitors • Straightforward primer design • Initial heating not needed	• Enzymes used are expensive • Difficult to amplify long fragment	[52]
LAMP	• Short run time, 30–60 min • Low cost • High specificity • High sensitivity (10-fold than PCR) • High tolerance to impurities • Easy to operate	• Complication of primer design • Tendency to form primer-dimers and non-specific amplicons • Requirement of electrophoresis post LAMP if not combined with suitable detection step	[36, 53]
NABSA	• Direct detection of viable bacteria • High specificity • Desirable sensitivity • Low cost	• Complicated reaction components • High cost • Equipment requirement of RNA detection	[17, 36]
PCR	• Desirable specificity • Desirable sensitivity • Reliable	• Longer run time, 2–3 h • Thermal cycler requirement • Relatively specificity (2 primers correspond to 2 target sites) • Low tolerance to inhibitors • Requirement of electrophoresis post PCR	[36]

TABLE 3.2 *(Continued)*

Method	Advantages	Disadvantages	References
qPCR	• Desirable specificity and sensitivity • Real-time result interpretation • Amenable for multiplexing	• Sophisticated instrument requirement • Skilled personnel • Expensive	[82]
RCA	• Sensitive • Simple and single primer design • Minimal assay optimization • Resistant to contaminants	• High cost of padlock probe • Longer run time • Susceptible to background signal interference	[6, 64]
RPA	• High specificity • Desirable sensitivity • Rapid • Low operating temperatures (37–42°C) • Robust to biological impurities	• Comparatively long chain of primers (>30) • Optimization complexities	[37]
SDA	• Specific and sensitive • High amplification efficiency • Instrument free	• Amplification of long fragment is tedious • Template with nicked end is needed	[77, 87]

- No influence on microbial growth;
- Effective isolation of target pathogen;
- Non-interference of the food components with pathogens due to enrichment;
- Concentrated amount should meet the minimum/maximum legal requirements of sample requirements as specified by ISO, USDA, or FDA.

3.4.1 MAGNETIC SEPARATION PROCESS

The separation relying on paramagnetic beads coated with a biorecognition element unique to the target food pathogen. Various specific receptors like antibodies, aptamers, functionalized proteins, phage endolysins, etc., can be coupled with magnetic beads. The complexed target cell along with these functionalized magnetic beads could be segregated from the food matrix using a magnet and thereafter, could be detected with a suitable detection technique.

Magnetic nanoparticles are simpler and effective analyte concentration techniques from the aqueous solutions without any complex instrumentation on account of their high surface area to volume proportion thus promoting microbial adherence efficiency, capacity, and rapid reaction kinetics of target pathogen with the receptors. Iron oxide micro or nanoparticles have been widely researched upon for pathogen detection. A method for separation of *E. coli* using magnetic glycol nanoparticles functionalized with D-mannose groups was reported, wherein 88% binding efficiency was achieved within an incubation period of 45 minutes [21].

Magnetic bioactive nanoparticles consisting of vancomycin-conjugated iron-platinum were able to capture and detect pathogens including *Enterococcus faecalis* and *Staphylococcus aureus* at a concentration of one log/cfu within an hour [27]. The non-specific adsorption and their potential interference in detection may be overcome by inert chemicals coating.

3.4.2 PRE-FILTRATION/FILTER MEMBRANES

Several researchers have used flat sheet and cross-flow filters to recover the pathogenic bacteria from egg whites, spinach, chicken leg, chicken carcass and ground turkey [12]. The common recommended procedure includes the following steps:

- Microbial shattering within the food matrix using the mechanical treatment such as stomaching, blending or rinsing;
- Pre-filtration by glass/nylon filters;
- Enzyme assisted breakdown of food particles;
- Concentration using microfiltration and microfugation.

Some of the major limitations of microfilters include: expensive, difficulty in reusability, removal of microorganisms during pre-filtration step and fouling of membranes.

3.5 ACOUSTOPHORESIS AND MAGNETOPHORESIS

These techniques can be integrated into microfluidic devices. These methods can be applied in a continuous flow fashion for pathogen enrichment. Acoustophoresis allows label free cell concentration according to differential size, density, and compressibility of analyte against surrounding medium. A piezo-ceramic transducer generates a resonating acoustic force acting differently on a pathogen and other food components, facilitating size-based separation of pathogen from food components. This results in quick separation of bigger size particles depending on the amount of acoustic power. This principle was applied to detect approximately 10^3 cfu/mL cells of *Salmonella* in artificial contaminated meat samples with a recovery rate of 60 to 90% [47].

Magnetophoresis is another technique involving the use of external magnetic fields for continual separation of magnetic beads from non-magnetic components. Magnetophoresis employs the application of magnetic field perpendicularly to fluid flow, guiding the deflection of pathogen-bead complexes from straight line according to their size, magnetic power applied and percentage of iron oxide in beads. This technique allows enrichment of different pathogens from food matrix [34]. The principle is exploited using simultaneous separation of *E. coli* and *Salmonella* using their respective functionalized hyglosbeads and dynabeads based upon the differences in iron oxide and magnetic properties [46]. The method can recover 62% and 72% of *E. coli* and *Salmonella* cells within 1.2 min from pathogen-contaminated chicken enrichment broths.

3.6 OTHER PROTOCOLS

Presently, low-cost mini centrifuges under the brand name of Mini Dancer Plus (Sango Biotech) have been commercialized. These mini centrifuges are used to extract nucleic acids in bacterial and viruses from blood plasma and culture media. Similarly, disposable Q-tips were used to directly extract DNA templates using sodium hydroxide from the spiked shrimp for detecting *Vibrio parahaemolyticus* within an hour. Being disposable in nature, the risk of carry over contamination during sample preparation was minimized significantly [71].

3.7 ON-SITE COLORIMETRIC DETECTION

The detection step subsequent to nucleic acid isothermal amplification is designed in such a way to enable instrument-free visual result interpretation. Colorimetric visual detection is growing in popularity as signal read-out can be monitored in a simplified manner by naked eyes without requiring skilled trained personnel for further data analysis. These methods offer potential for online detection in low resource facilities without using any complicated and costly instruments. Colorimetric detection works on the basis of color change due to biochemical or chemical reaction between target analytes and probes [64]. Visual detection can be applied either with direct detection of amplicons or indirect detection of amplification by-products [82]. The different visual endpoint detection techniques after isothermal amplification are explained further.

3.7.1 TURBIDITY

This method utilizes the release of pyrophosphate ions ($P_2O_7^{4-}$) from dNTPs as a by-product during isothermal amplification of target DNA to million-folds especially in LAMP. The ions generated bind with Mg^{2+} ions providing white insoluble precipitate ($Mg_2P_2O_7$), which can be observed with the naked eye [84]. This mode of detection minimizes contamination risk as reaction is completed in a single tube and applied on-site without extra instrument cost. However, its sensitivity is lower due to low degree of differentiation and stability of reaction for a short time.

3.7.2 COLORIMETRIC DYES

Colorimetric dyes are superior to fluorescent dyes like calcein as the latter requires a UV torch for signal read-out creating difficulty for direct detection via naked eye. The detection principle for colorimetric dyes can be based on pH change, ds DNA binding or metal ion binding occurring as the amplification proceeds without affecting amplification efficiency. Their sensitivity is significantly higher than turbidity-based methods. In addition to pyrophosphate (PPi) ions, hydrogen ions are also released as byproducts during DNA polymerization, creating a drop in pH [81]. pH-sensitive dyes such as neutral red, phenol red, etc., can be exploited to measure the change in pH from alkaline to acidic side which can be achieved within 30 min to 10 copies of DNA [56]. To facilitate the unambiguous distinction in color measurement, primary buffers used in DNA amplification have been replaced with $(NH_4)_2SO_4$ due to its low buffering capacity [48].

Metal ion indicators like eriochrome black T (EBT) and hydroxy naphthol blue (HNB), monitor the variation in Mg^{2+} ion concentration during amplification. As the amplification proceeds, metal ion indicators compete with the newly generated PPi ions for reacting with pre-added Mg^{2+} ions. HNB changes in color to blue from purple as the Mg^{2+} ion concentration gradually decreases with amplification [14]. The sensitivity of HNB-based method was found parallel to SYBR green-based method, and higher than calcein-based method [3, 41]. Acid chrome blue K (ACBK) is another metal indicator with principle same as that of HNB and changes its color from red to blue as the amplification proceeds [72].

DNA intercalating dyes, like SYTO-9 and SYBR Green I, are normally used in the detection phase. These dyes change color from orange to yellow-green enabling visualization by unaided eye [69]. However, high SYBR Green I concentration could lower the amplification efficiency and are usually preferred to be added after the amplification step. Moreover, the toxicity and frameshift mutation associated with these dyes limit their application. A new DNA intercalating dye called GeneFinder™ is found to be suitable in respect of its high sensitivity, differentiability, safety, low cost, and compatibility with addition prior to amplification [4]. It allows obvious color detection from red to green, however, is not suitable for color-blind persons.

Apart from the above, a simple visual detection method has been developed to detect trace phosphate (Pi) released after cleavage of PPi ions using added thermostable inorganic pyrophosphatase (PPase). The release Pi reacts with ammonium molybdate yielding blue phosphor-molybdenum complex in presence of potassium antimonyl tartrate, ascorbic acid, and sulfuric acid [80]. The use of colorimetric dye is simple with no additional equipment required. However, only a "trained-eye" is required to monitor and differentiate colors easily.

3.7.3 NANOPARTICLES

Nanoparticle based colorimetric detection is another approach, which can be integrated with an isothermal amplification process to detect the specific sequence of target DNA. Gold nanoparticles (AuNPs) exhibit the kaleidoscopic properties and bear simple structure, cheaper, and equipment free which can be exploited for rapid direct visual detection. AuNPs are frequently used in bio-sensing after being conjugated with suitable bio-receptors (or probes). They can be directly conjugated with probes specific to target DNA and visually detected by altering their dispersion (red color) or aggregation (blue color) in the presence of optimized salt concentration in a single tube [67].

Otherwise, AuNPs can be used in lateral flow assay by attaching them with a secondary biomolecule, e.g., rabbit anti-fluorescein isothiocyanate (FITC) which captures the corresponding FITC labeled probe hybridized to target DNA [70]. A LAMP based assay utilizing a complementary biotin labeled loop primers was used to bring together the target amplicons and streptavidin functionalized AuNPs [35]. This leads to clusterization of AuNPs, changing the color from red to blue within 5 minutes. This requires uncapping operation for adding the biotin labeled primers after amplification and may contribute towards contamination.

AuNPs coupled with ssDNA probe can be used for endpoint detection in case of RCA wherein once this probe is hybridized toss circular DNA leads to polymerization and subsequent aggregation of AuNPs during amplification leading color change to blue from red [30]. This method joins the high selectivity and sensitivity of RCA with simple, convenient, and rapid AuNPs based detection.

3.8 PORTABLE MICROFLUIDIC NUCLEIC ACID DEVICES

Microfluidics have paved the way in bringing the rapid diagnostic devices in the form of sample-to-answer miniaturized kits. Microfluidics is the technology, wherein the flow of nanoliters of the fluid is precisely manipulated through the microchannels. These channels can be embedded either in chip or paper form. These emerging devices can be used for rapid online quality monitoring due to following advantages-cost effective, high throughput, portability, ease of operation, low sample volume and short analysis time [74]. All the analysis operations ranging from sample pretreatment to detection can be performed on a single miniaturized platform.

Isothermal NABT can be easily commingled into microfluidic devices for on-site food safety investigation. They have been extensively incorporated into microfluidic paper-based analytical devices (μPAD) conferring higher sensitivity and specificity than immunoassays [1]. μPADs is considered as an attractive option than silicon, glass, and plastic due to its low cost, simplicity, availability, biodegradability, capillary action mediated fluid uptake, ease of fabrication and modification. Their universal properties make them suitable for application in-field, domestic, or resource-poor usage.

Further, all the operations of isothermal NABT, i.e., sample preconcentration and preparation; extraction, amplification, and measurable detection can be easily integrated onto μPADs [24]. Moreover, reaction mixtures of isothermal NABT (e.g., LAMP, and RPA) can be freeze-dried for longer shelf life and stored into these devices prior to use [31].

The limitations, which need to be overcome with μPADs, are their poor sensitivity and multiplexing. However, efforts have been made to address these issues using molecularly imprinted polymers (MIP) for improving specificity and sensitivity; and different channels coated with target analyte specific antibiotics for concurrent identification of multiple pathogens [54, 61]. Subsequently, some of the advancements combining these miniaturized devices with rapid isothermal NABT for the swift and cost-effective point-of-need detection of FBP will be discussed.

A multiplex LAMP detection of FBP was carried out using paper disks impregnated with reaction mix and target specific primers [58]. It uses Fisetin, a plant flavonoid derived eco-friendly dye, exhibiting strong fluorescence under UV when intercalated with amplified dsDNA for sensing of *E. coli* and *Salmonella* spp. directly through polycarbonate microdevice. Centrifugation is used to direct the fluid flow while polydopamine coated paper was used for purification of DNA from inhibiting components of artificially contaminated milk with *Salmonella* spp. The LOD of the method was found as 170 cfu/mL for spiked milk.

RCA based microdevice was developed for detection of pathogen *C. difficile* using glutamate dehydrogenase aptamer [32]. Two nitrocellulose disks connected with a paper bridge were used. In one disk, a fluorescently labeled DNA aptamer-graphene oxide was used as a biorecognition element which after hybridization with target analyte emits fluorescence and moves to the second disk via a bridge where DNA is amplified through RCA. After addition of colorimetric substrates, signals can be visualized by the naked eye. Another study used RPA based paper chips for diagnosis of *Salmonella typhimurium*, *E. coli* and *S. aureus* in a single step [2]. A membrane made up of ether sulfone polymer was coated with dry reagents and fluorescent probes and results can be achieved within 20 min at 37°C with LOD of 10^2 cfu/mL.

A handheld double-strip rotary detection device was developed containing four test lines for capturing pathogen specific antibodies and their detection using AuNP tagged secondary antibodies for multiple diagnosis of *B. cereus*, *S. aureus* and *E. coli* O157:H7 [54]. The assay has a sensitivity of 1×10^4 cfu/g of contaminated lettuce without any external enrichment step. Further, a fully automated sample-to-result paper device was fabricated using flinders technology associates (FTA) card, glass fibers and polyvinyl chloride for substrates of nucleic acid extraction, amplification, and transport of samples to different zones, respectively [13]. This device has been tested on *E. coli* spiked contaminated spinach and milk with LOD of 10–1,000 cfu/mL. Recently, a completed automated and disposable paper-based platform consisting of sponge-based reservoirs for reagent storage, DNA extraction using glass fibers and paper values, heater for HDA followed by lateral flow detection for tracking *S. typhimurium*.

The cellulose paper microchip integrated both the steps of sample preparation and amplification using LAMP was developed for *E. coli*, wherein end-point was detected using SYBR Green I having LOD of 5 cells/reaction within 60 min [16]. Similarly, cellulose paper-based LAMP immobilized by polyethylenimine (PEI) yielded a semi-quantitative diagnosis of *E. coli* having LOD of 4.14×10^3 copies/μL within 60 min using HNB dye as an indicator [28]. A microspot paper assay utilizing wax printing was demonstrated for concurrent sensing of *E. coli*, *Salmonella typhimurium*, and *L. monocytogenes* at a sensitive level of 10 cfu/cm^2 in spiked ready-to-eat meat after enrichment of 12 hours, to be used as screening medium in food industry [33].

NASBA-based microfluidic devices have also addressed issues of on-site RNA detection. A single, monolithic, PDMS-based microfluidic device merged steps of tRNA (transfer ribonucleic acid) purification using silica bead immobilization, followed by NASBA mediated amplification and fluorescent detection [19]. All these efforts have significantly made advancements in fabricating devices for rapid on-site pathogen detection. However, still research is being conducted to further simply the operations from sample pretreatment to detection for practical applications.

3.9 CHALLENGES AND FUTURE PERSPECTIVES

Considerable advancements have been made in building rapid platform-based tests for detection of FBP in agriculture-based sectors as well as for food industry stakeholders. Among them, isothermal nucleic acid amplification techniques have offered an edge over others by displaying the properties of high sensitivity, selectivity, low cost, and thermal cycler free operation showing their applicability for on-spot detection especially in poor resource settings.

In the past two decades, several pathogen diagnostics tests using isothermal NABT, e.g., LAMP, RCA, NASBA, etc., have been developed by researchers throughout the globe. However, each of these methods has some associated challenges like complex primer designing, meticulous optimization of procedures and minimizing the cross-contamination, which once overcome can generate an effective technology. Since NABT requires steps of sample pretreatment/enrichment prior to nucleic acid extraction and amplification, tedious, and time-consuming protocols are required to segregate and selectively enrich the interested pathogen from the food matrix. Moreover, interference of inhibitors present in the food matrix is another bottleneck that needs to be addressed for avoiding the false negatives/positives and uncapping operations for swift method development.

Alternative separation cum concentration strategies (viz. immunomagnetic, microfiltration, portable centrifuges, etc.), can be integrated with NABT for in-field detection. Further, microfluidic platforms have also witnessed sample preparation and purification using techniques of acoustophoresis and magnetophoresis simplifying operations to a greater extent, lot needs to be done in this field.

Colorimetric visual detection is another important step providing the user with endpoint results without relying on further complicated data analysis. Colorimetric detection using indicator dyes and nanoparticles, e.g., AuNPs can be easily coupled with NABT to observe signal read-out by cutting the need of additional equipment such as UV torch required in case of fluorescent detection. Apart from the above, some recent detection tools using G-quadruplex DNAzyme and CRISPR/Cas12a system have also been exploited for rapid molecular diagnosis (for details, see Refs. [10, 39]).

Nowadays, microfluidics has emerged as an interdisciplinary field in developing point-of-care devices in various scientific and engineering fields including biomedicine, chemistry, biotechnology, and nanotechnology. Recent biosensor-based advancements are also directed to merge the NABT in portable microfluidic devices to widen its field of application. Several microfluidic devices are fully automated and integrated with isothermal NABT to enable swift sample-to-answer measurements. Paper-based microfluidic diagnostics further assisted in meeting the ASSURED requirements for foodborne pathogen detection. Still, research improvements in using µPADs with respect to sensitivity, functionality, and multiplex detection of pathogens are underway for better onboard storage of reagents, controlling fluid flow for physical and chemical modification of paper, etc.

Furthermore, existing microfluidic platforms are quite complex and expensive which needs to be replaced with cost-effective functional systems, convenient for use in food safety. In short, there is everlasting room for improvement for further simplifying the user steps combining sample concentration, separation, reagent storage, extracting nucleic acid, exponential amplification and direct visualization in a single device without any uncapping operations by integrating these NABT with microfluidics, nanoparticles, etc., for rapid, portable, and on-site pathogen tracking at various facets of food production, processing, and storage to minimize the foodborne outbreaks and early release of assured food products in market.

3.10 SUMMARY

The timely detection and containment of foodborne outbreaks warrants for rapid, simple, sensitive, robust, and portable detection tools for food safety analysis. The inherent limitations of traditional time-consuming culturing methods and sophisticated rapid instruments (viz. HPLC, qPCR, SPR (surface plasmon resonance), etc.), have paved the way for isothermal NABT in various clinical and environmental food safety analysis for their application in poor remote settings especially in developing countries. The chapter describes the role of various isothermal NABT (such as: LAMP, RCA, NASBA, HDA, and SDA) in prompt diagnosis of FBP. Since most of the molecular diagnostics are hampered by the requirement of long incubations for selective enrichment of target pathogens for improving sensitivity, efforts are being made to discuss strategies to bypass long enrichment steps and minimize matrix interference by concentrating the desired cells using immunomagnetic separation, microfiltration, acousto-phoresis, magnetophoresis, etc. Moreover, colorimetric visual detection has attracted a lot of attention for integration with these techniques due to portability, compatibility, cheap, and instrument free operations. The chapter mainly focused on on-site visual detection principles based on detection of amplicons or their by-products.

The chapter further summarizes the integration of all the analysis steps of pathogen detection using microfluidic devices for yielding direct sample-to-answer results with particular importance to paper based microfluidic devices. However, fabrication of these isothermal nucleic acid amplification-based microfluidics is still in its infancy and much need to be explored to refine and update the existing research to achieve wide-spread commercial acceptance. Nevertheless, significant advancements have been conducted for point-of-need detection of notorious microbes responsible for global outbreaks using the isothermal NABT coupled with microfluidics and research is continuously conducted to achieve sensitive and stand-alone nucleic acid devices of industrial importance.

KEYWORDS

- **colorimetric**
- **foodborne pathogens**
- **isothermal amplification**
- **microfluidics**
- **nucleic acid detection**
- **polymerase chain reaction**

REFERENCES

1. Adley, C. C., (2014). Past, present and future of sensors in food production. *Foods, 3*(3), 491–510.
2. Ahn, H., Batule, B. S., Seok, Y., & Kim, M. G., (2018). Single-step recombinase polymerase amplification assay based on a paper chip for simultaneous detection of multiple foodborne pathogens. *Analytical Chemistry, 90*(17), 10211–10216.
3. Ali, S. A., Kaur, G., Boby, N., & Sabarinath, T., (2017). Rapid and visual detection of *Leptospira* in urine by LigB-LAMP assay with pre-addition of dye. *Molecular and Cellular Probes, 36*, 29–35.
4. Almasi, M. A., Manesh, M. E., Jafary, H., & Dehabadi, S. M. H., (2013). Visual detection of potato leafroll virus by loop-mediated isothermal amplification of DNA with the GeneFinder™ dye. *Journal of Virological Methods, 192*(1, 2), 51–54.
5. Asiello, P. J., & Baeumner, A. J., (2011). Miniaturized Isothermal nucleic acid amplification: Review. *Lab on a Chip, 11*(8), 1420–1430.
6. Baner, J., Nilsson, M., Mendel-Hartvig, M., & Landegren, U., (1998). Signal amplification of padlock probes by rolling circle replication. *Nucleic Acids Research, 26*(22), 5073–5078.
7. Blais, B. W., Turner, G., Sooknanan, R., & Malek, L. T., (1997). Nucleic acid sequence-based amplification system for detection of *Listeria monocytogenes* hlyA sequences. *Applied and Environmental Microbiology, 63*(1), 310–313.
8. Cao, A., & Zhang, C. Y., (2012). Sensitive and label-free DNA methylation detection by ligation-mediated hyperbranched rolling circle amplification. *Analytical Chemistry, 84*(14), 6199–6205.
9. Chen, H., Zhao, Y. Y., & Shu, M., (2019). Detection and evaluation of viable but non-culturable *Escherichia coli* O157: H7 induced by low temperature with a BCAC-EMA-Rti-LAMP assay in chicken without enrichment. *Food Analytical Methods, 12*(2), 458–468.
10. Chen, J. S., Ma, E., & Harrington, L. B., (2018). CRISPR-Cas12a target binding unleashes indiscriminate single-stranded DNase activity. *Science, 360*(6387), 436–439.
11. Chen, X., Wu, X., Gan, M., Xu, F., & He, L., (2015). Rapid detection of *Staphylococcus aureus* in dairy and meat foods by combination of capture with silica-coated magnetic nanoparticles and thermophilic helicase-dependent isothermal amplification. *Journal of Dairy Science, 98*(3), 1563–1570.
12. Cho, I. H., & Ku, S., (2017). Current technical approaches for the early detection of foodborne pathogens: Challenges and opportunities. *International Journal of Molecular Sciences, 18*(10), 2078.
13. Choi, J. R., Hu, J., Tang, R., & Gong, Y., (2016). An integrated paper-based sample-to-answer biosensor for nucleic acid testing at the point of care. *Lab on a Chip, 16*(3), 611–621.
14. Choopara, I., Arunrut, N., Kiatpathomchai, W., Dean, D., & Somboonna, N., (2017). Rapid and visual *Chlamydia trachomatis* detection using loop-mediated isothermal amplification and hydroxy naphthol blue. *Letters in Applied Microbiology, 64*(1), 51–56.
15. Compton, J., (1991). Nucleic acid sequence-based amplification. *Nature, 350*, 91–92.

16. Connelly, J. T., Rolland, J. P., & Whitesides, G. M., (2015). Paper-machine for molecular diagnostics. *Analytical Chemistry, 87*(15), 7595–7601.
17. Cook, N., (2003). The use of NASBA for the detection of microbial pathogens in food and environmental samples. *Journal of Microbiological Methods, 53*(2), 165–174.
18. Daher, R. K., Stewart, G., Boissinot, M., & Bergeron, M. G., (2016). Recombinase polymerase amplification for diagnostic applications. *Clinical Chemistry, 62*(7), 947–958.
19. Dimov, I. K., Garcia-Cordero, J. L., & O'grady, J., (2008). Integrated microfluidic tmRNA purification and real-time NASBA device for molecular diagnostics. *Lab on a Chip, 8*(12), 2071–2078.
20. D'souza, D. H., & Jaykus, L. A., (2003). Nucleic acid sequence based amplification for the rapid and sensitive detection of *Salmonella enterica* from foods. *Journal of Applied Microbiology, 95*(6), 1343–1350.
21. El-Boubbou, K., Gruden, C., & Huang, X., (2007). Magnetic glyco-nanoparticles: A unique tool for rapid pathogen detection, decontamination, and strain differentiation. *Journal of the American Chemical Society, 129*(44), 13392, 13393.
22. Esmatabadi, M. D., & Bozorgmehr, A., (2015). Techniques for evaluation of LAMP amplicons and their applications in molecular biology. *Asian Pacific Journal of Cancer Prevention, 16*(17), 7409–7414.
23. Fang, Z., Wu, W., Lu, X., & Zeng, L., (2014). Lateral flow biosensor for DNA extraction-free detection of salmonella based on aptamer mediated strand displacement amplification. *Biosensors and Bioelectronics, 56*, 192–197.
24. Feng, S., Choi, J. R., Lu, T. J., & Xu, F., (2015). State-of-art advances in liquid penetration theory and flow control in paper for paper-based diagnosis. *Advances in Porous Flow, 5*, 16–29.
25. Gill, P., Alvandi, A. H., Abdul-Tehrani, H., & Sadeghizadeh, M., (2008). Colorimetric detection of *Helicobacter pylori* DNA using isothermal helicase-dependent amplification and gold nanoparticle probes. *Diagnostic Microbiology and Infectious Disease, 62*(2), 119–124.
26. Gootenberg, J. S., & Abudayyeh, O. O., (2017). Nucleic acid detection with CRISPR-Cas13a/C2c2. *Science, 356*(6336), 438–442.
27. Gu, H., Ho, P. L., Tsang, K. W., Wang, L., & Xu, B., (2003). Using biofunctional magnetic nanoparticles to capture vancomycin-resistant enterococci and other gram-positive bacteria at ultralow concentration. *Journal of the American Chemical Society, 125*(51), 15702–15703.
28. Hongwarittorrn, I., & Chaichanawongsaroj, N., (2017). Semi-quantitative visual detection of loop mediated isothermal amplification (LAMP)-generated DNA by distance-based measurement on a paper device. *Talanta, 175*, 135–142.
29. Hsieh, K., Patterson, A. S., & Ferguson, B. S., (2012). Rapid, sensitive, and quantitative detection of pathogenic DNA at the point of care through microfluidic electrochemical quantitative loop-mediated isothermal amplification. *Angewandte Chemie International Edition, 51*(20), 4896–4900.
30. Hu, B., Guo, J., Xu, Y., Wei, H., Zhao, G., & Guan, Y., (2017). Sensitive colorimetric assay system for nucleic acid detection based on isothermal signal amplification technology. *Analytical and Bioanalytical Chemistry, 409*(20), 4819–4825.

31. Hu, J., Wang, S., Wang, L., & Li, F., (2014). Advances in paper-based point-of-care diagnostics. *Biosensors and Bioelectronics, 54*, 585–597.
32. Hui, C. Y., Liu, M., Li, Y., & Brennan, J. D., (2018). Paper sensor printed with multifunctional bio/nano materials. *Angewandte Chemie International Edition, 57*(17), 4549–4553.
33. Jokerst, J. C., & Adkins, J., (2012). Development of a paper-based analytical device for colorimetric detection of select foodborne pathogens. *Analytical Chemistry, 84*(6), 2900–2907.
34. Kant, K., & Shahbazi, M. A., (2018). Microfluidic devices for sample preparation and rapid detection of foodborne pathogens. *Biotechnology Advances, 36*(4), 1003–1024.
35. Kong, C., Wang, Y., Fodjo, E. K., Yang, G. X., Han, F., & Shen, X. S., (2018). Loop-mediated isothermal amplification for visual detection of *Vibrio parahaemolyticus* using gold nanoparticles. *Microchimica Acta, 185*(1), 35.
36. Law, J. W. F., Ab Mutalib, N. S., Chan, K. G., & Lee, L. H., (2015). Rapid methods for the detection of foodborne bacterial pathogens: Principles, applications, advantages and limitations. *Frontiers in Microbiology, 5*, 770–775.
37. Lillis, L., Siverson, J., & Lee, A., (2016). Factors influencing recombinase polymerase amplification (RPA) assay outcomes at point of care. *Molecular and Cellular Probes, 30*(2), 74–78.
38. Long, Y., Zhou, X., & Xing, D., (2011). Sensitive and isothermal electrochemiluminescence gene-sensing of *Listeria monocytogenes* with hyperbranching rolling circle amplification technology. *Biosensors and Bioelectronics, 26*(6), 2897–2904.
39. Lu, X., Shi, X., Wu, G., Wu, T., Qin, R., & Wang, Y., (2017). Visual detection and differentiation of classic swine fever virus strains using nucleic acid sequence-based amplification (NASBA) and G-quadruplex DNAzyme assay. *Scientific Reports, 7*. Article ID: 44211.
40. Mangal, M., & Bansal, S., (2016). Molecular detection of foodborne pathogens: A rapid and accurate answer to food safety. *Critical Reviews in Food Science and Nutrition, 56*(9), 1568–1584.
41. Manjunatha, C., & Sharma, S., (2018). Rapid detection of *Puccinia triticina* causing leaf rust of wheat by PCR and loop-mediated isothermal amplification. *PloS One, 13*(4). Article ID: e0196409.
42. Maruyama, F., Kenzaka, T., Yamaguchi, N., Tani, K., & Nasu, M., (2003). Detection of bacteria carrying the stx2 gene by *in situ* loop-mediated isothermal amplification. *Applied and Environmental Microbiology, 69*(8), 5023–5028.
43. Min, J., & Baeumner, A. J., (2002). Highly sensitive and specific detection of viable *Escherichia coli* in drinking water. *Analytical Biochemistry, 303*(2), 186–193.
44. Mollasalehi, H., & Yazdanparast, R., (2013). Development and evaluation of a novel nucleic acid sequence-based amplification method using one specific primer and one degenerate primer for simultaneous detection of *Salmonella enteritidis* and *Salmonella typhimurium*. *Analytica Chimica Acta, 770*, 169–174.
45. Murinda, S. E., Ibekwe, A. M., & Zulkaffly, S., (2014). Real-time isothermal detection of Shiga toxin-producing *Escherichia coli* using recombinase polymerase amplification. *Foodborne Pathogens and Diseases, 11*, 529–536.
46. Ngamsom, B., & Esfahani, M. M., (2016). Multiplex sorting of foodborne pathogens by on-chip free-flow magnetophoresis. *Analytica Chimica Acta, 918*, 69–76.

47. Ngamsom, B., & Lopez-Martinez, M. J., (2016). On-chip acoustophoretic isolation of microflora including *S. typhimurium* from raw chicken, beef and blood samples. *Journal of Microbiological Methods, 123*, 79–86.
48. Niessen, L., Bechtner, J., Fodil, S., Taniwaki, M. H., & Vogel, R. F., (2018). LAMP-based group specific detection of aflatoxin producers within *Aspergillus* section *Flavi* in food raw materials, spices, and dried fruit using neutral red for visible-light signal detection. *International Journal of Food Microbiology, 266*, 241–250.
49. Notomi, T., Okayama, H., & Masubuchi, H., (2000). Loop-mediated isothermal amplification of DNA. *Nucleic Acids Research, 28*(12), e63–e63.
50. Park, J. W., Louzada, E. S., & Braswell, W. E., (2018). New diagnostic real-time PCR method for huanglongbing detection in citrus root tissue. *Journal of General Plant Pathology, 84*(5), 359–367.
51. Piepenburg, O., Williams, C. H., Stemple, D. L., & Armes, N. A., (2006). DNA detection using recombination proteins. *PLoS Biology, 4*(7), e204–e210.
52. Ramalingam, N., San, T. C., Kai, T. J., Mak, M. Y. M., & Gong, H. Q., (2009). Microfluidic devices harboring unsealed reactors for real-time isothermal helicase-dependent amplification. *Microfluidics and Nanofluidics, 7*(3), 325–330.
53. Safavieh, M., & Kanakasabapathy, M. K., (2016). Emerging loop-mediated isothermal amplification-based microchip and microdevice technologies for nucleic acid detection. *ACS Biomaterials Science and Engineering, 2*(3), 278–294.
54. Shin, J. H., Hong, J., Go, H., & Park, J., (2018). Multiplexed detection of foodborne pathogens from contaminated lettuces using a handheld multistep lateral flow assay device. *Journal of Agricultural and Food Chemistry, 66*(1), 290–297.
55. Simpkins, S. A., & Chan, A. B., (2000). RNA transcription-based amplification technique (NASBA) for the detection of viable *Salmonella enterica. Letters in Applied Microbiology, 30*(1), 75–79.
56. Tanner, N. A., Zhang, Y., & Evans, Jr. T. C., (2015). Visual detection of isothermal nucleic acid amplification using pH-sensitive dyes. *Biotechniques, 58*(2), 59–68.
57. Tomita, N., Mori, Y., Kanda, H., & Notomi, T., (2008). Loop-mediated isothermal amplification (LAMP) of gene sequences and simple visual detection of products. *Nature Protocols, 3*(5), 877–882.
58. Trinh, T. N. D., & Lee, N. Y., (2018). A rapid and eco-friendly isothermal amplification microdevice for multiplex detection of foodborne pathogens. *Lab on a Chip, 18*(16), 2369–2377.
59. Tsaloglou, M. N., & Nemiroski, A., (2018). Handheld isothermal amplification and electrochemical detection of DNA in resource-limited settings. *Analytical Biochemistry, 543*, 116–121.
60. Umesha, S., & Manukumar, H. M., (2018). Advanced molecular diagnostic techniques for detection of foodborne pathogens: Current applications and future challenges. *Critical Reviews in Food Science and Nutrition, 58*(1), 84–104.
61. Uzun, L., & Turner, A. P., (2016). Molecularly-imprinted polymer sensors: Realizing their potential. *Biosensors and Bioelectronics, 76*, 131–144.
62. Valderrama, W. B., Dudley, E. G., Doores, S., & Cutter, C. N., (2016). Commercially available rapid methods for detection of selected foodborne pathogens. *Critical Reviews in Food Science and Nutrition, 56*(9), 1519–1531.

63. Vidic, J., Manzano, M., Chang, C. M., & Jaffrezic, R. N., (2017). Advanced biosensors for detection of pathogens related to livestock and poultry. *Veterinary Research, 48*(1), 1–22.
64. Vidic, J., Vizzini, P., & Manzano, M., (2019). Point-of-need DNA testing for detection of foodborne pathogenic bacteria. *Sensors, 19*(5), 1100–1110.
65. Vilela, D., González, M. C., & Escarpa, A., (2012). Sensing colorimetric approaches based on gold and silver nanoparticles aggregation: Chemical creativity behind the assay: A review. *Analytica Chimica Acta, 751*, 24–43.
66. Vincent, M., Xu, Y., & Kong, H., (2004). Helicase-dependent isothermal DNA amplification. *EMBO Reports, 5*(8), 795–800.
67. Wachiralurpan, S., Sriyapai, T., & Areekit, S., (2018). Rapid colorimetric assay for detection of *Listeria monocytogenes* in food samples using LAMP formation of DNA concatemers and gold nanoparticle-DNA probe complex. *Frontiers in Chemistry, 6*, 90–96.
68. Walker, G. T., (1993). Empirical aspects of strand displacement amplification. *Genome Research, 3*(1), 1–6.
69. Wang, C., Li, R., Quan, S., Shen, P., Zhang, D., Shi, J., & Yang, L., (2015). GMO detection in food and feed through screening by visual loop-mediated isothermal amplification assays. *Analytical and Bioanalytical Chemistry, 407*(16), 4829–4834.
70. Wang, L., Qian, C., Qian, W., Wang, R., Wu, J., & Ying, Y., (2017). A highly specific strategy for in suit detection of DNA with nicking enzyme assisted amplification and lateral flow. *Sensors and Actuators B: Chemical, 253*, 258–265.
71. Wang, R., Xiao, X., Chen, Y., Wu, J., & Qian, W., (2017). Loop-mediated, isothermal amplification-based method for visual detection of *Vibrio parahaemolyticus* within only 1 h, from shrimp sampling to results. *Analytical Methods, 9*(11), 1695–1701.
72. Wang, X., Fu, Z., Chen, X., Peng, C., Xu, X., Wei, W., Li, F., & Xu, J., (2017). Use of a novel metal indicator to judge loop-mediated isothermal amplification for detecting the 35s promoter. *Analytical and Bioanalytical Chemistry, 409*(4), 881–889.
73. Wang, Y., & Salazar, J. K., (2016). Culture-independent rapid detection methods for bacterial pathogens and toxins in food matrices. *Comprehensive Reviews in Food Science and Food Safety, 15*(1), 183–205.
74. Weng, X., & Neethirajan, S., (2017). Ensuring food safety: Quality monitoring using microfluidics. *Trends in Food Science & Technology, 65*, 10–22.
75. Wong, Y. P., Othman, S., Lau, Y. L., Radu, S., & Chee, H. Y., (2018). Loop-mediated isothermal amplification (LAMP): A versatile technique for detection of microorganisms. *Journal of Applied Microbiology, 124*(3), 626–643.
76. Wu, M. Y. C., Hsu, M. Y., & Chen, S. J., (2017). Point-of-care detection devices for food safety monitoring: Proactive disease prevention. *Trends in Biotechnology, 35*(4), 288–300.
77. Wu, W., Zhao, S., & Mao, Y., (2015). Sensitive lateral flow biosensor for *Escherichia coli* O157:H7 detection based on aptamer mediated strand displacement amplification. *Analytica Chimica Acta, 861*, 62–68.
78. Yan, L., Zhou, J., & Zheng, Y., (2014). Isothermal amplified detection of DNA and RNA. *Molecular BioSystems, 10*(5), 970–1003.

79. Yang, W., Song, X., Wang, J., Li, Z., Ji, M., & Li, Y., (2014). Detection methods for milk pathogenic bacteria by loop-mediated isothermal amplification. *Bioscience Trends, 8*(6), 316–321.
80. Zhang, F., Wang, R., Wang, L., Wu, J., & Ying, Y., (2014). Tracing phosphate ions generated during DNA Amplification and its simple use for visual detection of isothermal amplified products. *Chemical Communications, 50*(92), 14382–14385.
81. Zhang, F., Wu, J., Wang, R., Wang, L., & Ying, Y., (2014). Portable pH-inspired electrochemical detection of DNA amplification. *Chemical Communications, 50*(61), 8416–8419.
82. Zhang, M., Ye, J., He, J. S., Zhang, F., Ping, J., Qian, C., & Wu, J., (2020). Visual detection for nucleic acid-based techniques as potential on-site detection methods: A review. *Analytica Chimica Acta, 1099,* 1–15.
83. Zhang, X., Guo, L., Ma, R., & Cong, L., (2017). Rapid detection of *salmonella* with recombinase - aided amplification. *Journal of Microbiological Methods, 139*, 202–204.
84. Zhang, X., Lowe, S. B., & Gooding, J. J., (2014). Review of monitoring methods for loop-mediated isothermal amplification (LAMP). *Biosensors and Bioelectronics, 61*, 491–499.
85. Zhao, X., Lin, C. W., Wang, J., & Oh, D. H., (2014). Advances in rapid detection methods for foodborne pathogens. *Journal of Microbiology and. Biotechnology, 24*(3), 297–312.
86. Zhao, Y., Chen, F., Li, Q., Wang, L., & Fan, C., (2015). Isothermal amplification of nucleic acids. *Chemical Reviews, 115*(22), 12491–12545.
87. Zhong, J., & Zhao, X., (2018). Isothermal amplification technologies for the detection of foodborne pathogens. *Food Analytical Methods, 11*(6), 1543–1560.
88. Zhong, Q., Tian, J., Wang, B., & Wang, L., (2016). PMA based real-time fluorescent lamp for detection of *Vibrio parahaemolyticus* in viable but nonculturable state. *Food Control, 63*, 230–238.
89. Zhu, D., Yan, Y., Lei, P., Shen, B., & Cheng, W., (2016). Novel electrochemical sensing strategy for rapid and ultrasensitive detection of *Salmonella* by rolling circle amplification and DNA–AuNPs probe. *Analytica Chimica Acta, 846*, 44–50.

CHAPTER 4

ANTIMICROBIAL RESISTANCE IN FOOD GRADE LACTIC ACID BACTERIA

SHRUTI NAYYAR, SAURABH KADYAN, HEENA PARVEEN, RASHMI H. MALLAPPA, and DIWAS PRADHAN

ABSTRACT

Antimicrobial resistance (AMR) in bacteria has emerged as one of the major public health threats in the 21st century. An estimated 700,000 deaths occur annually due to AMR bacteria worldwide. In addition to the use of antimicrobial agents (AMAs) in humans, livestock have also been regarded as a crucial contributor in the overall emergence of AMR problem. However, most of the national and international programs are focused mainly on human pathogens, leaving behind the commensal and beneficial bacteria out of the purview. Besides the development of resistance in common pathogens, AMR in commensal and beneficial bacteria is also worrisome because of its ability to spread the resistant determinants horizontally to other infective bacteria. Lactic acid bacteria (LAB) are integral components of fermented foods, many of which are now increasingly finding use as beneficial probiotics. The role of LAB as reservoir of resistant determinants with transmission potential to pathogenic species is now increasingly acknowledged, thus representing a potential health risk. Various *in vitro* and *in vivo* experiments have shown that LAB can play an active role in disseminating resistant determinants across bacterial genus and species. Moreover, it has now been demonstrated that different fermented foods harbors resistance genes, hinting at a potential role these food items may play as a carrier of AMR genes. The recent metagenomic approaches have been instrumental in this regard to decipher the antibiotic resistance (AR) genes or the resistome present in these common foods. Shotgun metagenomics is an emerging tool that describes AMR abundance in bacterial communities more accurately than commonly used culture techniques on selected indicator organisms.

4.1 INTRODUCTION

Antimicrobial resistance (AMR) occurs when microorganisms are continuously exposed to antimicrobial agents (AMAs) and as a result of some mechanisms, they can grow and survive in presence of AMAs, which in normal conditions would generally inhibit or kill them. AMR is not a new phenomenon, but nowadays it has become a complex global issue that has serious implications on public health and is growing rapidly worldwide. Due to the growing AMR in common pathogens, it has become increasingly difficult to treat even common infections.

The main factor of AMR in bacteria is over prescribing and misuse of antibiotics in animal feed, agriculture, medicine, and aquaculture. Many studies report that the AMR genes or bacteria are widely distributed among the environment and spread through horizontal gene transfer (HGT) but few literatures cited the distribution pattern and origin of AMR. In India, the situation of AMR is much more serious since many of the multi-drug resistance traits such as extended spectrum beta-lactams (ESBLs), Metallo-beta-lactamases (MBLs), New Delhi, metallo-β-lactamase, etc., have already been documented in Indian isolates and now are found worldwide. A recent study was conducted in 2010, reported that India was the world's largest user of antibiotics for human health [20].

However, the major focus of most of the studies and schemes has been mainly on the pathogens or clinically relevant bacteria, leaving behind the commensal and beneficial bacteria out of the scope, though, the fact remains that AMR is not only a problem in pathogen or clinically relevant bacteria but also in commensals and dairy beneficial bacteria like LAB. Indeed, LAB represents the group of most important microbes that are present in large numbers in various habitats including gastrointestinal tract (GI) of animals and humans and also widely exploited as starter and probiotic culture in various fermented food products.

The role of LAB in transferring AR genes to pathogens and gut microbiota through HGT or acting as a reservoir of these genes is now increasingly acknowledged, hence, imposing a direct risk to consumer's health. The prevalence of AMR genes in specific niches such as fermented foods or GI tract which hosts myriad of LABs, can now be comprehensively studied employing the modern approach of metagenomics. These tools are culture independent and do not rely on isolating individual microbes for studying AMR profile, rather the technique employs the analysis of whole DNA directly extracted from the sample and studying the DNA for the presence of AMR genes by DNA sequencing and sequence analysis. The metagenomic sequencing technique has now emerged as a suggested tool to screen samples including fermented foods for the presence of multitude of genes that are involved in AMR and its movement across various bacterial species.

4.2 MECHANISMS OF ANTIBIOTIC RESISTANCE (AR) IN LAB

The use of lactic acid bacteria (LAB) in various fermentation processes was known to be healthy and safe. Their immense application as starter culture or probiotics in development of techno-functional fermented products is well recognized. Due to improper use of antibiotics in the past in agriculture and clinical settings, LAB poses a potential reservoir for transmitting antibiotic resistant determinants to commensal or pathogenic microorganisms as they are frequently isolated from the microbiome of both animals and humans.

Repeated antibiotic exposure has necessitated bacteria to adapt novel mechanisms to counteract the bactericidal effect of antibiotics. Broadly, the counteracting mechanisms evolved by bacteria are classified as intrinsic (or innate) resistance due to inherent structural or functional attributes of these species; and extrinsic (or acquired) resistance due to random mutations in chromosomal DNA or possession of exogenous DNA [29]. The tendency of which mechanism prevails over another is governed by antibiotic type, its site of action, the bacterial species and whether resistance resulted from plasmid or chromosomal mutation [35].

4.2.1 GENERAL MECHANISMS OF ANTIBIOTIC RESISTANCE (AR)

Differential uptake of antibiotics among bacterial species is considered as a general mechanism of resistance against multiple drugs [8]. The resistance mechanisms can be broadly sub-grouped into four categories (Figure 4.1):

- Minimization of antibiotic concentration either by restricting its entry inside cell or by pumping it out from the cell;
- Genetic mutation or post translational alteration in antibiotic's target site;
- Deactivation of antibiotic by its intracellular or extracellular modification/hydrolysis; and
- Acquisition of alternative route for resuming the metabolic processes blocked by the drug.

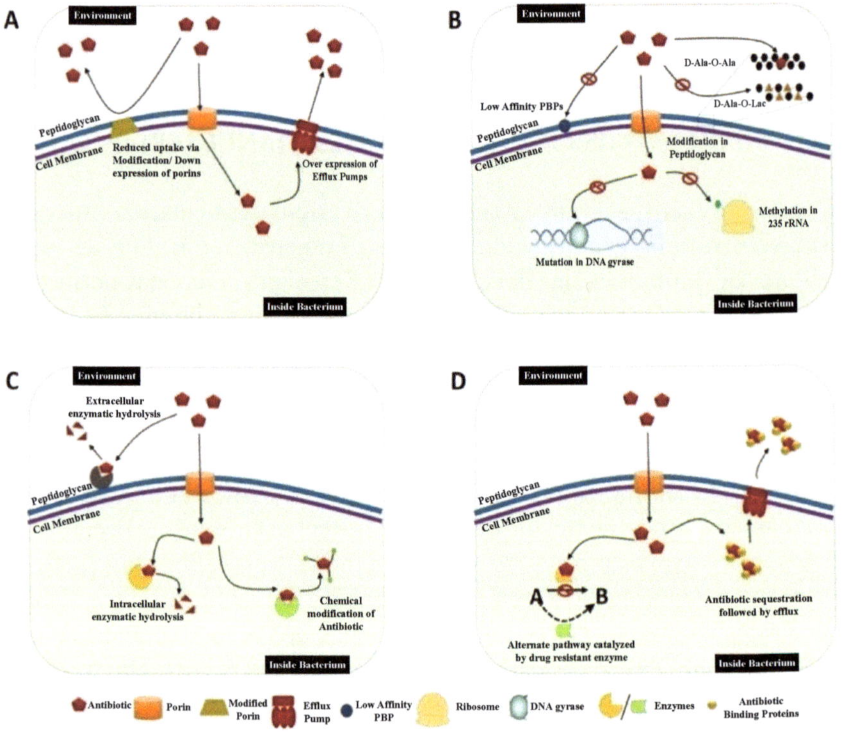

FIGURE 4.1 Overview of resistance mechanisms in LAB: (A) efflux pumps and membrane impermeability; (B) alteration in antibiotic target sites; (C) modification or hydrolysis of antibiotic; and (D) metabolic bypass and target protection.

4.2.1.1 *EFFLUX PUMPS AND MEMBRANE IMPERMEABILITY*

Decreasing the concentration of drugs within the cytoplasm of a cell either by active efflux pumps or through change in membrane permeability can be accomplished by some bacteria as a coping up mechanism of AR [22]. Efflux pumps play a pivotal role in underpinning the innate resistance of a bacterial species to a particular antibiotic. These active efflux pumps transport antibiotics away from bacteria. While some efflux pumps are antibiotic specific such as Tet pumps encoding for tetracycline resistance, others are non-specific conferring resistance to multiple antibiotics (multi-drug) [31]. Multidrug efflux pumps mainly fall under five categories [8]:

- ATP-binding cassette (ABC);
- Resistance-nodulation-division (RND);
- Small multidrug resistance (SMR);
- Major facilitator superfamily (MFS); and
- Multidrug and toxic compound extrusion (MATE) transporters.

These efflux systems when over expressed expel multiple antibiotics in a unidirectional manner and convert potential energy for power transport and majorly contribute AR in LAB by expulsion of structurally unrelated compounds [25]. Example includes two proton motive force driven hop excretion transporter proteins-HorA and HorC*in Lactobacillus* genus [12].

Membrane impermeability helps in reducing the intracellular concentration of antibiotics in bacteria. Some LAB species use this mechanism against aminoglycosides as they are deprived of cytochrome-mediated electron transport to facilitate drug intake [10]. Furthermore, many antibiotics diffuse across the outer membrane through large, non-selective channels called porins. Membrane can be made impermeable to antibiotics by down regulation of porins, replacement with small channeled porins or mutations in porin genes. However, control of resistance through these mechanisms is less favored in Gram positive bacteria as compared to Gram negative bacteria because of the absence of outer membrane.

4.2.1.2 ALTERATION IN ANTIBIOTIC TARGET SITES

Several antibiotics require specific target binding sites to exert their bactericidal effects. Changes in target structure to evade efficient antibiotic binding and at same time enable the target to continue its normal function is used as self-resistance mechanism by bacteria. Modification in cell-wall components such as penicillin binding proteins (PBPs) or D-alanine residues of muramyl pentapeptide can bring resistance to β-lactam group of antibiotics (e.g., penicillin), glycopeptides group (e.g., teicoplanin, and vancomycin) and non-ribosomal (bacitracin) antibiotics [1].

Enterococcus faecium and *Lactobacillus plantarum* exhibits intrinsic resistant to vancomycin due to swapping of D-alanine remnants in N-acetyl muramic acid (NAM) pentapeptide of peptidoglycan with D-serine (low-level resistance) or D-lactate (high-level resistance) by ligase-related enzymes, i.e., D-alanyl D-alanine ligase [28, 43]. This change in structural configuration bypasses binding of vancomycin to D-alanine terminus of pentapeptide and promotes cell wall synthesis of resistant species. However, *L. acidophilus* despite belonging to the same genus is vancomycin sensitive might be due to loss of its plasmid resistance gene during the transfer process [13].

AR development through mutation in a target gene is a common phenomenon especially, where enzymatic resistance mechanisms are absent. Pronounced fluoroquinolone resistance in most bacteria occurs via these genetic variations in quinolone resistance determining regions of gyrA subunit and parC subunit of topoisomerase II and DNA gyrase enzyme respectively, as both enzymes facilitate negative supercoiling of DNA [33]. Resistance to other antibiotics viz. Macrolides, streptogramin, and lincosamides can be brought about by post-translational methylation of distinct adenine residue in 23S ribosomal RNA. The process is further catalyzed by erythromycin resistance methylase (erm) protein. 23S ribosome methylation predominates in *Lactobacillus* genus due to presence of several *erm* encoding proteins [35].

4.2.1.3 MODIFICATION OR HYDROLYSIS OF ANTIBIOTIC

The resistance process by preventing specific binding of an antibiotic to a target can be achieved either by hydrolysis of structural features of an antibiotic or by chemical modification in the structure of antibiotics contains vulnerable hydroxyl or amine groups. Aminoglycosides (kanamycin, gentamicin, and streptomycin) are diverse molecules containing an aminocylitol nucleus attached to different sugar amino groups via glycosidic bonds. Modification in their structure by the enzymes (such as: AAC (N-acetyl transferases), ANT (O-adenyltransferases) and APH (O-phosphotransferases) lead to phosphorylation, acetylation, and adenylylation of antibiotic respectively thereby preventing binding to 30S ribosomal unit of 16SrRNA [30]. In contrast, certain hydrolyzing enzymes produced by bacteria are capable of either inactivating or destroying an antibiotic molecule as defense mechanism to protect them.

Moreover, these hydrolytic enzymes require only water as co-substrate and function well by degrading antibiotics in an extracellular environment. Such enzymes can be effective against either broad or narrow range of antibiotics [7]. The classic example is β-lactamase superfamily of proteins that hydrolyze the β-lactam ring of carbapenems, penicillin, cephalosporins, and monobactams antibiotics. Another not so common example includes macrolide esterase enzymes which target the ester bond of ring structure of these cyclic molecules [2].

4.2.1.4 METABOLIC BYPASS AND TARGET PROTECTION

Another possible mechanism whereby bacteria expand their AR portfolio is via metabolic bypass resulting in internal changes.

Bacteria can also adapt to various phenotypic changes failing antibiotics contact with target sites thus, rendering them worthless against the bacterial species [25]. One such example includes bacterial resistance to sulfonamide and trimethoprim class of antibiotics which are responsible for inhibiting enzyme necessary for biosynthesis of nutrient folate. Bacteria counteract by acquiring a drug resistant version of this enzyme allowing folate production in presence of these drugs [11]. Secretion of certain proteins, such as, Qnr, and ribosomal protection proteins (RPPs) reported to provide protection against fluoroquinolones and tetracyclines [38, 41].

Antibiotic sequestration is a phenomenon usually found in producer soil bacteria wherein target protection is assured by expression of certain drug binding proteins by bacteria responsible for capturing metal bound or metal free antibiotics followed by their efflux through related ABC transporting systems [30].

4.3 INTRINSIC AND EXTRINSIC RESISTANCE IN LAB

4.3.1 INNATE (OR INTRINSIC) RESISTANCE

Innate resistance is the inherent trait of a bacterium to survive in the existence of antibacterial agents. Generally, a high level of endurance to an antibiotic is common within members of a particular bacterial species. For instance, *Enterococcus* spp. is intrinsically immune to clindamycin, cephalosporins, and aminoglycosides [24, 47]. Pediococci, *Leuconostoc* spp. and *Lactobacilli* exhibited intrinsic resistance to vancomycin, a valuable trait for their selective isolation from other Gram-positive bacteria. *Bifidobacterium* spp. universally employed as probiotics, carry resistance to quinolones, mupirocin, tetracyclines, and aminoglycosides [16]. Normally, intrinsic AR genes are chromosomally encoded and hence pose minimum risk of spreading to other bacteria unless or until it is flanked by insertion sequences.

4.3.2 ACQUIRED (OR EXTRINSIC) RESISTANCE

Acquired resistance refers to incorporation of exogenous mobile genetic material by a bacterium having the property of delivering resistance for some antibiotics. Contrary to innate resistance, the extrinsic resistance is associated with some bacterial strains [1]. The transfer of AR within intra- and inter-bacterial species is achieved through the horizontal gene transfer (HGT) mechanism. According to WHO, HGT occurs when the bacteria enhance its antibiotic portfolio through acquisition of new genes and carry the potential to transfer such resistance determinants among other microorganisms or straight to humans and animals thereby posing a greater health risk. Mutations in indigenous DNA of bacteria and acquisition of antimicrobial resistant genes from exogenous DNA contribute to this type of resistance, albeit mutations play a minor role in resistance development. However, mutation becomes prominent for those bacteria, which do not participate in HGT, e.g., *Mycobacterium tuberculosis* as well as for novel resistant genes lacking natural reservoir, e.g., mutation of topoisomerase genes against fluoroquinolones [6, 42].

HGT over genus and species borders occurs mainly through [16]: (i) transformation: intake of naked exogenous DNA (from dead cells) by competent cells (ii) conjugation: AR genes transfer through conjugative plasmids and transposons; and (iii) transduction: transfer via lytic and temperate bacteriophages (Figure 4.2). Incorporation of DNA directly in the genomes of naturally competent bacteria is confined to closely related species only as homologous recombination occurs among specific regions of high sequence similarities. Hence, majority of HGT mechanisms among distantly related species are dependent on the conjugation or direct cell-to-cell contact especially where resistant determinant genes are encoded on mobile genetic elements (MGEs), such as, integrons, transposons, and plasmids [17].

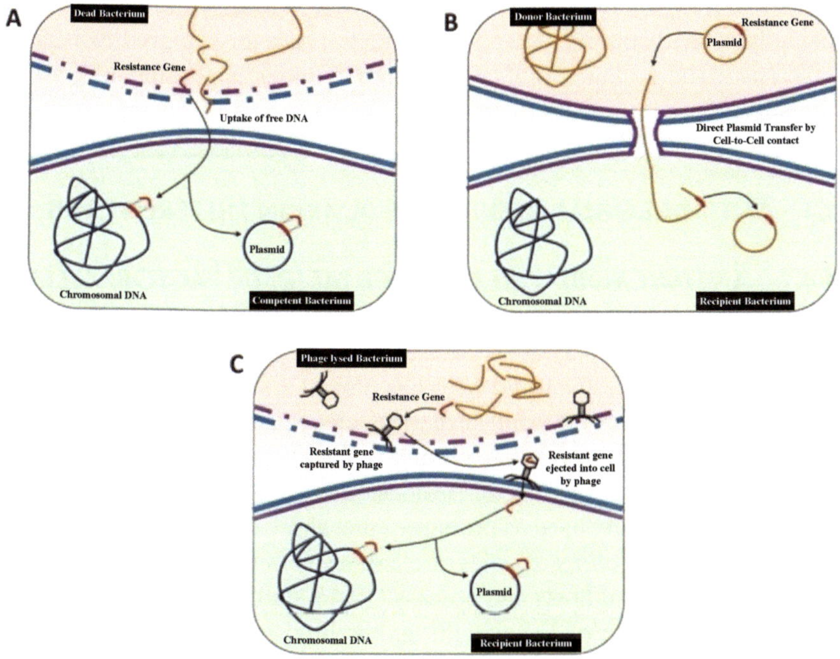

FIGURE 4.2 Horizontal gene transfer mechanism. (A) transformation, (B) conjugation; and (C) transduction.

Some authors have highlighted the role of plasmids in contributing the genetic variation of resistance among bacteria. Higher the plasmids present in the environment, more will be the AR diversity as transposons and integrons are unable to self-replicate and require plasmids/phages for their transportation [17, 57]. Transition from chromosomal to plasmid mediated resistance mechanisms warrants close attention for assuring 'One Health' due to their enhanced expression and easy dissemination in the environment.

Lactococcus lactis subsp. *lactis* was reported with a plasmid encoded efflux protein (Mdt(A), that involved in the resistance mechanism towards various antibiotics such as erythromycin, spiramycin, tetracycline, clindamycin, lincomycin, minocycline, and azithromycin [44]. Several conjugative transposons described in LAB covers Tn (916, 918, 920, 925, 2702) in *E. faecalis*, Tn5233 in *E. faecium*, Tn (5276, 5301) in *Lactococcus lactis* [75]. In addition to atypical resistance, intrinsic resistance related genes if encoded on plasmids, or flanked by transposons or integrons followed by their successful expression and dissemination among distantly related species also pose a potential risk for acquired resistance which needs careful attention and research to understand safety of animal, human, and environment health.

4.4 ANTIMICROBIAL RESISTANCE (AMR) IN DIFFERENT LAB

4.4.1 ANTIMICROBIAL RESISTANCE (AMR) IN LACTOBACILLUS SPECIES

For several centuries, *Lactobacillus* has played an integral role in various food fermentations. At present, several species of *Lactobacillus* are widely used as starter and/or protective culture in fermentation of dairy products, vegetables, meat, sausages, etc. Besides, several strains of *Lactobacillus* are also commercially used as probiotics that show various health benefits. Although, *Lactobacillus* species are common members of indigenous microbiota of several hosts and possess GRAS status, several reports have documented the role of lactobacilli as potential reservoirs of AR genes, thus, the safety aspects of *Lactobacillus* regarding transferable AR genes should be considered as a matter of concern.

The antibiotic susceptibility profile of lactobacilli has been reported as early as by Sozzi et al. [79] who reported susceptibility of lactobacilli towards cloxacillin, novobiocin, doxycycline, oxacillin, and streptomycin, whereas, variable resistance had been observed in case of kanamycin and streptomycin antibiotic. Teuber et al. [83] also reported that *Lactobacillus* species were susceptible to nalidixic acid, mycostatin, polymyxin B, neomycin, colimycin, sulphonamides. Zhou et al. [99], later observed that *Lactobacillus* species are generally susceptible to those classes of antibiotics that inhibit the protein synthesis such as erythromycin, tetracycline followed by clindamycin and chloramphenicol. Subsequent studies by Ammor et al. [6]; Luh et al. [56]; Patel et al. [67] also confirmed that *Lactobacillus* species showed susceptibility towards penicillin (penicillin G and ampicillin), tetracycline, streptomycin, erthyromycin, clindamycin, and quinupristin/ dalfopristin.

In terms of intrinsic resistance, *Lactobacillus* are classical vancomycin (Glycopeptides) resistant along with demonstrating resistance against aminoglycosides class of antibiotics such as neomycin, kanamycin, streptomycin, and gentamycin and other classes of antibiotics that inhibit nucleic acid synthesis and folic acid synthesis (sulfamethoxazole, trimethoprim, and co-trimoxazole) (Table 4.1). The vancomycin resistance in lactobacilli is due to a change in vancomycin target, i.e., D-alanine/D-alanine terminus of the pentapeptide of the peptidoglycan precursor replaced by D-lactate or D-serine in the muramyl pentapeptide, thereby, preventing vancomycin binding [72]. The mechanism describing the aminoglycoside resistance is lactobacilli has been attributed to the absence of cytochrome-mediated electron transport, which mediates drug uptake.

TABLE 4.1 *Lactobacillus* Intrinsic Resistance against Antibiotic Groups and its Mechanisms

Antibiotic	Antibiotic Class	Mechanism of Resistance	References
Kanamycin, neomycin	Amino glycosides	Active efflux pump, decreases, cell membrane permeability (porin loss), alteration of ribosome, inactivation of aminoglycosides modifying enzymes	[62]
Vancomycin, teicoplanin	Glycopeptides	Vancomycin resistance resulted from production of modified peptidoglycan precursors, low binding affinities	
Ciprofloxacin, levofloxacin, norfloxacin	Quinolones	Decreases cell membrane permeability, over expression of naturally occurring efflux pump or mutation of molecular targets such as DNA gyrase	
Trimethoprim		Resistance generally achieved by bypass mechanisms	

The wild types of *Lactobacillus* species are normally susceptible to inhibitors of cell wall synthesis and β-lactams such as penicillin, ampicillin, oxacillin, piperacillin, etc., however, frequent studies have reported resistance towards these otherwise generally susceptible antibiotics in *Lactobacillus*. Zhou et al. [99] documented *L. casei* from *Dahi* that showed resistance towards penicillin groups, especially Penicillin G. Apart from penicillin, some reports also suggest *L. casei* to show resistance against ampicillin.

Belleti et al. [14] have also reported *Lactobacillus* (especially *L. rhamnosus*) from certain cheese showing resistance towards oxacillin and cephalosporins. Hu et al. [39] also reported a number of *Lactobacillus* from fermented foods to be resistant towards erythromycin; chloramphenicol, clindamycin, and tetracycline. Additionally, a number of other studies have reported unnatural resistance in lactobacilli. Hence, it can be understood that lactobacilli evolve resistance to the earlier sensitive antibiotics which may be attributed to either genomic mutations or HGT by various mechanisms.

Many researchers have also tried to study AMR in lactobacilli at gene level in order to decipher the genetic basis of unnatural resistance against susceptible antibiotics. Devirgiliis et al. [25], documented lactobacilli from various fermented food that showed acquired resistance to clindamycin, erythromycin, tetracycline, and chloramphenicol and which was mediated by genes encoded in plasmids along with another plasmid encoding resistant *cat* genes. It is interesting to see that genes against these antibiotics are the most commonly reported and studied resistance gene observed in different species of lactobacilli isolated from various food and dairy products.

The presence of AR genes such as *tet(M), tet(W), tet(S)* coding for tetracycline resistance and *erm(B)* and *erm(C)* for erythromycin resistance as the most commonly detected AR genes in different species of lactobacilli have been validated independently by different researchers [6, 24, 27, 36]. Thuma and Halami [86] have reported frequent incidence of resistance against tetracycline in lactobacilli which have been attributed due to the presence of genes, *tetM, tetW, tetO, tetK, and tetL* alone or in combination. In addition to tetracycline and erythromycin resistance genes, other genes such as *aphA3* and *aadE* genes that conferred acquired resistance against aminoglycoside.

Han et al. [38] have reported that *L. casei* and *L. pentosus* strains isolated from dairy products were found to carry *aph(3')-IIIA*, *aadA* and *aadE* genes. Although the presence and transferability of multi-resistant gene in lactobacilli is rare, yet Gfeller et al. [35] showed that *erm-vat-tet* resistance gene that provided resistance against tetracycline and erythromycin in *L. fermentum* strain isolated from cheese could be transferred to other species.

4.4.2 ANTIMICROBIAL RESISTANCE (AMR) IN *STREPTOCOCCUS THERMOPHILUS*

Streptococcus thermophilus was thought to have been used since animals were domesticated and milk availability was more consistent (for some 10,000 years). Among LAB, *S. thermophilus* is considered as the second most important economic species after *Lactococcus*. It is mainly employed as starter culture in fermentation of various milk products, predominantly involved in the manufacture of fermented milk, yogurt, various types of cheeses such as Grana Padano, Mozzarella, Emmental, etc.

Some strains of *S. thermophilus* are generally paired with *Lactobacillus* species during fermentation and synergistically enhance the growth of lactobacilli by releasing formate as a by-product. Since *S. thermophilus* has a long history of healthy use as fermenting natural bacteria, it has also been given the GRAS status by the U.S. FDA (United States Food and Drug Administration). Over the last decade, several researchers have worked on detecting and characterizing the resistance among *S. thermophilus* species, and there are quite a few data available regarding this species pattern of 'susceptibility and resistance.' It has just been discovered that *S. thermophilus* act as a carrier of various AR factors in retail cheeses, suggesting that fermentation of cheese is a susceptible method by which antibiotic resistant bacteria can develop and thrive [95].

Sozzi et al. [79] observed that *S. thermophilus* strains were resistant to ampicillin, aureomycin, carbenicillin, cephaloridine, cephalothin, clindamycin, dicloxacillin, doxycycline, erythromycin, mandelamine, rifampin. Further studies by Katla et al. [45]; Temmerman et al. [82]; Aslim and Beyatli [8] have shown that *S. thermophilus* is generally susceptible to cephalothinquinupristin/dalfopristin, chloramphenicol, tetracycline, erythromycin, and ciprofloxacin, whereas it intrinsically shows resistance towards aminoglycoside group of antibiotics (streptomycin, trimethoprim, gentamicin, kanamycin, and sulfadiazine) (Table 4.2).

TABLE 4.2 Intrinsic Resistance in *Streptococcus thermophilus*

Antibiotic	Antibiotic Class	Mechanism of Resistance	References
Gentamicin, kanamycin, and streptomycin	Aminoglycosides	Active efflux pump, decreases, cell membrane permeability (porin loss), alteration of ribosome, inactivation of aminoglycosides modifying enzymes	[6]
Trimethoprim		Resistance generally achieved by bypass mechanisms	
Sulphadiazine	Sulphonamides		

Aslim et al. [8] found that certain strains of *S. thermophilus* isolated from Turkish yogurt showed resistance towards vancomycin which may be an acquired characteristic. Karapetkov et al. [47] also observed that some of the *S. thermophilus* strains demonstrated acquired resistance towards erythromycin, streptomycin, and tetracycline. At the genetic level AR genes (*tetM* and *ermB*) have also been reported in *S. thermophilus* strains from certain pharmaceutical and dairy products.

In a study by Flórez et al. [31], AMR genes were reported in *S. thermophilus* spp. isolated from raw milk samples against tetracycline, erythromycin, clindamycin, streptomycin, and neomycin, which hints the ability of *S. thermophilus* to acquire resistance against otherwise sensitive antimicrobials. Yang et al. [95] have also documented *S. thermophilus* strains from various fermented dairy products of China which showed high level of resistance towards ciprofloxacin (79.5%), streptomycin (92.3%), chloramphenicol (71.8%) and gentamycin (87.2%) and showed lower resistance towards erythromycin (7.7%).

Hence, it is highly probable that *S. thermophilus* can acquire AMR genes and develop resistance to resistant antibiotics, however, the chances that *S. thermophilus* may pass AMR genes to pathogenic bacteria present either in the GI tract or in food and pose a potential clinical threat needs to be properly assessed.

4.4.3 ANTIMICROBIAL RESISTANCE (AMR) IN LACTOCOCCUS SPECIES

Lactococcus constitutes one of the important genera that form the LAB family, used primarily as a starter culture in the fermentation process of various dairy products. According to Odamaki et al. [64], *Lactococcus* genera consists of total of seven species; and among all, *L. lactis* is commercially important due to its excessive use in dairy and food processing and in food preservative nisin. Two major subspecies of *Lactococcus*, i.e., *L. lactis* ssp. *lactis* and *L. lactis* ssp. *cremoris* strains are generally robust species and mostly associated as a milk-souring isolate and in cheddar cheese production. Devirgiliis et al. [25] reported that lactococci, like other LAB, can also acquire AMR under selective pressure, withstand specific AMAs and serve as a reservoir for transmissible AR genes. Although it is not yet recognized as probiotic strains, yet there is increasing interest in AMR issue in this genus due to the successful use of different *Lactococcus* strains as a dairy starter for fermentation of various dairy products.

Number of reports have documented the AMR profile of lactococci from various sources. Florez et al. [29] studied the susceptible nature of *L. lactis* towards lincomycin, erythromycin, novobiocin, teicoplanin, vancomycin, chloramphenicol, rifampicin, spectinomycin, and beta-lactams group of antibiotics. Khemariya et al. [48] also documented susceptibility of *L. lactis* subsp. *lactis* isolated from dairy sources towards ampicillin, spiramycin, erythromycin, spectinomycin, rifamycin, ciprofloxin, and trimethoprim. Singh et al. [78] also observed ampicillin and vancomycin to be highly effective against lactococci strains from dairy sources. Thokchom and Joshi [85] documented strains of *Lactococcus* spp. was sensitive towards different classes of antibiotics, most commonly aminoglycosides, β-lactam, glycopeptides, cephalosporins, chloramphenicol, and lincosamides. Variable responses were observed in lactococci against streptomycin and gentamycin.

In terms of intrinsic resistance, Devirgillis et al. [25] stated that species of lactococci show resistance towards aminoglycosides class of antibiotics (such as kanamycin and gentamycin), trimethoprim, metronidazole, and cefoxitin. However, variable resistance against a number of other antibiotics have been observed in lactococci (Table 4.3). Florez et al. [30] documented a single strain of *L. lactis*, which showed resistance towards chloramphenicol, clindamycin, erythromycin, streptomycin, and tetracycline. Similarly, Tommey et al. [87] reported *L. lactis* with phenotypic resistance against six antibiotics, i.e., ampicillin, erythromycin, streptomycin, chloramphenicol, and tetracycline.

Thokchom and Joshi [85] also reported *Lactococcus* spp. from traditionally fermented Indian foods, resistant against cephalexin and cloxacillin. Hummel et al. [40] stated that like other LAB, *tetracycline* resistance gene is not frequently documented in lactococci, however, Florez et al. [30], reported two strains of *L. lactis* from artisanal starter free cheese with *tet*(M) gene carried by Tn916 transposon. Ammor et al. [6] also reported *L. lactis* K214 strain from raw milk soft cheese with plasmid-mediated AR determinants for chloramphenicol, streptomycin, and *tet(S)* for tetracycline. Further, Devirgiliis et al. [25] reported the availability of erythromycin and tetracycline resistance genes in *L. lactis* strains from Italian traditional cheese (*Mozzarella di Bufala Campana*).

Li et al. [54] reported the existence of *dfrA* gene in lactococci carried by Tn4003 transposons described in *S. aureus*, illustrating the possible transmission route of the AR genes. The *dfrA* gene encodes DHFR (dihydro-folatereductase), a drug resistant enzyme that is related with trimethoprim resistance. Toomey et al. [87] revealed the recipient ability of *Lactococcus* for *tet*(M) gene in the conjugative transfer experiment by using *Lactobacillus* as donor.

TABLE 4.3 Intrinsic AMR in *Lactococcus lactis* and the Mechanism of Resistance

Antibiotic	Antibiotic Class	Mechanism of Resistance	References
Colistin	Polymyxins	Resistance gene encoded in plasmid	[62]
Pipemidic acid	Quinolones	Resistance typically arises due to alteration in the target enzymes (DNA gyrase), additional mutation, it can also be mediated by plasmids.	
Rifamycin	Rifampin	Rifampicin resistance occurs from mutations that modify rifampicin binding site residues on RNA polymerase, resulting in diminished affinity for rifampicin.	

4.4.4 ANTIMICROBIAL RESISTANCE (AMR) IN ENTEROCOCCUS SPECIES

The *Enterococcus* genus, like all other LAB, is also being used in the dairy fermentation industry for a long time because of its valuable role in the production of (artisanal) cheeses 'organoleptic quality and the development of enterococci (enterocin) bacteriocins. Significant function of *E. feacium*, *E. feacalis* and *E. durans* in the maturation of traditional cheeses has shown that enterococci play a vital role in the maturation of these cheeses, thus leading to their characteristic taste and flavor. Due to these technological applications, enterococci have also been used as an adjunct or protective culture in the manufacture of fermented foods.

In addition, enterococci are nowadays marketed as probiotics, believed to preserve natural intestinal microflora, stimulate the immune system, and enhance food nutritional value. Besides these characteristics, enterococci present to humans an evolving array of emerging pathogens as they cause disease, act as a reservoir of AR agents, and are often armed with possible virulence factors. *E. feacalis* or *E. feacium* strains are mainly involved in infection processes. The use of enterococci remains a debatable problem because of this 'dualistic' character.

Enterococci virulence is greatly improved by their repeated tolerance to widely used antibiotics, making them an important opportunist for nosocomial infections. Enterococci possess both innate resistance against β-lactams, cephalosporins, sulphonamides, and low levels of clindamycin and aminoglycosides and transferable resistance against fluoroquinolones, chloramphenicol, erythromycin, and glycopeptides, such as vancomycin (Table 4.4) [33]; and Penicillin and ampicillin, both active β-lactams work against enterococci by inhibiting its peptidoglycan synthesis. Enterococci produce five types of PBPs.

TABLE 4.4 Intrinsic AMR Profiling in *Enterococcus* Species

Antibiotic	Antibiotic Class	Mechanism of Resistance	References
Cephalosporin	β-lactams	By expressing β lactamase, plasmid mediated AmpC enzymes	[62]
Gentamycin	Aminoglycosides	Active efflux pump, decreased permeability, ribosome alteration and by inactivation of drug by aminoglycoside modifying enzyme	
Sulfonamide	Sulfonamide	Chromosomally encoded, through mutation, etc.	

E. faecalis and *E. faecium* shows intrinsic resistance towards β-lactams (ampicillin and cephalosporins) and this resistance was associated with the presence of class B penicillin binding protein, encoded by a species-specific chromosomal gene, *pbp5* [76]. Enterococci with resistance genes against various antibiotics can easily spread to other non-pathogenic/pathogenic cultures by HGT mechanism. Various studies reported the transfer ability and diversity of resistance determinants in various organisms.

Rizzotti et al. [73] studied 20 species of enterococci originating from swine meat to examine the diversity and transfer ability of tetracycline gene *tet(M)* belonging to *E. faecium* (4 = no. of species), *E. hirae* (1), *E. faecalis* (12), *E. durans* (2) and *E. mundtii* (1). It was observed that 50% strains possessed *tet(L)* and tet(M) genes and were found correlated with the Tn916-1545 family transposon. Transposon regulated genetic transfer between species has recently been identified as being the most prevalent route in spreading of AR in bacteria. The transposons of this family generally carry *tet(M)* gene and exhibit broad host range towards Gram positive/negative bacteria. The 50% enterococci strains displaying the transferability of *tet(M)* gene to other *E. faecalis* and *L. innocua* strains showing the role of enterococci in spreading of tetracycline resistance to other pathogenic bacteria. Jaimee et al. [42]; Cisneros et al. [4] also observed multidrug resistance mechanisms in enterococci species that include enzyme modification, inactivation of therapeutic agents and overexpression of efflux pump.

In addition, some strains of enterococci have developed intrinsic resistance towards vancomycin by changing the D-alanine residues of NAM either by D-serine or D-lactate group in the peptidoglycan structure. Vancomycin resistance is of particular interest, as the prevalence of enterococci resistant to vancomycin in hospitals has led to serious infections that cannot be treated with traditional antibiotic therapy. Enterococci genus exhibiting 6 distinct gene clusters that mediate resistance to glycopeptides, and from these five clusters are acquired traits including *van* (*A, B, D, E, and G*), whereas *vanC* is inherently present in motile enterococci. The most significant form of enterococcal glycopeptide resistance is *vanA* and its principal reservoir is *E. faecium* [50]. Both inherent and acquired resistance in species of enterococci possesses a major factor of food safety as they can cause infections as well as spread AR to other harmful bacteria.

4.4.5 ANTIMICROBIAL RESISTANCE (AMR) IN LEUCONOSTOC SPECIES

Leuconostoc is another genus under the LAB group involved in the fermentation of numerous dairy and meat products and also in the production of sauerkraut and pickles. *Leuconostoc* spp. is widely used as an adjunct culture for aroma production in various fermented food products. The *L. dextranicum* and *L. cremoris* are two important species that belong to this genus. Very few studies acknowledge the presence of AMR in *Leuconostoc* species, hence, detailed studies are needed to check the existence and transferability of AR genes in *Leuconostoc* species, in particular in those strains that are employed in fermentation of various foods and dairy products.

Swenson et al. [80]; and Florez et al. [30] have reported the susceptibility of *Leuconostoc* to clindamycin, erythromycin, rifampicin, chloramphenicol, and tetracycline. Some other studies have described the presence of innate resistance (Table 4.5) in *Leuconostoc* species against various antibiotics such as cefoxitin, metronidazole, glycopeptides (vancomycin), streptomycin, nitrofurantoin, nalidixic acid, gentamycin, kanamycin, and trimethoprim [30].

TABLE 4.5 Intrinsic AMR Profiling in *Leuconostoc* Species

Antibiotic	Antibiotic Class	Mechanism of Resistance	References
Vancomycin	Glycopeptides	Vancomycin resistance resulted from production of modified peptidoglycan precursors, low binding affinities	[62]
Kanamycin and gentamycin	Aminoglycosides	Active efflux pump, decreases, cell membrane permeability (porin loss), alteration of ribosome	
Trimethoprim and sulphadiazine	Sulfonamides	Resistance generally achieved by bypass mechanisms	

Zdolec et al. [96] also reported *L. mesenteroides* strain from fermented sausages to be resistant against streptomycin, enrofloxacin, nalidxic acid, metronidazole, and kanamycin. More recently, Florez et al. [31] revealed that *Leuconostoc* species harbor resistant determinants that can be transferred to other species. It was demonstrated that the *erm*(B) gene conferring resistance to erythromycin, could be transferable from *Enterococcus faecalis* to *Leuconostoc* spp. both under wet lab conditions as well as in the cheese environment, which hints the possible role *Leuconostocs* could play in disseminating AMR genes in the food environment.

4.4.6 ANTIMICROBIAL RESISTANCE (AMR) IN PEDIOCOCCUS SPECIES

Pediococcus is also an industrially important LAB involved in the production of various fermented dairy and food products. Among different species of *Pediococcus*; *P. acidilactici* and *P. pentosaceus* are the two species that most commonly employed in the development of various fermented food and dairy goods. Due to the presence of techno functionality in *Pediococcus* species, it is commonly used in milk and milk products, meat, vegetables, sausages, silage crops, etc. The *Pediococcus* species help to control the succession of microbes and inhibit the spoilage microbes during fermentation by releasing organic acids, hydrogen peroxide and bacteriocins that cause in the decrement of the pH [1].

Singla et al. [77] documented that some *Pediococcus* strains may also act as a safe probiotic. While some literature has already recognized the presence of AMR in the *Pediococcus* species [62], the issue of AMR in these genera has not received much attention compared to other LAB genera such as *Lactobacillus*, *Enterococcus*, etc. Therefore, by considering all the mentioned facts that emerge from the AMR problem in the *Pediococcus* genus, it is important to concentrate or draw attention to the safety use of this genus as an adjunct or starter culture and also as a probiotic.

Tankovic et al. [81] found *P. acidilactici* strains susceptible to novobiocin. Danielsen et al. [22] also suggested that pediococci is typically susceptible towards Penicillin G, gentamycin, erythromycin, netimicin, clindamycin, rifampin, and ramoplanin. Further, Uymaz et al. [92] documented that the BH105 strain of *P. pentosaceus* was sensitive to chloramphenicol and penicillin. In some studies, *P. acidilactici* has also been reported to be more susceptible to trovafloxacin than *P. pentosaceus* [22].

Singla et al. [77] recorded that most of the *Pediococcus* isolated in their research showed sensitivity to various antibiotics β-lactams, *viz.*, Penicillin-100%, amoxycillin-92.5%, cloxacillin-90%, ampicillin-67.5% and protein synthesis inhibitors, such as, aminoglycosides (netilmicin-90%, tobramycin-82.5%, gentamicin-87.5% and amikacin-72.5%).

In terms of intrinsic resistance, Danielsen et al. [22] reported that *Pediococcus* species showed intrinsic resistance towards vancomycin, teicoplanin, kanamycin, streptomycin, and tetracycline (especially *P. acidilacti*) along with ciprofloxin, doxycycline, sulfamethoxazole, and trimethoprim (Table 4.6).

TABLE 4.6 Intrinsic AMR Profiling in *Pediococcus* Species

Antibiotic	Antibiotic Class	Mechanism of Resistance	References
Vancomycin and teicoplanin)	Glycopeptides	Intrinsic resistance to vancomycin in *P. pentosaceus* is caused by a transformed peptidoglycan precursor ending in d-Ala-dlactate	[62]
Kanamycin, neomycin	Aminoglycosides	Active efflux pump, decreases, cell membrane permeability (porin loss), alteration of ribosome, inactivation of aminoglycosides modifying enzymes	

Federici et al. [28] documented that *P. pentosaceus* isolated from fermented sausages was found to carry AR gene for tetracycline *tet*(M)

and *erm*(B) for erythromycin. Mandal et al. [59] reported that *Pediococcus* strains from buttermilk showed resistance towards co-trimoxazole and also strains were highly resistant towards vancomycin. Toomey et al. [87]; and Banwo et al. [9] stated that *Pediococcus* species favored widespread AR towards vancomycin but it is non-transferable.

4.5 METAGENOMICS APPROACH FOR STUDYING ANTIMICROBIAL RESISTANCE (AMR) IN FOOD CHAIN

Majority of the previously described studies on the identification or presence of AMR genes in food systems has been carried out using traditional culture-based techniques. However, in these conventional culturing techniques, many of the bacterial isolates and therefore their resistance genes are left out or remain undetectable since it is not possible to culture all the microorganisms in a sample. Hence, in order to overcome this problem, recently culture independent techniques using metagenomic DNA sequencing approach are increasingly being used to identify or monitor the AMR genes in different environments or ecosystems. This approach required DNA isolation directly from a sample and analyzed for presence of genes rather than isolating and analyzing individual microorganisms. The metagenomics approach is able to provide a holistic insight on the microbial composition of specific communities, the metabolic activities, and functional roles of the different microbes in various environmental samples, in addition to providing information about their relative abundance, specific contribution in fermentation, virulence, AR profile, etc. (Table 4.7).

The resistome analysis of a sample type by metagenomic technique typically utilizes the shotgun metagenomics approach. Shotgun metagenomic sequencing tool is an emerging strategy to study the resistome profile of specific niches along with identifying majority of organisms (cultivable as well as non-cultivable bacteria) and revealing many other functional features. In this technique typically utilizing the sequence-based screening approach, the DNA is extracted directly from the sample and the extracted DNA is fragmented into smaller components, which are independently sequenced. In order to determine the AMR genes, the metagenomic sequences are then aligned with the reference sequences that have been collected in data banks over the years.

TABLE 4.7 Studies Utilizing Shotgun Metagenomic Sequencing to Check the Prevalence of AMR Genes in Different Sample Types

Sample Type	AMR Genes Detected	Bacterial Flora	References
	Non-Food Samples		
Marine microbial community	Fluoroquinolone-resistance genes	Coral, *Porites astreoides*	[93]
Agroecosystem (soil, animal feces and manure samples)	Multidrug resistance transporters (44.75%) were found most abundant. Bacitracin, vancomycin, beta lactam, tetracycline resistance and MLS efflux pump resistance genes are also determined.	Animal faces, manure, and soil microflora	[70]
Coastal beach and sewage water	Two clinically relevant AR genes, i.e., *TEM-4* and *TEM-33* betalactamase found to be most frequent. Virulence factor-availability of shET2 enterotoxin (*sen*B) in a single sewage sample.	Bacterial community of coastal beach and sewage water	[34]
	Food Samples		
Mastitis milk (Krankej and Gir cattle breed)	Reported resistance genes towards fluoroquinolones and methicillin which, along with other antibiotic resistance sequences, indicated the presence of potential multidrug resistance	*E. coli, Ps. aeruginosa*	[15]
Mozzarella cheese	Ampicillin and kanamycin resistance gene	*S. thermophilus* and *L. helveticus*	[26]
Blue veined cheese and raw milk	Tetracycline resistance *tet*(A), *tet*(L), *tet*(M) and *tet*(S)	Cheese microflora and Starter bacteria	[32]
Red pepper	Identified predominant ARG types as multidrug resistance genes (60.6%), which were: *TolC*, *omp*R, *mdt*C, *mdf*A, *emr*D, *mdt*H, *emr*A, *emr*B, *acr*B, *mdt*B, *mdt*K, *mdt*G, *omp*F, *mdt*E, *mex*E, *mdt*A, and *acr*A	*Enterobacteriaceae* including genera *Enterobacter*, *Citrobacter*, *E. coli*, *Salmonella*, and *Klebsiella*.	[55]

Various computational tools, such as, IMGIMER, MGRAST tools for functional annotation and IMGIMER, CAMERA (community cyber infrastructure for advanced marine microbial ecology research and analysis), MG-RAST (metagenomic rapid annotations using subsystem technology), and EBI (European Bioinformatics Institute) metagenomics, SILVANGS for data management, storage, and sharing also play a crucial role to gather information from the bulk metagenome sequences [53]. A single metagenome can hence, predict the resistance potential against multiple antibiotics using a wide range of known resistance genes. Many studies have already utilized this approach to check the prevalence of AMR genes in water, sewage, soil, etc. [3, 39, 79].

Understanding the recent concern placed on the role of the food chain in the transmission of AR genes from bacteria to humans, shotgun metagenome sequencing approach is now highly sought to better detect and understand the transmission and selection process of AMR genes in the food chain. Although, various scientific reports and publications have already highlighted the potential of genomic approach in studying AMR, the application of whole metagenome shotgun sequencing to determine or survey the AMR in food and food related products have not been as extensive.

As reported by Oniciuc et al. [65], only 13% of all whole metagenome sequencing related publications or reports deal with AMR. In food science, the first reported studies of use of shotgun metagenome sequencing for checking of AMR genes prevalence have come up from studies conducted in the last five years and most studies have been focused mainly on traditional foodborne pathogens, such as, *Salmonella* [2, 98], *Campylobacter* spp. [18], Shiga toxin-producing *E. coli* [21], *L. monocytogenes* and *S. aureus* [37].

The metagenomics approach is now increasingly being adopted to analyze the persistence of AMR genes in various microbial ecosystems. As most of the environment, associated microorganisms are yet not fully cultivable, metagenomics represents a powerful culture-independent tool to study AMR in any environment. In addition, the application of Shotgun metagenomic sequencing can not only promote the future identification of novel genes of resistance, but can also be used to predict the future evolution of AR and enable us to further research on genetic elements that participate in the transfer of AR genes.

4.6 SUMMARY

AMR is a threat not only in classical pathogens but also in commensals and food bacteria since considerable evidence have demonstrated the vital role these microbes play in disseminating the AMR genes in the food chain. Early on, the topic has been introduced and an explanation has been made as to why AMR is important in the LAB group of microorganisms. In the next section, the probable mechanisms of action as to how these microbes evade the antimicrobial action have been elaborated. Further, the next section attempts to comprehensively summarize the prevalent status of AMR in six major LAB genera. Finally, a brief reference to the latest metagenomic approach by shotgun sequencing has been made to study the overall resistome pool in any sample type.

KEYWORDS

- **antibiotics**
- **antimicrobial resistance**
- **extended-spectrum beta-lactams**
- **lactic acid bacteria**
- **metagenomics**
- **resistance-nodulation-division**

REFERENCES

1. Albano, H., Todorov, S. D., & Van, R. C. A., (2007). Characterization of two bacteriocins produced by *Pediococcus acidilactici* isolated from alheira (a fermented sausage) traditionally produced in Portugal. *International Journal of Food Microbiology, 116*(2), 239–247.
2. Allard, M. W., Luo, Y., & Strain, E., (2012). High resolution clustering of *Salmonella enterica* serovar *Montevideo* strains using a next-generation sequencing approach. *BMC Genomics, 13*(2), 32–39.
3. Allen, H. K., Moe, L. A., Rodbumrer, J., Gaarder, A., & Handelsman, J., (2009). Functional metagenomics reveals diverse β-lactamases in a remote alaskan soil. *The ISME Journal, 3*(2), 243–251.
4. Alvarez, C. Y. M., & Fernández, F. J., (2017). Assessment of virulence factors, antibiotic resistance and amino decarboxylase activity in *Enterococcus faecium* MXVK 29 isolated from Mexican Chorizo. *Letters in Applied Microbiology, 64*(2), 171–176.
5. Álvarez-Cisneros, Y. M., & Ponce-Alquicira, E., (2018). Antibiotic resistance in lactic acid bacteria. In: *Antimicrobial Resistance: Global Threat* (p. 23). Intech Open, online.

6. Ammor, M. S., Flórez, A. B., & Mayo, B., (2007). Antibiotic resistance in non-enterococcal lactic acid bacteria and *Bifidobacteria*. *Food Microbiology, 24*(6), 559–570.
7. Ammor, M. S., Flórez, A. B., & Van, H. A. H., (2008). Molecular characterization of intrinsic and acquired antibiotic resistance in lactic acid bacteria and *Bifidobacteria*. *Journal of Molecular Microbiology and Biotechnology, 14*(1–3), 6–15.
8. Aslim, B., & Beyatli, Y., (2004). Antibiotic resistance and plasmid DNA contents of *Streptococcus thermophilus* strains isolated from Turkish yogurts. *Turkish Journal of Veterinary and Animal Sciences, 28*(2), 257–263.
9. Banwo, K., Sanni, A., & Tan, H., (2013). Functional properties of *Pediococcus* species isolated from traditional fermented cereal gruel and milk in Nigeria. *Food Biotechnology, 27*(1), 14–38.
10. Barthelemy, P., Autissier, D., Gerbaud, G., & Courvalin, P., (1984). Enzymic hydrolysis of erythromycin by a strain of *Escherichia coli*. *The Journal of Antibiotics, 37*(12), 1692–1696.
11. Blair, J. M., Webber, M. A., Baylay, A. J., Ogbolu, D. O., & Piddock, L. J., (2015). Molecular mechanisms of antibiotic resistance. *Nature Reviews Microbiology, 13*(1), 42–51.
12. Behr, M. A., (2013). Evolution of *Mycobacterium tuberculosis*. In: *The New Paradigm of Immunity to Tuberculosis* (pp. 81–91) Springer, New York, NY.
13. Belletti, N., Gatti, M., & Bottari, B., (2009). Antibiotic resistance of lactobacilli isolated from two Italian hard cheeses. *Journal of Food Protection, 72*(10), 2162–2169.
14. Bhatt, V. D., Patel, M. S., Joshi, C. G., & Kunjadia, A., (2011). Identification and antibiogram of microbes associated with bovine mastitis. *Animal Biotechnology, 22*(3), 163–169.
15. Bhatt, V. D., Ahir, V. B., Koringa, P. G., Jakhesara, S. J., Rank, D. N., Nauriyal, D. S., Kunjadia, A. P., & Joshi, C. G., (2012). Milk microbiome signatures of subclinical mastitis affected cattle analyzed by shotgun sequencing. *Journal of Applied Microbiology, 112*(4), 639–650.
16. Blair, J. M., Webber, M. A., Baylay, A. J., Ogbolu, D. O., & Piddock, L. J., (2015). Molecular mechanisms of antibiotic resistance. *Nature Reviews Microbiology, 13*(1), 42–51.
17. Brown-Jaque, M., Calero-Cáceres, W., & Muniesa, M., (2015). Transfer of antibiotic resistance genes via phage-related mobile elements. *Plasmid, 79*, 1–7.
18. Chen, Y., Mukherjee, S., & Hoffmann, M., (2013). Whole-genome sequencing of gentamicin-resistant *Campylobacter coli* isolated from U.S. Retail meats reveals novel plasmid-mediated aminoglycoside resistance genes. *Antimicrob. Agents Chemother, 57*, 5398–5405.
19. Condon, S., (1983). Aerobic metabolism of lactic acid bacteria. *Irish Journal of Food Science and Technology, 7*(1), 15–25.
20. Dale, G. E., Broger, C., & D'Arcy, A., (1997). Single amino acid substitution in *Staphylococcus aureus* dihydrofolate reductase determines trimethoprim resistance. *Journal of Molecular Biology, 266*(1), 23–30.
21. Dallman, T. J., Byrne, L., & Ashton, P. M., (2015). Whole-genome sequencing for national surveillance of Shiga toxin-producing *Escherichia coli* O157. *Clinical Infectious Diseases, 61*(3), 305–312.

22. Danielsen, M., Simpson, P. J., O'connor, E. B., Ross, R. P., & Stanton, C., (2007). Susceptibility of *Pediococcus* spp. to antimicrobial agents. *Journal of Applied Microbiology, 102*(2), 384–389.
23. Das, D. J., Shankar, A., Johnson, J. B., & Thomas, S., (2020). Critical insights into antibiotic resistance transferability in probiotic *Lactobacillus*. *Nutrition, 69*, 110567.
24. Devirgiliis, C., Coppola, D., Barile, S., Colonna, B., & Perozzi, G., (2009). Characterization of the Tn916 conjugative transposon in a food-borne strain of *Lactobacillus paracasei*. *Appl. Environ. Microbiol, 75*(2), 3866–3871.
25. Devirgiliis, C., Zinno, P., & Perozzi, G., (2013). Update on antibiotic resistance in food borne *Lactobacillus and Lactococcus* species. *Frontiers in Microbiology, 4*, 301–316.
26. Devirgiliis, C., Zinno, P., Stirpe, M., Barile, S., & Perozzi, G., (2014). Functional screening of antibiotic resistance genes from a representative metagenomic library of food fermenting microbiota. *BioMed Research International*, 2014,
27. Egervärn, M., Roos, S., & Lindmark, H., (2009). Identification and characterization of antibiotic resistance genes in *Lactobacillus reuteri* and *Lactobacillus plantarum*. *Journal of Applied Microbiology, 107*(5), 1658–1668.
28. Federici, S., Ciarrocchi, F., & Campana, R., (2014). Identification and functional traits of lactic acid bacteria isolated from ciauscolo salami produced in central Italy. *Meat Science, 98*(4), 575–584.
29. Flórez, A. B., Delgado, S., & Mayo, B., (2005). Antimicrobial susceptibility of lactic acid bacteria isolated from a cheese environment. *Canadian Journal of Microbiology, 51*(1), 51–58.
30. Flórez, A. B., Ammor, M. S., & Mayo, B., (2008). Identification of Tet (M) *in two Lactococcus lactis* strains isolated from a Spanish traditional starter-free cheese made of raw milk and conjugative transfer of tetracycline resistance to *Lactococci* and *Enterococci*. *International Journal of Food Microbiology, 121*(2), 189–194.
31. Flórez, A. B., Campedelli, I., & Delgado, S., (2016). Antibiotic susceptibility profiles of dairy *Leuconostoc*, analysis of the genetic basis of atypical resistances and transfer of genes *in vitro* and in a food matrix. *PLoS One, 11*(1), 8.
32. Flórez, A. B., Vázquez, L., & Mayo, B., (2017). A functional metagenomic analysis of tetracycline resistance in cheese bacteria. *Frontiers in Microbiology, 8*, 907–914.
33. Franz, C. M., Stiles, M. E., Schleifer, K. H., & Holzapfel, W. H., (2003). Enterococci in foods—a conundrum for food safety. *International Journal of Food Microbiology, 88*(2, 3), 105–122.
34. Fresia, P., Antelo, V., & Salazar, C., (2019). Urban metagenomics uncover antibiotic resistance reservoirs in coastal beach and sewage waters. *Microbiome, 7*(1), 1–9.
35. Gfeller, K. Y., Roth, M., Meile, L., & Teuber, M., (2003). Sequence and genetic organization of the 19.3-kb erythromycin- and dalfopristin-resistance plasmid pLME300 from *Lactobacillus fermentum* ROT1. *Plasmid, 50*(3), 190–201.
36. González-Rodríguez, I., Ruiz, L., Gueimonde, M., Margolles, A., & Sánchez, B., (2013). Factors involved in the colonization and survival of *Bifidobacteria* in the gastrointestinal tract. *FEMS Microbiology Letters, 340*(1), 1–10.
37. Gordon, N. C., Price, J. R., Cole, K., & Everitt, R., (2014). Prediction of *Staphylococcus aureus* antimicrobial resistance by whole-genome sequencing. *Journal of Clinical Microbiology, 52*(4), 1182–1191.

38. Han, J. H., Li, X. F., & Gao, W. H., (2013). Susceptibility of *Lactobacillus pentosus* strains isolated from fermented products to streptomycin and kanamycin. *International Food Research Journal, 20*(4), 1927.
39. Hu, L. H., Wu, F. Y., & Lin, C. T., (2013). Graphene. Modified $LiFePO_4$ cathode for lithium ion battery beyond theoretical capacity. *Nature Communications, 4*(1), 1–7.
40. Hummel, A., Holzapfel, W. H., & Franz, C. M., (2007). Characterization and transfer of antibiotic resistance genes from enterococci isolated from food. *Systematic and Applied Microbiology, 30*(1), 1–7.
41. Imperial, I. C., & Ibana, J. A., (2016). Addressing the antibiotic resistance problem with probiotics: Reducing the risk of its double-edged sword effect. *Frontiers in Microbiology,* 7, 1983–1990.
42. Jaimee, G., & Halami, P. M., (2016). Emerging resistance to aminoglycosides in lactic acid bacteria of food origin: Impending menace. *Applied Microbiology and Biotechnology, 100*(3), 1137–1151.
43. Johnson, A. P., Uttley, A. H., Woodford, N., & George, R. C., (1990). Resistance to vancomycin and teicoplanin: An emerging clinical problem. *Clinical Microbiology Reviews, 3*(3), 280–291.
44. Jose, N. M., Bunt, C. R., & Hussain, M. A., (2015). Implications of antibiotic resistance in probiotics. *Food Reviews International, 31*(1), 52–62.
45. Katla, A. K., Kruse, H., Johnsen, G., & Herikstad, H., (2001). Antimicrobial susceptibility of starter culture bacteria used in Norwegian dairy products. *International Journal of Food Microbiology, 67*(1, 2), 147–152.
46. Kapoor, G., Saigal, S., & Elongavan, A., (2017). Action and resistance mechanisms of antibiotics: A guide for clinicians. *Journal of Anesthesiology, Clinical Pharmacology, 33, 33*(3), 300.
47. Karapetkov, N., Georgieva, R., Rumyan, N., & Karaivanova, E., (2011). Antibiotic susceptibility of different lactic acid bacteria strains. *Beneficial Microbes, 2*(4), 335–339.
48. Khemariya, P., Singh, S., Nath, G., & Gulati, A. K., (2013). Isolation, identification and antibiotic susceptibility of nis+ *Lactococcus lactis* from dairy and non-dairy sources. *Czech Journal of Food Sciences, 31*(4), 323–331.
49. Khemariya, P., Singh, S., Nath, G., & Gulati, A. K., (2017). Probiotic *Lactococcus lactis*: A review. *Turkish Journal of Agriculture-Food Science and Technology, 5*(6), 556–562.
50. Klein, G., (2003). Taxonomy, ecology and antibiotic resistance of enterococci from food and the gastro-intestinal tract. *International Journal of Food Microbiology, 88*(3), 123–131.
51. Knudtson, L. M., & Hartman, P. A., (1993). Antibiotic resistance among enterococcal isolates from environmental and clinical sources. *Journal of Food Protection, 56*(6), 489–492.
52. Laxminarayan, R., & Chaudhury, R. R., (2016). Antibiotic resistance in India: Drivers and opportunities for action. *PLoS Medicine, 13*(3).
53. Lee, W. P., Stromberg, M. P., Ward, A., Stewart, C., Garrison, E. P., & Marth, G. T., (2014). MOSAIK: Hash-based algorithm for accurate next-generation sequencing short-read mapping. *PloS One, 9*(3), 7.

54. Li, X. Z., & Nikaido, H., (2009). Efflux-mediated drug resistance in bacteria. *Drugs, 69*(12), 1555–1623.
55. Li, Z., Dong, L., Zhao, C., & Zhu, Y., (2020). Metagenomic insights into the changes in microbial community and antimicrobial resistance genes associated with different salt content of red pepper (*Capsicum annuum L.)* sauce. *Food Microbiology,* 85. Article ID: 103295.
56. Luh, K. T., Hsueh, P. R., Teng, L. J., Pan, H. J., Chen, Y. C., Lu, J. J., Wu, J. J., & Ho, S. W., (2000). Quinupristin-dalfopristin resistance among gram-positive bacteria in Taiwan. *Antimicrobial Agents and Chemotherapy, 44*(12), 3374–3380.
57. Ma, L., Li, B., & Zhang, T., (2014). Abundant rifampin resistance genes and significant correlations of antibiotic resistance genes and plasmids in various environments revealed by metagenomic analysis. *Applied Microbiology and Biotechnology, 98*(11), 5195–5204.
58. Madhavan, H. N., & Murali, S., (2011). Mechanisms of development of antibiotic resistance in bacteria among clinical specimens. *Journal of Clinical and Biomedical Sciences, 1*(2), 42–48.
59. Mandal, V., Sen, S. K., & Mandal, N. C., (2011). Isolation and characterization of pediocin NV 5 producing *Pediococcus acidilactici* LAB 5 from vacuum-packed fermented meat product. *Indian Journal of Microbiology, 51*(1), 22–29.
60. Mazurkiewicz, P., Sakamoto, K., Poelarends, G. J., & Konings, W. N., (2005). Multidrug transporters in lactic acid bacteria. *Mini Reviews in Medicinal Chemistry, 5*(2), 173–181.
61. Miller, W. R., Munita, J. M., & Arias, C. A., (2014). Mechanisms of antibiotic resistance in enterococci. *Expert Review of Anti-Infective Therapy, 12*(10), 1221–1236.
62. Moračanin, S. V., Djukic, D., Zdolec, N., Milijašević, M., & Mašković, P., (2017). Antimicrobial resistance of lactic acid bacteria in fermented food. *Journal of Hygienic Engineering and Design, 18*, 25–35.
63. Nawaz, M., Wang, J., Zhou, A., Ma, C., Wu, X., Moore, J. E., Millar, B. C., & Xu, J., (2011). Characterization and transfer of antibiotic resistance in lactic acid bacteria from fermented food products. *Current Microbiology, 62*(3), 1081–1089.
64. Odamaki, T., Yonezawa, S., & Kitahara, M., (2011). Novel multiplex polymerase chain reaction primer set for identification of *Lactococcus* species. *Letters in Applied Microbiology, 52*(5), 491–496.
65. Oniciuc, E. A., & Likotrafiti, E., (2018). The present and future of whole genome sequencing (WGS) and whole metagenome sequencing (WMS) for surveillance of antimicrobial resistant microorganisms and antimicrobial resistance genes across the food chain. *Genes, 9*(5), 268–272.
66. Pan, L., Hu, X., & Wang, X., (2011). Assessment of antibiotic resistance of lactic acid bacteria in Chinese fermented foods. *Food Control, 22*(8), 1316–1321.
67. Patel, A. R., Shah, N. P., & Prajapati, J. B., (2012). Antibiotic resistance profile of lactic acid bacteria and their implications in food chain. *World J. Dairy Food Sci., 7*(2), 202–211.
68. Peterson, E., & Kaur, P., (2018). Antibiotic resistance mechanisms in bacteria: Relationships between resistance determinants of antibiotic producers, environmental bacteria, and clinical pathogens. *Frontiers in Microbiology, 9*, 2928–2933.

69. Piddock, L. J. V., (2006). Clinically relevant chromosomally encoded multidrug resistance efflux pumps in bacteria. *Clinical Microbiology Reviews, 19*(2), 382–402.
70. Pitta, D. W., Dou, Z., Sanjay, K., & Indugu, N., (2016). Metagenomic evidence of the prevalence and distribution patterns of antimicrobial resistance genes in dairy agroecosystems. *Foodborne Pathogens and Disease, 13*(6), 296–302.
71. Port, J. A., Wallace, J. C., Griffith, W. C., & Faustman, E. M., (2012). Metagenomic profiling of microbial composition antibiotic resistance determinants in Puget sound. *PloS One, 7*(10), 6.
72. Pradhan, D., Mallappa, R. H., & Grover, S., (2020). Comprehensive approaches for assessing the safety of probiotic bacteria. *Food Control, 108*. Article ID: 106872.
73. Rizzotti, L., Gioia, F. L., Dellaglio, F., & Torriani, S., (2009). molecular diversity and transferability of the tetracycline resistance gene Tet (M), carried on Tn916-1545 family transposons, in enterococci from a total food chain. *Antonie Van Leeuwenhoek, 96*(1), 43–52.
74. Ruiz, J., (2003). Mechanisms of resistance to quinolones: Target alterations, decreased accumulation and DNA gyrase protection. *Journal of Antimicrobial Chemotherapy, 51*(5), 1109–1117.
75. Sharma, P., Tomar, S. K., Goswami, P., Sangwan, V., & Singh, R., (2014). Antibiotic resistance among commercially available probiotics. *Food Research International, 57*, 176–195.
76. Signoretto, C., Boaretti, M., & Canepari, P., (1994). Cloning, sequencing and expression in *Escherichia Coli* of the low-affinity penicillin binding protein of *Enterococcus faecalis*. *FEMS Microbiology Letters, 123*(1), 99–106.
77. Singla, V., Mandal, S., Sharma, P., Anand, S., & Tomar, S. K., (2018). Antibiotic susceptibility profile of *Pediococcus* Spp. from diverse sources. *3-Biotech, 8*(12), 489–493.
78. Singh, A., Pradhan, D., & Goyal, C., (2017). Characterization of antibiotic resistance in dairy lactococci. *Indian Journal of Dairy Science, 70*(3), 24–33.
79. Sozzi, T., & Smiley, M. B., (1980). Antibiotic resistances of yogurt starter cultures *Streptococcus thermophilus* and *Lactobacillus bulgaricus*. *Appl. Environ. Microbiol, 40*(5), 862–865.
80. Swenson, J. M., Facklam, R. R., & Thornsberry, C., (1990). Antimicrobial susceptibility of vancomycin-resistant *Leuconostoc*, *Pediococcus*, and *Lactobacillus* species. *Antimicrobial Agents and Chemotherapy, 34*(4), 543–549.
81. Tankovic, J., Leclercq, R., & Duval, J., (1993). Antimicrobial susceptibility of *Pediococcus* Spp. and genetic basis of macrolide resistance in *Pediococcus acidilactici* HM3020. *Antimicrobial Agents and Chemotherapy, 37*(4), 789–792.
82. Temmerman, R., Scheirlinck, I., Huys, G., & Swings, J., (2003). Culture-independent analysis of probiotic products by denaturing gradient gel electrophoresis. *Appl. Environ. Microbiol, 69*(1), 220–226.
83. Teuber, M., Meile, L., & Schwarz, F., (1999). Acquired antibiotic resistance in lactic acid bacteria from food. In: *Lactic Acid Bacteria: Genetics, Metabolism and Applications* (pp. 115–137). Springer, Dordrecht.
84. Teuber, M., (1999). Spread of antibiotic resistance with food-borne pathogens. *Cellular and Molecular Life Sciences CMLS, 56*(9, 10), 755–763.

85. Thokchom, S., & Joshi, S. R., (2012). Antibiotic resistance and probiotic properties of dominant lactic microflora from tungrymbai (an ethnic fermented soybean food in India). *Journal of Microbiology, 50*(3), 535–539.
86. Thumu, S. C. R., & Halami, P. M., (2012). Presence of erythromycin and tetracycline resistance genes in lactic acid bacteria from fermented foods of Indian origin. *Antonie Van Leeuwenhoek, 102*(4), 541–551.
87. Toomey, N., Monaghan, A., Fanning, S., & Bolton, D., (2009). Transfer of antibiotic resistance marker genes between lactic acid bacteria in model rumen and plant environments. *Appl. Environ. Microbiol, 75*(10), 3146–3152.
88. Toomey, N., Bolton, D., & Fanning, S., (2010). Characterization and transferability of antibiotic resistance genes from lactic acid bacteria isolated from Irish pork and beef abattoirs. *Research in Microbiology, 161*(2), 127–135.
89. Tosi, L., Berruti, G., & Danielsen, M., (2007). Susceptibility of *Streptococcus thermophilus* to antibiotics. *Antonie Van Leeuwenhoek, 92*(1), 21–28.
90. Tran, J. H., & Jacoby, G. A., (2002). Mechanism of plasmid-mediated quinolone resistance. *Proceedings of the National Academy of Sciences, 99*(8), 5638–5642.
91. Trieber, C. A., Burkhardt, N., Nierhaus, K. H., & Taylor, D. E., (1998). Ribosomal protection from tetracycline mediated by Tet (O): Tet (O) interaction with ribosomes is GTP-dependent. *Biological Chemistry, 379*(7), 847–856.
92. Uymaz, B., Şimşek, O., Akkoç, N., Ataoğlu, H., & Akçelik, M., (2009). *In vitro* characterization of probiotic properties of *Pediococcus Pentosaceus* BH105 isolated from human feces. *Annals of Microbiology, 59*(3), 485–491.
93. Wegley, L., Edwards, R., Rodriguez-Brito, B., Liu, H., & Rohwer, F., (2007). Metagenomic analysis of the microbial community associated with the coral *Porites astreoides*. *Environmental Microbiology, 9*(11), 2707–2719.
94. Woodford, N., Reddy, S., & Fagan, E. J., (2007). Wide geographic spread of diverse acquired Ampc B-lactamases among *Escherichia coli* and *Klebsiella* spp. in the UK and Ireland. *Journal of Antimicrobial Chemotherapy, 59*(1), 102–105.
95. Yang, C., & Yu, T., (2019). Characterization and transfer of antimicrobial resistance in lactic acid bacteria from fermented dairy products in China. *The Journal of Infection in Developing Countries, 13*(2), 137–148.
96. Zdolec, N., Filipovic, I., & Fleck, Ž. C., (2011). Antimicrobial Susceptibility of lactic acid bacteria isolated from fermented sausages and raw cheese. *Veterinarski Arhiv, 81*(1), 133–141.
97. Zhang, S., Oh, J. H., & Alexander, L. M., (2018). D-alanyl-D-alanine ligase as a broad-host-range counter selection marker in vancomycin-resistant lactic acid bacteria. *Journal of Bacteriology, 200*(13), e00607–e00617.
98. Zhang, S., Yin, Y., Jones, M. B., & Zhang, Z., (2015). *Salmonella* serotype determination utilizing high-throughput genome sequencing data. *Journal of Clinical Microbiology, 53*(5), 1685–1692.
99. Zhou, J. S., Pillidge, C. J., Gopal, P. K., & Gill, H. S., (2005). Antibiotic susceptibility profiles of new probiotic *Lactobacillus* and *Bifidobacterium* strains. *International Journal of Food Microbiology, 98*(2), 211–217.

CHAPTER 5

ZOONOSES: INFECTIOUS DISEASES IN HUMANS

ARUN KHARATE, RAVINDRA DOMBAR, J. ADEPPA, and LINA DHOTE

ABSTRACT

Animals are part and parcel of human life that are dependent for various needs ranging from draft, meat, milk, animal byproducts, pet, human culture, and entertainment. During the interaction with animals and birds, including wildlife, humans can be encountered by pathogenic agents, such as viruses, bacteria, fungi, and parasitic organisms. Zoonoses are those diseases, which will transfer between animals/insects and humans naturally. More than 60% of the human diseases will come under the zoonotic diseases. The consequences can be direct in the form of morbidity and mortality or indirect in the form of loss of man-hours, loss of livestock, loss to the meat and milk industry, travel, and tourism, etc. Varying levels of monitoring and surveillance of zoonoses have complicated the quantification of outbreaks; and inadequate collaboration among the human and animal health sector is another added challenge. Systematic planning and organized approach in detection of zoonoses, cessation of transmission, preparedness, and continuous research can help in curtailing zoonoses.

5.1 INTRODUCTION

The 60% of infectious diseases and 75% of emerging diseases throughout the world are zoonotic in nature [91, 92]. It is believed that the density and growth of the human population, the richness of wildlife and low latitude are seers for the zoonotic emergence [88]. India has an encouraging atmosphere for the spreading of both familiar and emerging diseases among animals and humans [46, 58, 76, 88]. For instance, India loses thousands of human lives annually, suffering high rabies burden among the globe [60, 98]; hundreds of deaths reported during outbreaks of anthrax in domestic animals and wildlife [70]; the reports of emerging agents (like Nipah and Hendra) are increasing over a period of time [53]; and a good number of different endemic zoonoses have also been reported [49, 81]. Most of these diseases excessively attack India's poor and marginal population [5, 57].

According to the World Health Organization (WHO), microbial agents are responsible for 25% of annual global deaths and it is even more in developing countries [66]. Wool-house [83] has reported that 1,415 kinds of pathogens are known to be infectious to man, which also comprises 217 viruses and prions, 538 bacteria and rickettsia, 307 fungi, 66 protozoa and 287 helminths. The emerging zoonoses account for about 75% out of 175 zoonotic diseases.

Foods of animal origin causing zoonoses are prime source of zoonoses seriously impacting the developing countries. The transmission of zoonoses to animal keepers may occur via infection while handling and processing of animal products. Around 68% of the Indian working population is juxtaposition to livestock and their occupation [85]. The group of the population lives in an animal shed, they carry unsuitable dumping of refuge from livestock shed, slaughter, diseased animal, carry skinning of a diseased animal, dispose infected material from diseased animals and they also practice poor hygienic animal husbandry activities. Inadequate knowledge on zoonoses occurrence and their public health impacts have aggregated the condition and has become a major stumbling block in initiating effective and adequate control measures [74]. For developing and improving existing control measures, the assessment of perception, farming management, eating habits, and culture of dairy farmers is necessary [65].

In the food chain, humans can get contracted with the disease, when they come in contact with contaminated meat, fruits, vegetables; and water zoonotic diseases are an economic burden on production in the animal sector. Though large-scale use of antimicrobials is adopted in animal husbandry globally, non-therapeutic use has brought a possible risk in the surfacing of antimicrobial-resistant pathogens including the zoonotic agent. Unfortunately, the data on control effects of zoonoses (such as: estimates of disease burden) in India is unavailable.

Aims including, multi-sectoral and multidisciplinary approach is essential to recognize the risk factors for zoonoses in India and adopt the arbitration required to control zoonoses. These efforts are complicated by fragmented research and control programs in India. There is a need to collaborate among institutions of education, research, and policy organizations like the Ministry of Health and Family Welfare, Ministry of Agriculture, and Wildlife institutes that suffer many challenges at different levels. The interdisciplinary collaboration and communication of One-Health concept [44] has been extended to all sections of health care for humans, the environment, and animals. One-Health approach is the need of the hour to control zoonoses by adopting inter-sectoral coordination on a national level.

The effort for controlling zoonoses that prevail in developing countries is adequate in comparison to developed countries that have networked efforts adjoining One-Health concept. Though there are many zoonoses, some listed endemic zoonoses of concerns are: rabies, brucellosis, bovine tuberculosis, human African trypanosomiasis, echinococcosis, cysticercosis, and anthrax [19]. These zoonoses are a threat to the community health [80], especially for those groups that domesticate animals and breeds for their food and clothing. There is an increased risk of emerging infectious agents in India that has been known as one of the four important hotspots globally. The recent climate change forecast and scenario in India (providing a regional climate for impact studies) reports a rise of annual mean surface temperature by the end of this centenary period from 2.5–5°C with many getting heated and more marked in northern India. Summer and monsoon rainfall will be increasing 20% indicating the potential impact on zoonoses in days to come. Therefore, veterinary public health has opened an interesting field of inquiry in India.

This chapter reviews the incidence of zoonoses in India including: status, risk factors, control measures and health programs, impact of foods and food products on the spread of zoonoses in India.

5.2 ZOONOSES: AN EMERGING THREAT TO HEALTH

The zoonoses in India capture major public health significance because of multiple favorable factors, such as suitable climate abundance of vectors, lack of proper medical care, low socioeconomic condition; and more importantly, about 80% of the people reside in rural areas close to the domestic livestock. However, the existence of zoonoses is blurred because of improper coordination among physicians, veterinarians, epidemiologists; and the physician is unfamiliar with zoonoses besides inadequate diagnostic facilities. Since 1998, plague has killed almost 12 million persons; more than 1.7 million people are bitten by suspected rabid animals and obtain anti-rabies vaccination; and approximately 20,000 people die due to rabies annually. Death for typhus recorded was more than deaths in world war-I. Brucellosis is a huge economic burden causing an annual loss of Rs. 2,400 (US$ 400) million through brucellosis in cattle and buffaloes apart from human lives. Japanese encephalitis is also another emerging zoonoses causing many epidemics and substantial morbidity and mortality. Though epidemiological evidence suggests Kala-azar as zoonotic, yet it confirmed nonzoonotic in India.

5.3 FACTORS AFFECTING THE SPREAD OF ZOONOSES

5.3.1 OCCUPATIONAL HAZARDS: HANDLING ANIMAL BYPRODUCTS AND WASTES

The occurrence of disease incidences is significantly higher in those workers, who are involved in handling of animal by-products and wastes in comparison to those that are not involved in such activities, e.g., leptospirosis in paddy field workers, listeriosis in agricultural farmers, anthrax in livestock farmers, carpet weavers and workers Erysipeloid in fresh merchant and butchers, trypanosomiasis, and tularemia in hunters, creeping eruption trench diggers and plumber, etc. Some of the examples of occupational hazard-oriented zoonoses are: Q-fever in meat plant and rendering plant workers, jungle yellow fever in woodcutters, tick-borne disease, bovine tuberculosis in farmers and salmonellosis in food processing.

5.3.2 ECOLOGICAL CHANGES IN THE ENVIRONMENT

With the increase in human population, humans have exploited the untouched boundaries and natural resources, such as: exploiting the rivers, constructing pipelines and roads through sparsely populated areas carrying deforestation, irrigating, and cultivating new lands. These activities have led to the entering of humans into the virgin ecosystem in which infectious agents make part of the biotic community. Megaprojects (such as: irrigation scheme, construction of dams, an extension of agriculture and engineering resources, artificial lakes, and deforestation) led to a change of buying pattern of insects and change in reserve animal population, which has increased the spread of leptospira, tularemia, and helminthic infestation.

5.3.3 INCREASED TRADE IN ANIMAL PRODUCTS

Those centers that import wool, hides, meat, and bone from the regions where some of the zoonotic infections are endemic, deserve into their boundaries, e.g., foot, and mouth disease, salmonellosis, anthrax, and Newcastle disease.

5.3.4 INCREASED DENSITY OF ANIMAL POPULATION

The animal possesses the great risk of increased frequency of zoonotic agent in humans, e.g., tuberculosis, dermatophytosis, and brucellosis.

5.3.5 TRANSPORTATION OF VIRUS-INFECTED MOSQUITOES

Journeys carried in ship, cruise, aircraft, train, motor, and other vehicles can transport the viruses into a new region, e.g., yellow fever, chikungunya fever, dengue fever, etc.

5.3.6 CULTURAL ANTHROPOLOGICAL NORMS

People in Kenya allow the hyenas and dogs to consume dead animal bodies infected with hydatidosis that aids in the transmission cycle of hydatidosis.

5.3.7 INCREASED MOVEMENT OF MAN

Human activities (such as: pilgrimages, tourism, engineering project work, and land development) expose people to contaminated foods and water leading to infections, such as: colibacillosis, amoebiasis, salmonellosis, giardiasis, and shigellosis, etc.

5.4 SALIENT ZOONOTIC DISEASES

5.4.1 RABIES

Rabies is a dangerous disease that claims the lives of over 20,000 people in India and 59,000 people across the world every year [6], which is predominant in marginalized or rural populations [79]. Rabies is an ideal One-Health challenge [38], e.g.: death has been recorded to a rabid dog bite in 99% of cases [78]. Canine rabies also can be prevented using the vaccine, e.g., in humans that constitute the primary step in post-exposure prophylactic (PEP) procedure following a dog bite [93, 97]; nevertheless improper perception [34] complicated by irregular availability of PEP [33] has continued the existence of human rabies mortality, though rabies transmission to humans can be drastically reduced by canine vaccination that has been generally considered as prohibitively costly in India [43, 86]. The mortality in rabies is seen as characteristically furious in 80% of cases, and another 20% reported as paralytic form are wrongly diagnosed [45, 79] often as malaria [24] despite high priority efforts by zoonoses specialists in the Indian subcontinent [77].

Lawmakers have been unwilling to execute extensively spread canine rabies involvement lacking analysis of stability among effectiveness and expenditure [55]. Vaccination against canine rabies has been abolished it in Great Britain, the United States (US), and Western Europe, South America, and Asia. Tanzania retained at adequate quantity [72, 97], where PEP accessibility is doubtful campaigns. In such countries, common centers, where pet owners bring their dogs for vaccination, is the only cost-effective method for the prevention of human rabies death [47, 84].

However, for India, the cost-effective veneers of canine rabies vaccination have not been analyzed yet [96]. Campaigns of stray-dogs sterilization in combination with vaccination are advocated by an approach for rabies control [31]; and it is also argued by Animal Welfare Board of India as a proper approach for rabies reduction [31]. These types of approaches have been successful in only a few cities but have not been widely adopted. However, it is a cost-saving strategy to control rabies and has not been evaluated in any settings [43].

5.4.2 BRUCELLOSIS

Brucellosis is an important, neglected zoonotic disease. The disease is more prevalent in developing countries because of health issues, economic, and livelihood burden in the society. In cattle, the majority of the cases of brucellosis are caused by *B. abortus*, seldom by *B. melitensis* and sometimes by *B. suis* [62].

The major sources of infection are uterine discharge of infected animals, aborted fetuses, and infected semen samples. Contact with infected animals during artificial insemination and different veterinary activities at the workplace is the common way of spread. The disease is worldwide in animals and humans. The cases of brucellosis are increasing because of urbanization, lack of hygienic practices, expansion of animal industries and improper food handling practices, etc.

The *Brucella* sp. include: *B. neotomae, B. ovis, B. canis* again two species *B. ceti* and *B. pennipedialis* [37]. A new microbe named *B. microti* was detected from vole, foxes, and soil in central Europe [40].

Adult male cattle show orchitis and infertility in both sexes. Young animal and non-pregnant women are usually asymptomatic after the first abortion.

Brucella infection is transmissible to the human causing acute febrile illness, undulant fever and can also produce complications related to the musculoskeletal, cardiovascular, and central nervous symptoms. There is an occupational risk to the workers, veterinarians, farmers, who handle the susceptible animals. Infection can be acquired in the laboratory also.

5.4.3 LISTERIOSIS

Listeriosis is caused by the genus listeria. The disease is often related with the silage feed (the disease is also called silage disease). The disease is found in a variety of animals including man. It affects all ages and both sexes. The recovery rate up to 30% is possible in sheep and goat with prompt and aggressive therapy. In sheep, the disease is also called circling disease. The other clinical symptoms include: depression, twisting, and turning of the head towards the lateral side, walking in a circle, dropping of eyelids and ear [62]. The most common finding is focal liver necrosis while iritis and keratoconjunctivitis [54]. Listeriosis is one of the bacterial foodborne diseases having zoonotic importance. The increased movement of live animals and their product internationally increases the risk of zoonotic disease [61].

Total 17 spp. of genus listeriosis are responsible for causing listeriosis. Out of all these, only two are pathogenic, viz., *L. monoctogenes* and *L. ivanovii* [23]. *L. monocytogenes* is infectious for both animal and human while *L. ivanovii* is infectious to animals only.

The *L. monocytogenes* having 13 serovars [i.e., 1/2a, 1/2b, 1/2c, 3a, 3b, 3c, 4a, 4b, 4c, 4d, and 4e] all are responsible to cause listeriosis in humans but under Indian scenario only 1/2a, 1/2b, and 4b serovars have clinical importance [63].

The possible explanation for the emergence of human foodborne listeriosis is a public health issue because of the changing pattern of food production, processing, transportation, distribution, preservation techniques, changes in feeding habits of people and more preference to the ready-to-eat food. The people at high risk include: newborn, immune-compromised persons, elders, and pregnant women.

5.4.4 TUBERCULOSIS

Bovine tuberculosis is most significant economically notable zoonotic disease, causing leading deaths in comparison to other infectious diseases [82]. WHO reported that bovine tuberculosis having the ability to infect humans is one of the neglected zoonotic diseases [27]. Currently, tuberculosis has importance in developing nations, as humans spend most of their time in close association with animals. It has been reported that new tuberculosis cases in developing countries are increasing at a higher rate (10–15%); India is reported of having the greatest rate of infection in the world. It has been mentioned in the "WTO report on tuberculosis statistics for 2015 for India," 2.2 million cases were in India out of 9.6 million cases in the world, and 40% of the Indian people are affected with tuberculosis. Africa reported the highest cases of tuberculosis followed by South-East Asia.

Mycobacterium tuberculosis, the main etiological agent causing human tuberculosis, comes under the order *Actinomycetales,* family *Mycobacteriaceae* and the genus Mycobacterium. *M. bovis* is the causative agent for Bovine tuberculosis that has the largest host range among known pathogenic microbes and is distributed worldwide. Cattles are known to be the common and true host [94]. However, both ruminants and non-ruminants of wild and domestic animals including primates are reservoir for tuberculosis. Infection in domestic animals is increasing due to the movement of wild animals to cities.

Bovine tuberculosis can be transmitted among animals, between animals to humans and rarely has been observed among humans [39]. Spread is mainly through the aerosol route and other possible routes including congenital, genital, and cutaneous [25]. Calves get infected through suckling of infected mother milk. Humans get infected from cattle through direct contact by aerosol routes or secretions of infected animals [10, 90] and indirectly by consumption of raw milk and other by-products of diseased animals [50]. Individuals working in farms, abattoir, zoo, and laboratory are at higher risk including veterinarians and who are in close association with infected animals.

The common clinical symptoms in infected individuals include: cough, intermittent fever, diarrhea, low appetite, weight loss, and weakness and so on. Clinical infestation in humans requires a longer incubation period to appear, usually months or years. Abdominal infection is common in children affected with *M. bovis*, whereas neck lymph glands are inflamed and ulcerated in adult patients [12]. Respiratory illness is most usual in an individual with revived infections [73].

In several countries in the world, tuberculosis in individuals is diagnosed based on sputum analysis and other rapid tests are also used, such as: Xpert MTB/RIF. PCR and gene sequencing is used to identify the *M. tuberculosis* isolates, which can help in differentiating the complex of *M. tuberculosis*. Mycobacterium culturing method is routinely used to confirm infection in both animals and humans. Tuberculin test is the simplest method but is not routinely used method in live animals to diagnose bovine tuberculosis, which is based on delayed hypersensitivity reactions. In this method, bovine tuberculin is injected intra-dermal and swelling thickness on the site of injection is measured after 72 h. Digital Vernier caliper is used to calibrate the thickness of the skin before and after tuberculin injection.

Zoonotic tuberculosis is the most neglected disease in humans; and it cannot be controlled completely without considering the animal reservoir. It is tough to control without the involvement of humans, food safety and animal health sector involvement in addressing the part of animals in acquiring and transmission of bovine tuberculosis. Therefore, in October

2017, WHO, FAO, and the World Organization for Animal Health together launched the ground plan for zoonotic tuberculosis in humans and animals depending on the "One-Health" approach. The Revised National Tuberculosis Control Program (RNTPC) was started by the Government of India (GOI) in 1997 with the objective of early detection of disease for controlling the transmission rather than eradication [69].

Ideal solution for prevention of transmission is through isolation or culling of infected animals. Other strategies include: improving the awareness in public about the transmission of disease, impact on socio-cultural and economic factors through health camps, blocking the risk pathways and improving the food safety measures. Test and slaughter program by extending remuneration to the farmer can help in controlling the transmission. The involvement of public and private stakeholders can help in the control and eradication of bovine tuberculosis. Food safety measures like routine inspection of carcass at the abattoir, removal of contaminated products from the food chain and preventing the use of raw milk consumption can help in controlling the disease. Use of the vaccine in animals against bovine tuberculosis should be strengthened as in humans as part of a preventive measure.

In developing countries like India, 70% of the people are living with agricultural and livestock activities and livestock serves as a reservoir for bovine tuberculosis disease having public health significance [16]. Therefore, the development of a vaccine against tuberculosis for animals is a successful strategy in controlling the disease. At the same time, it is essential to develop a rapid screening method for detection of bacteria within a short period [64]. Disease surveillance programs are highly essential in risky areas to mitigate the diseases. Bovine tuberculosis has been largely eradicated by strict control measures and elimination programs, such as, test-and slaughter [3, 68]. Similar program is in urgent need for developing countries, where control measures are not followed regularly [7].

5.4.5 SALMONELLOSIS

Salmonellosis is the most common and important bacterial disease that affects the intestinal tract of humans and animals. The Salmonella organisms are: Gram-positive, relatively anaerobic, non-spore-forming, straight rods classified under the family *Enterobacteriaceae* [20, 29, 52, 58] and the genus *Salmonella*. The *Salmonella* spp. are able to survive variable environmental conditions at a temperature (optimum temperature 37°C and pH 6.5–7.0) between 8 and 45°C at pH 4.0 to 9.5 (even can grow at 54 or less at 2–4°C]. Because of these reasons, salmonella organisms can pose pronounced warning signals to the food industry [52].

Seven subspecies of Salmonella have been identified on the basis of some criteria, viz., common habitat, biochemical character, and genetic makeup [14, 18, 32] and more than 2,500 serotypes differing in cell-wall structure [18, 26]. Most of the serotypes classified under the *S. enterica* subsp. *enterica* are mainly responsible for salmonellosis disease (99%) in all animals including humans and birds [15, 32, 51]. *S. paratyphi* A and B, *S. typhi* are pathogenic to humans but symptomless for animals. *S. choleraesuis, S. enteritidis* and *S. typhimurium* are infectious to humans, pigs, and other infected animals [21, 87].

Virulence of the disease is connected with plasmid and chromosomal genes [17]. In the world, over 90 million diarrhea cases are caused by *Salmonella* spp. per year, among 85% of those cases are linked to food origin [41]. More than 200 million to over 1 billion cases of salmonellosis were reported annually in the whole world [11]. The disease is highly fatal in children and deaths have been observed in children below four years old infected with serotype enteritidis/typhimurium [28]. The typhoid fever mainly spreads between humans with no mediation of animals as intermediate host/vector [59], which is mainly associated with food/water, unhygienic, high population in areas of poor sanitary conditions/environment [48].

Central and Southeast Asia and Africa have reported the highest cases of typhoid fever [4]. Common symptoms of the disease are: headache, fever, constipation, or diarrhea, stomach ache, and respiratory problems, loss of appetite, lethal neurological and hepatic changes and even intestine, and spleen damage also observed [29].

Most often food poisoning is caused by *S. typhimurium, S. enteritidis, S. newport* and *S. heidelberg* [4, 51]. The incubation period of the disease varies between 6 h to 72 h and it depends on the dose of infection and host immunity [29]. The salmonella organism resides for 4 weeks even after symptoms disappear from adult patients and whereas up to 7 weeks in children. Humans get infected with salmonella mainly by consumption of infected eggs and their products [42]. Pig and pork serve as an important reservoir of salmonella spp. A high number of bacteria found in the intestine can infect humans involved in carcass processing. Fresh meat is also another source of infection of pathogenic Salmonella species [56]. Lastly, fruits and vegetables contaminated with animal food materials can also be a common source of infection.

5.4.6 YERSINIOSIS

Yersiniosis belongs to the *Enterobacteriaceae* family and it has 17 species. The *Y. pestis*, *Y. enterocolitica* and *Y. pseudotuberculosis* are infectious to the community. Yersinia bacteria are cocci or rod ranging from 1–3 µm in length and 0.5–0.8 µm in width, non-sporulating. Motility is by polar flagellum seen between a temperature range of 22 and 30°C [30]. Yersiniosis is a zoonotic disease caused by *Y. enterocolitica*. This bacterium is a Gram-negative group, catalase-positive, relatively anaerobic, psychrophilic, and non-sporulating.

Six biotypes of *Y. enterocolitica* have been reported, namely, 1A, 1B, 2, 3, 4 and 5 depending on the biochemical reactions. Acids are produced from xylose, trehalose, sorbitol, and saccharine, decarboxylation of ornithine, O-nitrophenyl-β-D galactopiranoide and salicin, hydrolysis of esculin, Voges Proskauer reaction and reduction of nitrates [88]. Biotype A is common in the environment and it is considered as non-pathogenic due to absence of major virulence factor [75]. *Y. enterocolitica* serotypes are grouped based on Lipopolysaccharide O-antigen structure [36]. The serotype/biotypes, such as, 1B/O: 82/O: 5, 27, 2/O:9, 3/O:3 and 4/O:3 is considered as most pathogenic [67, 71].

Y. entorecolitica has virulence genes located on chromosomes and on plasmids, which are approximately 70 kbp [8, 9]. On ingestion of contaminated water/food, *Y. Entericolitica* bacteria colonize at the upper aspect of the large intestine and begins to multiply and cause infection [9].

Y. enterocolotica has a higher infective dose in comparison to other disease-causing bacteria in food and is 10^8–10^9 cells. The 3 to 7 days is the common incubation period but it may be as wide as 1 to 11 days. Clinical signs of Yersiniosis may be mild but severe gastritis and enteritis form may also be noticed [22]. Though this disease affects all age groups, yet children below five years old age group and low immune individuals are at high risk. The common symptoms of diseases are: fever, diarrhea, and stomach ache [9, 67]. Complications of disease may be erythema, bacteremia, nodosum, spleenitis, myocarditis, purulent hepatitis, and nephritis [13, 22].

Many cases of yersiniosis are misdiagnosed because of non-specific symptoms, which resembled disease of other etiology. Antibiotic therapy is recommended in case of infected and patients with reduced immunity. WHO recommends treatment with chloramphenicol, tetracycline, cotrimoxazole or gentamicin. Antibiotic resistance (AR) because of mutation on the chromosome has been observed not only to fluoroquinolone but also to ticarcillin, ampicillin, amoxicillin/clavulanate, cefalorine, and cefazoline [35]. The major source of yersiniosis is undercooked or raw pork but fresh/pasteurized milk and mother milk products, seafood, contaminated plants and drinking water. Though pigs are a potent reservoir of yersiniosis infection, yet bacteria may also be detected from other animals (such as: sheep, cattle, goat, dog, cat, deer, rodents), including many birds [95].

5.5 PREVENTION AND CONTROL MEASURES OF ZOONOSES

- Estimation of zoonoses burden among all species of animals and human beings, continuous active surveillance, health-based facility for a set of specific diseases.
- Improvement in diagnostic procedures that should be specific, rapid, affordable and simple.
- Efficient risk dissemination across agency and disciplinary boundaries.
- Preparedness for emergency and swift response.
- Enhancing risk-free control activities like use of alternative to antibiotics to prevent antimicrobial resistance (AMR).
- Vector control programs.
- Regular immunization to achieve satisfactory herd immunity.
- Awareness creation and public participation.

5.6 SUMMARY

Zoonoses appear like a complicated inter-relation of factors that specifically affect the poor group of population and therefore is an important barrier for successful prevention of zoonoses in under developed countries. Zoonoses and biodiversity 'hot spots' in developing countries like India are complicated by restricted resources to control. In this direction, gathering of resources in research, collaboration among different sectors and strong policy implementation will enable in prevention and control of zoonoses.

KEYWORDS

- **brucellosis**
- **listeriosis**
- **one-health**
- **rabies**
- **tuberculosis**
- **zoonoses**

REFERENCES

1. Abbas, S. S., Kakkar, M., & Rogawski, E. T., (2014). Roadmap to combat zoonoses in India (RCZI) initiative: Costs analysis of a population level rabies control program in Tamil Nadu, India. *PLOS Neglected Tropical Diseases, 8*(2), 721–728.
2. Abbas, S. S., Venkataramanan, V., Pathak, G., & Kakkar, M., (2011). Roadmap to combat zoonoses in India initiative - rabies control initiative in Tamil Nadu, India: Test case for the one-health approach. *International Health, 3*(4), 231–239.
3. Al-Maaly, N. M. H., & Abbas, M. S., (2015). Identification of the *Mycobacterium* spp. isolated from cows milk samples by using PCR technique. *Journal of Veterinary Science & Animal, 2,* 28–31.
4. Andino, A., & Hanning, I., (2015). *Salmonella enterica*: Survival, colonization, and virulence differences among serovars. *The Science World Journal.* Article ID: 520179.
5. Arit, A. K., Ichhpujani, R. L., Gupta, S., Gill, K. S., & Lal, S., (2006). Nipah/ hendra virus outbreak in Siliguri, West Bengal, India in 2001. *The Indian Journal of Medical Research, 123,* 553–560.
6. Asokan, G., Vanitha, A., & Prathap, T., (2011). One-health national program across species on zoonoses: A call to the developing world. *Infection Ecology & Epidemiology, 1,* 8293–8299.
7. Bakshi, C. S., Shah, D. H., Verma, R., Singh, R. K., & Malik, M., (2005). Rapid differentiation of *Mycobacterium bovis* and *Mycobacterium tuberculosis* based on a 12.7-kb fragment by a single tube multiplex-PCR. *Veterinary Microbiology, 109,* 211–216.
8. Bancerz-Kisiel, A., & Szewda, W., (2015). Yersiniosis. Zoonotic foodborne disease of relevance to public health. *Annals of Agricultural and Environmental Medicine, 22,* 397–402.
9. Bari, M. L., Hossain, M. A., Isshiki, K., & Ukuku, D., (2011). Behavior of *Yersinia enterocolitica* in foods. *Journal of Pathogens, 2011.* Article ID: 420732.

10. Beals, T., (2008). *The Risk of Bovine TB from Raw Milk Consumption with a Focus on Michigan.* https://www.realmilk.com/safety/risk-of-bovine-tb-from-raw-milk-consumption/(accessed on 21 September 2021).
11. Bierschenk, D., Boucher, D., & Schroder, K., (2017). Salmonella-induced inflammasome activation in humans. *Molecular Immunology, 86*, 38–43.
12. Bolognesi, N., (2007). *TB or Not TB: The Threat of Bovine Tuberculosis*. https://www.scidev.net/global/disease/feature/tb-or-not-tb-the-threat-of-bovine-tuberculosis.html (accessed on 21 September 2021).
13. Bottone, E. J., (2015). Yersinia enterocolitica: Revisitation of an enduring human pathogen. *Clinical Microbiology Newsletters, 37*, 1–8.
14. Brenner, F. W., Villar, R. G., Angulo, F. J., Tauxe, R., & Swaminathan, B., (2000). Salmonella nomenclature. *Journal of Clinical Microbiol*ogy, *38*, 2465–2467.
15. Bugarel, M., Tudor, A., Loneragan, G. H., & Nightingale, K. K., (2017). Molecular detection assay of five *Salmonella* serotypes of public interest: Typhimurium, enteritidis, Newport, Heidelberg, and Hadar. *Journal of Microbiological Methods, 134*, 14–20.
16. Cadmus, S. I. B., Adesokan, H. K., Adejuwon, T. A., & Adeyemi, M. O., (2008). Retrospective study of bovine tuberculosis and other diseases of public health importance in oko-oba abattoir, Lagos state. *Book of Abstracts, 89*, 72, 73.
17. Chaudhary, J. H., Nayak, J. B., Brahmbhatt, M. N., & Makwana, P. P., (2015). Virulence genes detection of salmonella serovars isolated from pork and slaughterhouse environment in Ahmedabad, Gujarat. *Veterinary World, 8*, 121–124.
18. Chen, H. M., Wang, Y., Su, L. H., & Chiu, C. H., (2013). Nontyphoid infection: Microbiology, clinical features, and antimicrobial therapy. *Pediatrics and Neonatology, 54*, 147–152.
19. Chugh, T. D., (2008). Emerging and re-emerging bacterial diseases. *Journal of Bioscience, 33*, 549–555.
20. Cianflone, N. F. C., (2008). Salmonellosis and the GI tract: More than just peanut butter. *Current Gastroenterology, 10*, 424–431.
21. Costa, L. F., Paixão, T. A., Tsolis, R., Bäumler, A. J., & Santos, R., (2012). Salmonellosis in cattle: Advantages of an experimental model. *Research in Veterinary Science, 93*, 1–6.
22. Dekker, J., & Frank, K., (2015). Salmonella, shigella and yersinia. *Clinics in Laboratory Medicine, 35*, 225–246.
23. Denbakker, H. C., & Warchocki, E. M., (2014). *Listeria floridensis sp. nov., Listeria aquatica sp. nov., Listeria cornellensis sp. nov., Listeria riparia sp. nov. and Listeria grandensis sp. nov.,* from agricultural and natural environments. *International Journal of Systematic and Evolutionary Microbiology, 64*, 1882–1889.
24. DMI (Director of Medical Education), (2015). *Medical Education Department - Rabies Control - Report of Dog Biter Cases, Rabies Treated and Rabies Death for the Period from 2010 to 2014 in Respect of Directorate of Medical Education*. Government of Tamil Nadu, Chennai, India. Ref 51023/Statistics/2015.
25. Elias, K. H. D., Asseged, B., Wondwossen, T., & Gebeyehu, M., (2008). Status of bovine tuberculosis in Addis Ababa dairy farms. *Scientific and Technical Review of OIE, 27*, 915–923.

26. El-Tayeb, M., & Ibrahim, A. S. S., (2017). Prevalence, serotyping and antimicrobials resistance mechanism of *Salmonella enterica* isolated from clinical and environmental samples in Saudi Arabia. *Brazilian Journal of Microbiology, 48*, 499–508.
27. Ereqat, S., Nasereddin, A., Levine, H., Azmi, K., & Al-Jawabreh, A., (2013). First-time detection of *Mycobacterium bovis* in livestock tissues and milk in the west bank, Palestinian territories. *PLoS Neglected Tropical Diseases, 7*(9). Article ID: e2417.
28. Evangelopoulou, G., Kritas, S., Christodoulopoulos, G., & Burriel, A. R., (2015). The commercial impact of pig *Salmonella* spp. infections in border-free markets during an economic recession. *Veterinary World, 8*, 257–272.
29. Fabrega, A., & Vila, J., (2013). *Salmonella enterica,* serovar typhimurium skills to succeed in the host: Virulence and regulation. *Clinical Microbiology Review, 26*, 308–341.
30. Fabrega, A., & Vila, J., (2012). *Yersinia enterocolitica*: Pathogenesis, virulence and antimicrobial resistance. *Enfermedades Infecciosas y Microbiología Clínica, 30*, 24–32.
31. Fitzpatrick, M. C., (2014). Cost-effectiveness of canine vaccination to prevent human rabies in rural Tanzania. *Annals of Internal Medicine, 160*(2), 91–100.
32. Foley, S. L., Johnson, T. J., Ricke, S. C., Nayak, R., & Danzeisen, J., (2013). *Salmonella* pathogenicity and host adaptation in chicken-associated serovars. *Microbiology and Molecular Biology Reviews, 77*, 582–607.
33. Fooks, A. R., (2014). Current status of rabies and prospects for elimination. *Lancet, 9951*, 1389–1399.
34. Franka, R., (2013). Current and future tools for global canine rabies elimination. *Antiviral Research, 1*, 220–225.
35. Frazão, M. R., Andrade, L. N., Darini, A. L. C., & Falcão, J. P., (2017). Antimicrobial resistance and plasmid replicons in *Yersinia enterocolitica* strains isolated in Brazil in 30 years. *Brazilian Journal of Infectious Disease, 21*, 477–480.
36. Garzetti, D., Susen, R., Fruth, A., & Tietze, E., (2014). Molecular scheme for *Yersinia enterocolitica* Patho-serotyping derived from genome-wide analysis. *International Journal of Medical Microbiology, 304*, 275–283.
37. Geoffrey, F., Bjorn, S. O., Jacques, G., Isabelle, J., & Axel, C., (2007). *Brucella ceti* sp. nov. and *Brucella pinnipedialis* sp. nov. for brucella strains with cetaceans and seals as their preferred hosts. *International Journal of Systematic and Evolutionary Microbiology, 57*, 2688–2693.
38. Hampson, K., Dushoff, J., & Cleaveland, S., (2009). Transmission dynamics and prospects for the elimination of canine rabies. *PLoS Biology, 7*(3). Article ID: e1000053.
39. Hampson, K., Coudeville, L., Lembo, T., Sambo, M., & Kieffer, A., (2015). Global alliance for rabies control partners for rabies prevention. estimating the global burden of endemic canine rabies. *PLoS Neglected Tropical Diseases, 9*(5). Article ID: e0003786.
40. Holger, C. S., Zdenek, H., Ivo, S., & Gilles, V., (2008). *Brucella microti* sp. nov., isolated from the common vole *Microtus arvalis*. *International Journal of Systematic and Evolutionary Microbiology, 58*, 375–382.
41. Hung, Y. T., Lay, C. J., Wang, C. L., & Koo, M., (2017). Characteristics of nontyphoidal gastroenteritis in Taiwanese children: 9-year period retrospective medical record review. *Journal of Infection and Public Health, 10*, 518–521.

42. Hurtado, A., Ocejo, M., & Oporto, B., (2017). *Salmonella* spp. and *Listeria monocytogenes* shedding in domestic ruminants and characterization of potentially pathogenic strains. *Veterinary Microbiology, 210*, 71–76.
43. Ichhpujani, R. L., (2006). Knowledge, attitude and practices about animal bites and rabies in general community: Multi-centric study. *Journal of Communicable Diseases, 4*, 355–361.
44. Ichhpujani, R. L., Rajagopal, V., Bhattacharya, D., Rana, U. V., & Mittal, V., (2004). An outbreak of human anthrax in Mysore -India. *The Journal of Communicable Diseases, 36*, 199–204.
45. Joint Director (Vector Borne Disease Control Program), (2015). *Integrative Assessment of Cost and Effectiveness of Rabies Control Interventions Among Different Sectors in Tamil Nadu*. Government of Tamil Nadu, Chennai, India.
46. Jones, K. E., Patel, N. G., & Levy, M. A., (2008). Global trends in emerging infectious diseases. *Nature, 451*, 990–993.
47. Kakkar, M., Venkataramanan, V., & Krishnan, S., (2012). Roadmap to combat zoonoses in India initiative: Moving from rabies research to rabies control - lessons from India. *PLoS Neglected Tropical Diseases, 6*(8). Article ID: e1748.
48. Kanj, S. S., Kanafani, Z. A., & Shehab, M., (2015). Epidemiology, clinical manifestations, and molecular typing of *Salmonella typhi* isolated from patients with typhoid fever in Lebanon. *Journal of Epidemiology and Global Health, 5*, 159–165.
49. Knobel, D. L., Cleaveland, S., & Coleman, P. G., (2005). Re-evaluating the burden of rabies in Africa and Asia. *Bulletin of the World Health Organization, 83*, 360–368.
50. Konhya, L. D., Himes, E. M., & Thoen, C. O., (1980). Mint: Bovine tuberculosis. In: Steele, J., (ed.), *Handbook Series in Zoonoses, Section A: Bacterial, Rickettsial and Mycotic Diseases* (Vol. 2, pp. 147–150). Boca Raton: CRC Press.
51. Kurtz, J. R., Goggins, J. A., & McLachlan, J. B., (2017). *Salmonella* infection: Interplay between the bacteria and host immune system. *Immunology Letters, 190*, 42–50.
52. Li, H. P., Wang, H., D'Aoust, J. Y., & Maurer, J., (2013). *Salmonella* species. Chapter 10. In: Doyle, M. P., & Buchanan, R., (eds.), *Food Microbiology: Fundamentals and Frontiers* (4th edn., pp. 225–261)Washington, DC: ASM Press.
53. Low, J. C., & Donachie, W., (1991). Clinical and serum antibody responses of lambs to infection by *Listeria monocytogenes*. *Research in Veterinary Science, 51*(2), 185–192.
54. Low, J. C., & Donachie, W., (1999). Review of *Listeria monocytogenes* and listeriosis. *The Veterinary Journal, 153*, 9–29.
55. Mallewa, M., (2007). Rabies encephalitis in Malaria - endemic Area, Malawi, Africa. *Emerging Infectious Diseases, 13*(1), 136–139.
56. Mani-Lopez, E., Garcia, H. S., & Lopez-Malo, A., (2012). Organic acids as antimicrobials to control salmonella in meat and poultry products. *Food Research International, 45*, 173–721.
57. Mantur, B. G., & Amarnath, S. K., (2008). Brucellosis in India- review. *Journal of Bioscience, 33*(4), 539–547.
58. Marchese, C., (2015). Biodiversity hotspots: Shortcut for a more complicated concept. *Global Ecology and Conservation, 3*, 297–309.
59. McSorley, S. J., (2014). Immunity to intestinal pathogens: Lessons learned from *Salmonella*. *Immunological Reviews, 260*, 168–182.

60. Nandakumar, T., (2015). *Drastic Increase in Incidence of Rabies Reported in State.* https://www.thehindu.com/news/cities/Thiruvananthapuram/drastic-increase-in-incidence-of-rabies-reported-in-state/article7745681.ece (accessed on 21 September 2021).
61. Nayakwadi, S., & Gupta, V. K., (2014). Isolation and characterization of *Listeria monocytogenes* from goat brain. *Advances in Animal and Veterinary Sciences, 2*(3S), 12–15.
62. Office International des Epizooties, (2019). Brucellosis (*Brucella abortus, B. melitensis and B. suis) (infection with B. abortus, B. melitensis* and *B. suis*). Chapter 3.1.4. *Manual of Diagnostic Tests and Vaccines for Terrestrial Animal*, pp. 355–398.
63. Office International des Epizooties, (2019). *Listeria monocytogenes*. Chapter 3.9.6. In: *Manual of Diagnostic Tests and Vaccines for Terrestrial Animals* (pp. 1705–1722).
64. Ozeki, Y., Igarashi, M., Doe, M., Tamaru, A., Kinoshita, N., & Ogura, Y., (2015). New screen for tuberculosis drug candidates utilizing a luciferase-expressing recombinant *Mycobacterium bovis Bacillus* Calmette-Guéren. *PLoS One, 10*(11). Article ID: 141658.
65. Pavani, G., (2014). Zoonotic diseases with special reference to India. *International Journal of Applied and Basic Medical Science, 4,* 73–87.
66. Prasad, H. K., Singhal, A., Mishra, A., Shah, N. P., & Katoch, V. M., (2010). Bovine tuberculosis in India: Potential basis for zoonosis. *Tuberculosis, 85*, 421–428.
67. Rahman, A., Bonny, T. S., Stonsaovapak, S., & Ananchaipattana, C., (2011). *Yersinia enterocolitica*: Epidemiological studies and outbreaks. *Journal of Pathology, 2011.* Article ID: 239391.
68. *Report of a Joint WHO/DFID-AHP Meeting with the participation of FAO and OIE, Geneva*, (2005). pp. 1–65. https://apps.who.int/iris/handle/10665/43485 (accessed on 21 September 2021).
69. *Revised National Tuberculosis Control Program [RNTCP]*, (2017). Guidelines for TB control in India. https://www.nhp.gov.in/revised-national-tuberculosis-control-programme_pg (accessed on 21 September 2021).
70. Rubin, C., Dunham, B., & Sleeman, J., (2014). Making one-health a reality - crossing bureaucratic boundaries. *Microbiology Spectrum, 2*(1), 269–283.
71. Schaake, J., Drees, A., & Grüning, P., (2014). Essential role of invasin for colonization and persistence of *Yersinia enterocolitica* in its natural reservoir host, the pig. *Infection and Immunity, 82*, 960–969.
72. Sekar, N., Shah, N. K., Abbas, S. S., & Kakkar, M., (2011). Roadmap to combat zoonoses in India initiative: Research options for controlling zoonotic disease in India, 2010–2015. *PLoS One, 6*(2). Article ID: 17120.
73. Shitaye, J. E., Getahun, B., & Alemayehu, T., (*2006*). Prevalence study of bovine tuberculosis by using abattoir meat inspection and tuberculin skin testing data, histopathological and IS6110 PCR examination of tissues with tuberculous lesions in cattle in Ethiopia. *Veterinary Medicine, 51*, 512–522.
74. Singh, B. B., Sharma, R., Gill, J. P. S., Aulakh, R. S., & Banga, H., (2011). Climate change, zoonoses: India. *International Office of Epizootics, 30,* 779–788.
75. Soderqvist, K., Boqvist, S., Wauters, G., Vagsholm, I., & Thisted-Lambertz, S., (2012). *Yersinia enterocolitica* in sheep-a high frequency of biotype 1A. *Acta Veterinaria Scandinavica, 54*, 39–45.

76. Stephen, C., & Stemshorn, B., (2016). Leadership governance and partnerships are essential: One-health competencies. *One-Health, 2*, 161–163.
77. Sudarshan, M. K., (2007). Assessing the burden of human rabies in India: Results of a national multi-center epidemiological survey. *International Journal of Infectious Diseases, 11*(1), 29–35.
78. Suraweera, W., & Million Death Study Collaborators, (2012). Deaths from symptomatically identifiable furious rabies in India: Nationally representative mortality survey. *PLoS Neglected Tropical Diseases, 10*, 1847–1852.
79. Swai, E. S., & Schoonman, L., (2010). The use of rose Bengal plate test to assess cattle exposure to Brucella infection in traditional and smallholder dairy production systems of tanga region of Tanzania. *Veterinary Medicine International, 2010*. Article ID: 837950.
80. Taylor, L. H., Latham, S. M., & Woolhouse, M. E. J., (2001). Risk factors for human disease emergence. *Philosophical Transactions of the Royal Society of London. Series B, Biological Sciences, 356*(1411), 983–989.
81. Thappa, D. M., & Karthikeyan, K., (2001). Anthrax: Overview within the Indian subcontinent. *International Journal of Dermatology, 40*, 216–222.
82. Theon, C., LoBue, P., Enarson, D., Kaneene, J., & De Kantor, I., (2009). Mint: Tuberculosis is a re-emerging disease in animals and humans. *Veterinaria Italiana, 45*, 135–181.
83. Traub, R. J., Robertson, I. D., Irwin, P. J., Mencke, N., & Thompson, R. C., (2005). Canine gastrointestinal parasitic zoonoses in India. *Trends in Parasitology, 21*, 42–48.
84. Velasco-Villa, A., (2008). Enzootic Rabies elimination from dogs and reemergence in wild terrestrial carnivores, United States. *Emerging Infectious Diseases, 14*(12), 1849–1854.
85. Wastling, J. M., Akanmori, B. D., & Williams, D. J. L., (1999). Zoonoses in West Africa: Impact and control. *Parasitology Today, 15*, 309–311.
86. WHO, (2014). *Human Rabies in India: Problem Needing More Attention*. www.who.int/bulletin/volumes/92/4/14-136044/en/ (accessed on 21 September 2021).
87. Wiedemann, A., & Virlogeux-Payant, I., (2015). Interactions of *Salmonella* with animals and plants. *Frontiers in Microbiology, 5*, 791–799.
88. Woolhouse, M., & Gaunt, E., (2007). Ecological origins of novel human pathogens. *Critical Reviews in Microbiology, 33*, 231–242.
89. World Health Organization (WHO), (2006). *The Control of Neglected Zoonotic Diseases: A Route to Poverty Alleviation* (p. 32). Geneva Report of a Joint WHO/DFID-AHP Meeting with the participation of FAO and OIE, Geneva.
90. World Health Organization, (1994). Zoonotic tuberculosis [*M. bovis*]: Memoranda from WHO meeting [with participation of FAO]. *Bulletin of the WHO, 72*(6), 851–857.
91. World Health Organization (WHO). (2020). *Zoonoses*. https://www.who.int/topics/zoonoses/en/ (accessed on 21 September 2021).
92. World Health Organization, (2013). *WHO Expert Consultation on Rabies: Second Report*. https://apps.who.int/iris/handle/10665/85346 (accessed on 21 September 2021).
93. World Organization for Animal Health [OIE]. (2020). *Bovine Tuberculosis*. https://www.oie.int/en/animal-health-in-the-world/animal-diseases/Bovine-tuber-

culosis/#:~:text=Bovine%20tuberculosis%20is%20an%20OIE,a%20closely%20related%20species%2C%20M (accessed on 21 September 2021).

94. Ye, Q., Wu, Q., Hu, H., Zhang, J., & Huang, H., (2016). Prevalence and characterization of *Yersinia enterocolitica* isolated from retail foods in China. *Food Control, 61*, 20–27.
95. Zinsstag, J., (2009). Transmission dynamics and economics of rabies control in dogs and humans in an African City. *Proceedings of the National Academy of Sciences of the United States of America, 106*(35), 14996–15001.
96. Zinsstag, J., Schelling, E., Waltner-Toews, D., Whittaker, M., & Tanner, M., (2015). *One-Health: The Theory and Practice of Integrated Health Approaches* (p. 480). UK: Cab International.
97. Zinsstag, J., Schelling, E., Waltner-Toews, D., & Tanner, M., (2011). From one medicine to one-health and systemic approaches to health and well-being. *Preventive Veterinary Medicine, 101*, 148–156.

PART II

Chemical Hazards in Milk and Milk Products

CHAPTER 6

MILK ADULTERATION: CURRENT SCENARIO AND CHALLENGES

TANMAY HAZRA, ROHIT G. SINDHAV, CH. V. K. SUDHEENDRA, and VIMAL M. RAMANI

ABSTRACT

Several hazardous and non-hazardous adulterants are used in milk to increase the margin of profit and also to dilute nutrients present in milk. Adulteration in milk and dairy products in any form (intentional or unintentional), should be stopped. Government and food manufacturers (small or large) should take necessary steps to solve this issue, which causes several health hazards in humans. Government should also promote the growth of the dairy business to strengthen the analysis of milk or dairy products at the rural level, where most of the Indian milk procurement has been conducted.

6.1 INTRODUCTION

Food is the fuel of human civilization; apart from providing basic nutrition; and different physiological and immunological functions are directed by milk products. The human body needs 10,500 kilojoules (KJ) of energy that is provided by basic ingredients of foods, such as carbohydrates, fats, proteins, vitamins, and minerals [86]. The food habit in human life is not only associated with normal biological activities but it has an integral relationship with the cultural emblem of any human society. Hence, safe food is a need of any society for development. Although due to huge population growth throughout the world, it has less food production; yet some time the food in use is adulterated or food is manipulated by different means. Adulteration of food is a curse to modern human society.

- **What is a Food Adulteration?** The history of food fraud is as parallel to the history of human civilization [43]. Adulteration of food is the degradation of food quality by adding with some inferior foreign components / components or by substituting the component from the actual food materials. An adulterant is defined by FSSAI (Food Safety and Standards Authority of India) that substance, which will degrade the quality of food [28].

  Food is categorized as an adulterated food when any substance that is harmful to health or low-priced or lower quality substances are added, or when the quality is below the prescribed legal standards, the principal component/components is/are removed from main food [12]. Adulterated food can cause different adverse health effects [66, 68] even death.

This chapter highlights the current status, issues, and challenges in milk adulteration.

6.2 CLASSIFICATION OF ADULTERATION OF FOODS [12]

1. **Intentional or Economical Motivated Adulteration (EMA):** Addition of foreign substances having resemblance to the original food; mainly for economic growth to the traders or farmers. Mixing of water in milk is a live example of this category of food adulteration.
2. **Unintentional Adulteration:** Entrance or inclusion of unwanted substances in food during pre and post-processing of food. Pesticide contamination is one of the live examples of this type of food adulteration.
3. **Microbial Contamination:** Inclusion of undesired micro-organism due to pre and post-processing activities of food.

6.2.1 HISTORY OF FOOD ADULTERATION OR FOOD FRAUD

Adulteration of foods dates back to ancient times, i.e., since food was sold on shelves. Even across the world, massive sale of adulterated foods or meats have been reported in the 1800 century. Some globally popular food scandals in the past are:

- In the year 1820, German chemist Frederick identified hazardous food colors in various foods and beverages; and in 1836, many foods were rejected from the German market [3].
- In the year 1850, thousands of children died in the USA due to consumption of poisonous milk and this incident was popular as Swill Milk Scandal. It was reported that milk was adulterated with plaster of Paris and starch [11].
- In the year 2003, carcinogenic Sudan dye was found in red chili [62].
- In the year 2008, melamine scandal in China snatched the life of many young children [32].
- The Irish pork crisis of 2008 was popular as dioxin contamination was detected in pork meat [83].
- The 2013 horse meat scandal was popular as beef meat was contaminated with horse meat [84].

Both developed and under-developed countries are under threat of adulteration [11, 72]. The foods most commonly adulterated are meat and meat products, poultry, food grains, spices, milk, and milk products [1, 64]. However, the types and nature of adulterants may change from time to time. For example, individual oils are being used to adulterate milk, but the recent trend is being followed by the unscrupulous trader to adulterate milk with designer oils or fats that cannot be detected by simple analysis [63, 64]. It is accounted that milk is one of the highest and soft targets of adulteration throughout the world [11, 50].

6.3 MILK: UNIQUE AND ALMOST COMPLETE FOOD

Milk is a natural ambrosia that provides superior quality of almost all macro and micronutrients for physiological development and growth; it can be readily consumed by infants, nursing women, children, and elderly people [64]. It is considered as nearly complete food. It is a very concentrated and basic source of macro and micro-elements of bioactive components, which are supportive to the development and growth of human body. Almost 260 million Cows throughout the world produce about 730 million tons of milk in the world [5].

Majority of world's milk production is contributed by cow (83%), buffalo (14%), goat (2%) and sheep (1%) and other milch animals (such as: camel, horse, donkey, yak (1% almost)) [26]. The growth of milk production in Asia has increased almost 283%, whereas in the USA it has increased by 19% [50]. The dairy sector holds a significant share in the global food market.

However, the image of milk as white ambrosia to human being has been considerably smudged due to the curse of adulteration or mixing of harmful chemicals to milk [19].

6.3.1 WHY MILK ADULTERATION IS EASY?

Under the Food Safety Regulations [28] by the Government of India (GOI), milk is a complex biological fluid from healthy mammals. Adulteration of milk is defined as addition or removal of some components with cheaper matters [19, 28].

The first and foremost reason for the widespread adulteration of milk is its chemical and physical nature. Chemically, milk is an extremely complex mixture comprising of lipids (mixture of mixed triglycerides, phospholipids, and other fat-soluble substances), proteins (caseins, whey proteins and other minor proteins), carbohydrates (mainly lactose to gather with oligosaccharides), minerals, and minor constituents (enzymes, pigments, NPN compounds, trace elements, gases, etc.), [42]. Moreover, physically milk is a highly heterogeneous system containing different components in emulsion, colloidal dispersion, true solution or partly in colloidal dispersion and partly in a true solution state. Due to this extremely complex chemical make-up and highly heterogeneous physical structure, milk can very easily hide many foreign substances when mixed with it [8].

6.4 ADULTERANTS IN MILK [8]

- Substances whose purpose is to increase the volume of milk with a view to getting more profit;
- Group of adulterants in milk consists of those substances, which are used for delaying spoilage of milk. These types of adulterants are also termed as preservatives and neutralizers;
- These groups of adulterants in milk include substances, whose purpose is to conceal watering of milk or that substance, which constitute synthetic milk.

The adulterants employed in EMA are cheaper fluids, such as, water, whey, skim milk and artificial fluid resembling physically and chemically to milk (so-called synthetic milk). Among the adulterants for EMA of milk, mixing of water is the most common practice [23, 35, 68] and now synthetic milk is infiltrating in the field. Apart from these adulterants, Table 6.1 indicates other chemical components, such as, synthetic fatty acids, color, emulsifier, gelatin that are frequently added in milk.

TABLE 6.1 Effects of Different Adulterants in Milk

Adulterants	Effects on Milk
Ammonium sulfate	Increases thickness of milk.
Benzoic acid formalin, hydrogen peroxide and salicylic acid	Preservative action and enhances the shelf life of the milk.
Detergents	It is generally use to emulsify oil; during oil adulteration in milk.
Low valued milk	The main aim of this type of adulteration to increase more profit. Addition of cow milk in goat milk. However, these types of adulterations cause allergic reaction to certain population.
Melamine	Increases non-protein nitrogen in milk apparently increasing protein in milk or milk powder.
Milk powder	Milk powder mixes with fresh milk to increase SNF however milk powder often mixes with milk when shortage of milk occurred.
Neutralizers	It is used for decreasing acidity in milk; also, neutralizers are the key ingredients of synthetic milk as well.
Non-edible colors	To improve the esthetic quality specially appearance of milk. Sometime in buffalo milk yellow color is added to represent it as cow milk.
Oil/fat	Increases fat in milk.
Salt	SNF increases. Also acts as a preservative to delay the spoilage of milk.
Starch	Density or SNF of milk increases to balance the addition of water.
Sugar	Increase in the lactometer reading by increasing the SNF in milk.
Urea	SNF in milk increase. whiteness in milk increases; total nitrogen content in milk increases; synthetic milk can be prepared
Water	Volume of milk increases with the decrease in the nutritive quality

Milk adulteration is not a single type chemical component addition but it is a multiple chemical approach. For example, for masking added water in milk or preparing of a synthetic variety of chemical substances are used. To make up the fat content, cheaper vegetable oil/fat is admixed. For incorporation of fat, the emulsion is prepared by employing surfactant, such as, detergent. Similarly, a decrease in density due to watering is compensated with the help of salt of sodium, potassium or ammonium with chloride, sulfate, or nitrate. For hiding the added water in milk or preparing of synthetic milk, selection of substances is planned and worked out so smartly that visual appearance, common physical properties, and gross chemical composition are simulated to match with those of the genuine milk.

6.5 SYNTHETIC MILK: AN EVIL FACE OF MILK

Synthetic milk is formulated using chemical substances (urea, oil, detergents, or non-edible color, etc.). Milk fat is replaced with foreign fat or vegetable oil. The nitrogen component in milk is replaced by urea or non-milk protein; and detergents are added to make foamy acceptable appearance [55, 73]. Physicochemical differences between synthetic and natural milk are shown in Table 6.2.

TABLE 6.2 Physicochemical Differences between Synthetic and Natural Milk

Characteristics	Synthetic Milk	Fresh Natural Milk
Color	Generally white	White to light yellow
Color during heating	Yellowish	No color change
Color during storage	Yellow	No color change
Flowability	Readily flows	Slowly flows
Odor or aroma	Becomes soapy and will be distinct while boiling	Typical milky aroma
pH	Alkaline in nature	Slightly acidic
Taste	Bitter (slightly)	Slight sweet and salty
Texture	Soapy texture	Not soapy texture however for high fat milk sometime gives soapy texture

6.6 HEALTH EFFECTS OF ADULTERATED MILK

Adulteration in milk with chemical substances not only affects the economic balance of a country but it directly affects the health of the consumers [55, 68]. WHO advised GOI that non-restricted consumption of adulterated milk can be fatal and may massacre Indian population by the 2025 [19]. Health hazards due to consumption of adulterated milk are listed in Table 6.3.

TABLE 6.3 Diseases Due to Consumption of Adulterated Milk

Adulterant	Diseases	References
Chlorine	Heart problem	[34]
Foreign fat/oil	Heart diseases	[35]
Hydrogen Peroxide	Inflammation in the intestine and disturbing the natural immunity in human body	[68]
Melamine	Renal failure and deaths	[20]
Neutralizers	Gastric ulcer, diarrhea	[19, 68]
Preservatives	Cancer and other poison-related symptoms	[68]
Starch	Diarrhea	[77]
Urea	Harmful to heart, liver especially for kidneys	[45]
Water	Decrease the nutritive quality of milk; however infected water my cause diarrhea	[15]

Antibiotics are generally considered as contaminants that are used for the treatment of Mastitis disease in milch animals. These are transferred to milk through ill-animals, which are given dose of antibiotics to cure a certain disease. Presence of Antibiotics causes serious health hazards including cancer and urticarial [23]. Pesticides are generally considered as contaminants in milk or foods. Presence of pesticides causes serious health hazards including cancer [6, 68]. Negative effects of adulteration of milk are summarized in Figure 6.1.

6.7 DETERMINATION OF AN ADULTERANT IN MILK BY DIFFERENT ANALYTICAL METHODS

Different analytical approaches are being adopted to determine (qualitative and quantitative) the adulterants added to milk or dairy products. The adoption of analytical methods mainly depends on different characteristics of adulterants. However, isolation techniques act as a pre-requisite step to proceed further for instrument-based detection analytical protocol. Color-based test protocols are performed for qualitative detection of adulterants in milk. These simple tests are generally used to detect common adulterants (starch, urea, sugar, etc.), in milk [29, 75]. Table 6.4 describes rapid qualitative color-based detection tests of seven common different adulterants in milk.

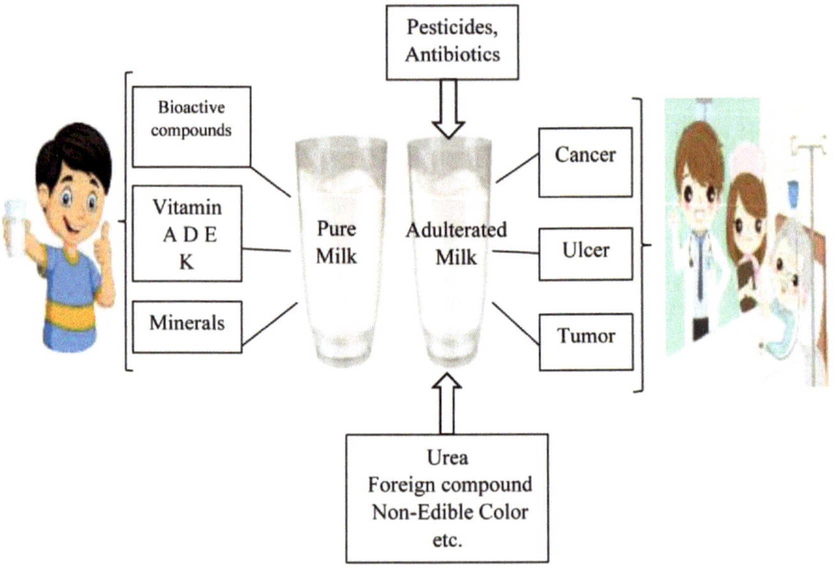

FIGURE 6.1 Evil effects of milk adulteration (created by authors).

TABLE 6.4 Protocols for Detection of Different Adulterants in Milk

Adulterant	Protocol	Observation
Cane sugar	Resorcinol test-5 ml milk + 1 ml HCl + 0.1 g of resorcinol powder. Red color during heating that indicate presence of cane sugar.	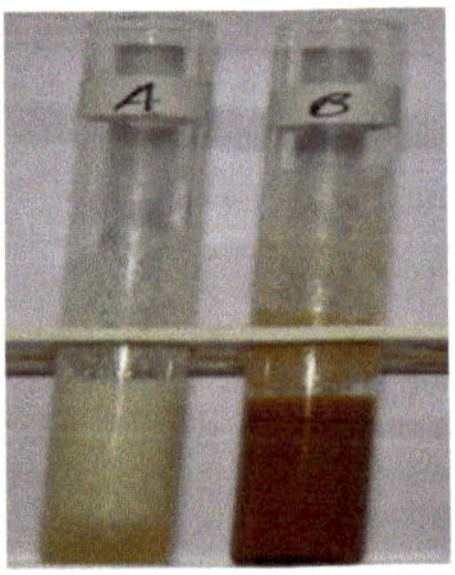

TABLE 6.4 *(Continued)*

Adulterant	Protocol	Observation
Common salt	Milk (1 ml) + 0.1 N of 5 ml $AgNO_3$ solution + 2–3 drops of potassium chromate (10%). Formation of Yellow color confirms the added salt. Control sample will remain red color as color of silver nitrate.	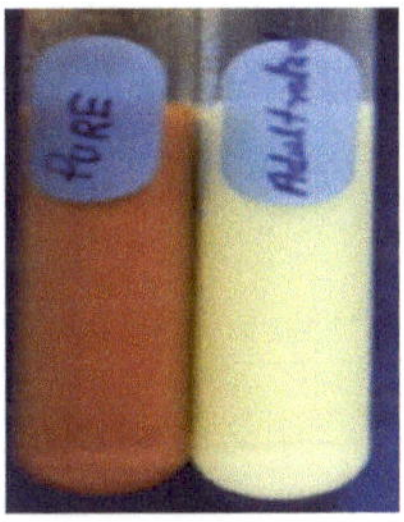
Formalin	Chronotropic acid-1 ml milk + chronotropic acid reagent (72%). Presence of starch turns yellow color but no color changed observed for pure milk.	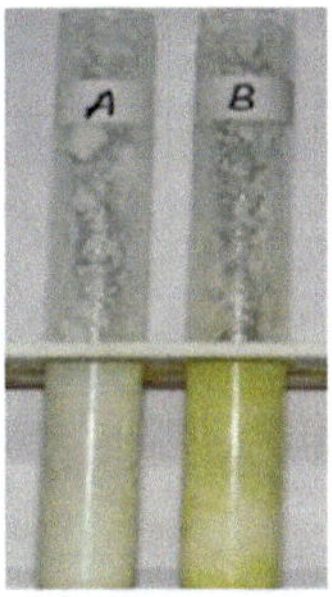
Glucose	Barfoed test-1 ml Milk + 1 ml Barfoed's reagent heat 3–4 minutes. Deep blue color complex indicates the addition of glucose. Pure milk gives faint bluish color due to Barfoed's reagent.	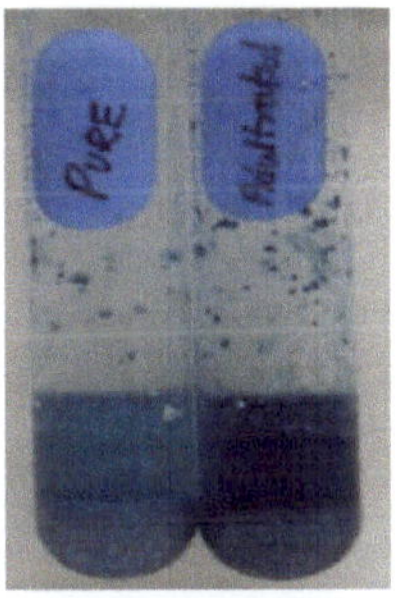
Neutralizers	Rosalic acid test-5 ml Milk + 5 ml ethanol+ 2 drop rosalic acid. Red rose color indicates the occurrence of alkali as neutralizer. Yellow color indicates pure milk.	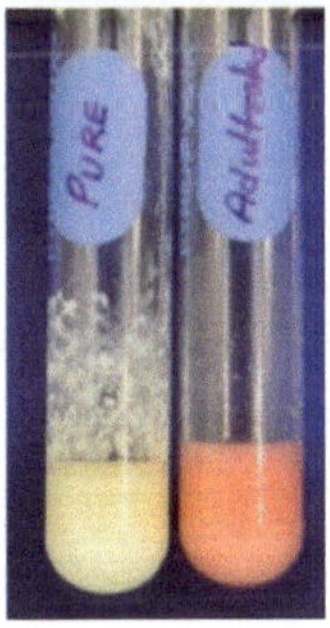

TABLE 6.4 *(Continued)*

Adulterant	Protocol	Observation
Starch	3 ml Milk + heat + Few drops of iodine solution (1%). Presence of starch confirmed by Blue-violet. Control sample will remain lite yellow color.	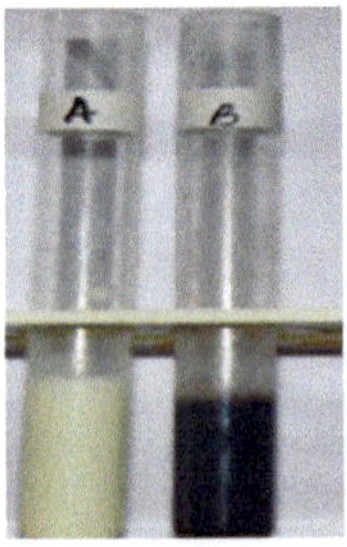
Urea	DMAB test reagent: 1 ml Milk + 1 ml 1.6% (w/v) alcoholic DMAB. Deep yellow color confirms the presence of added urea. Light yellow color indicates pure milk due to the presence of natural urea in milk.	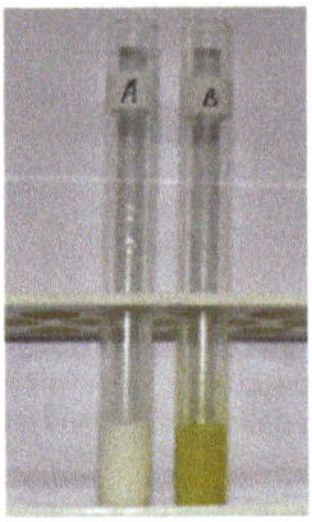

Legend: A: Control milk; B: adulterated milk.

6.7.1 ADVANTAGES OF QUALITATIVE TESTS

There are several important advantages in the use of qualitative tests (described in Table 6.3) in detecting several adulterants in milk [11]:

- Require very low initial investment and does not need any special trained manpower;
- Easily affordable even by small scale dairy plants.

6.7.2 DISADVANTAGES OF QUALITATIVE TESTS

Certain limitations are encountered while detecting some common adulterants added to milk using qualitative tests:

- Targeted analysis-individual test to be performed for each adulterant;
- Slow, time consuming and labor-intensive;
- The possibility of inter-adulterant interference on the performance of some test;

- Low sensitivity of some of the tests;
- Poor repeatability/reproducibility of some tests.

6.8 RECENT ADVANCEMENTS

Several sophisticated instrumental methods including chromatography (GLC (gas-liquid chromatography), HPLC, UPLC (ultra-performance liquid chromatography), etc.), spectroscopy (UV, Visible, IR, Raman, and NMR), mass spectrometry (MS) electrophoresis (PAGE), immunology (ELISA and RIA) and thermography (DTA and DSC) techniques have been adopted by the researchers [23, 64, 68]. Even combinations of different instrumental methods (GC-MS, LC-MS, CE-MS, MS-MS, etc.), are also employed as a hyphenated-techniques in milk forensic laboratory. In this section, authors of this chapter have developed some novel, innovative, and instrumental based protocols for detection of common adulterants in milk.

6.8.1 WATER

Freezing point depression method by Cryoscope is a well-known technique for detection of added water in milk up to 3% [76]. Water is the most common adulterants in milk. An Impedance based sensor is able to check the ionic property of milk and it is used to detect added water in milk up to 10% [24]. Another sensor-based method was developed, which uses nitrite sensor for detection of added water in milk [51].

6.8.2 DETERGENT

Detergent is mainly added to milk as an emulsifying agent during adulteration of foreign fats or oils in milk. Hence, it is an important component of synthetic milk. Several innovative methodologies have been adopted to detect detergents in milk.

Addition of anionic detergent in milk was detected using a spectrophotometric method [13]. The Azure A dye and EDTA solution were used that formed blue color complex in the presence of detergent in milk. This method was able to detect even very low quantity of detergent in milk (10–20 mg/L).

Pradeep et al. [65] used multimeter to detect detergent in milk by electric conductivity of milk. Milk is able to conduct electricity due to the presence of certain ions in milk; however, the addition of detergents increases the number of ions in the solution due to its dissociation of protein; thus it increases conductivity. This is the principle of this said methodology. However, those researchers were not able to detect low amount of detergent in milk.

Kumar et al. [48] applied gold nanoparticles (AuNPs) to detect detergent in milk. Gold nanoparticle-based method (due to optical properties) was very efficient to differentiate pure and detergent added milk. Thus, this method was able to detect low level of detergent in milk without the interface of other adulterants. Detergent present in milk can be swiftly detected and quantified using FT-IR (Fourier-transform infrared) spectroscopy [39]. Researchers have identified that 4,000–500 cm^{-1} spectral difference range can be used to differentiate between non-adulterated milk and adulterated milk. Hence, researchers have reported that FT-IR could be a better alternative method in detecting detergent in milk.

6.8.3 MELAMINE

After China scandal of melamine adulteration in milk and milk powders, USFDA has fixed the limits of melamine in normal foods and baby foods as 2.5 mg/kg and 1.0 mg/kg, respectively [11]. Generally, melamine adulteration can directly increase the total protein content of milk. Consumption of milk adulterated with melamine has a serious impact on the health of consumers; therefore, worldwide researchers have tried to detect melamine in milk through different novel technologies. The HPLC, LC-MS, GC-MS, MALDI-MS, NMR spectroscopy have been successfully used to detect low level of melamine adulteration in milk [79]. Innovative technologies, such as, SB-ATR, FTIR have successfully detected even 0.00025% of melamine adulteration in milk within 1–2 min without the requirement of any preparation steps for the sample [40]. Application of NIR hyperspectral imaging method successfully detected even 1% of melamine in milk [31]. These researchers have reported that these methods are useful even in detecting other adulterants in milk.

AuNPs surface grafted with derivative of Cyanuric acid could detect as low as 2.5 ppb of melamine in infant formulations (food) [4]. A simple colorimetric based detection of melamine using citrate stabilizes gold nanoparticle-based methodology is very fast. The presence of melamine in milk causes the accumulation of AuNPs resulting in a coffee color (red) to purple-blue color that is visible by the naked eye itself [49].

6.8.4 UREA

Chemical-based methods are being successfully employed to detect adulteration of urea in milk. Novel instrumental or sensor-based methodologies are developed by researchers to detect urea in milk. Raman spectroscopy-based method is used to detect urea's presence in milk both qualitatively and quantitatively. Raman excitation set up of 785-nm diode laser is able to detect presence of urea in pure milk instantly [46]. Another study applying Raman spectroscopy was found to detect even minute quantities of nitrogen-rich compounds, such as urea, and differentiate pure milk from adulterated without any false results [59]. A simple opto-fluidic micro-viscometer based on the stereo lithography technique is useful for detection of urea in milk up to 1% [85].

A potentiometric biosensor using NH^{+4} ion-sensitive electrode as the transducer can confirm the lower level (2.5×10^{-5} mol/L) of externally added urea in milk [84]. Application of dielectric couple with computational technique was used to detect different adulterants in milk effectively including urea [33].

The wireless network running on Android platform (Jelly-bean) coupled with Enzyme thermistor (FIA-ET) urea biosensor could detect urea in milk [56]. Similar types of chip-based technology developed on a chemical chip-based pad was able to detect multiple adulterants in milk within 40 minutes [50].

6.8.5 FOREIGN FATS AND OILS

The application of fluorescence spectroscopy using 2- and 3-dimensional front-face method was assessed for detection of vegetable oil in milk fat. This method had shown high potential for confirming the presence of vegetable oil in pure milk [61]. Difference in the compositional analysis of milk fat components was used for the identification of foreign fat addition in milk. A change in the fatty acids (such as: oleic acid ($C_{18:1}$) and linoleic acid ($C_{18:2}$)), triglycerides are directly proportional to the adulteration ratios. Therefore, changes in fatty acids and triacylglycerols analysis could be an efficient marker in detecting foreign fat in milk [47]. Triacylglycerols can be a potent marker in detecting foreign fat adulteration in milk fat; but cannot confirm the type of adulteration in milk fat [35]. A new approach has been followed to detect adulteration of refined palm oil and other fats in milk using gas chromatography (GC), wherein selected peaks of chromatograms of triglycerides are zoomed in and superimposed [63].

Researchers have also used PCR based method to ascertain the presence of animal depot fat up to 10% level added to milk fat [35, 74].

Nedeljkovic et al. [57] used Raman spectroscopy coupled with linear discriminant analysis for confirming the presence of vegetable oil in milk fat. This methodology was very sensitive for the detection of a wide range of vegetable oils including sunflower, coconut, and palm oils in milk fat. Makhoul et al. [52] applied direct injection mass spectroscopy for efficiently detection of foreign fat adulteration in milk. Sukthija et al. [78] used RP-TLC based protocol for the detection of adulterant oils in milk to a minimum level of 1%.

Researchers used FT-NIR and FT-MIR spectroscopy to identify palm oil adulteration in ghee [9, 53]. However, FT-IR-based analytical approaches were not suitable to detect palm oil in ghee lower than 10% level [9]. Annasamy et al. [7] used plastic optical fiber probes (U-structure) based fat sensor to confirm the presence of added oil in milk.

Ramani et al. [66] claimed to detect palm oil in milk fat at low level by unique colorimetric test protocol. However, those authors did not claim the effect of that said protocol for other oils or fats.

It is very difficult to confirm the presence of foreign fat in milk or milk fat; therefore, several researchers applied different analytical approaches (DSC, TLC, MS) to prove the originality of milk fat. Apart from these common adulterants, certain preservatives and low valued milk, whey, nondairy protein used to be adulterated with milk. Instrument-based approaches for detection of different adulterants in milk are listed in Table 6.5.

6.8.6 DISADVANTAGES

- High cost because of huge initial investment; Expensive operating software; the requirement of highly skilled and trained manpower; Very costly service, maintenance, and repairing/ replacement of equipments or parts.

- The lengthy operating procedure, particularly if the target adulterant necessitates isolation, purification, concentration, and/or derivatization. This renders the technique highly time-consuming very laborious and extremely tedious. Therefore, very hard to use in routine quality analysis especially rural-based dairy industry [50].

TABLE 6.5 Instrument-Based Approaches for Detection of Different Adulterants in Milk

Adulteration	Applied Analytical Techniques	References
Detergent	FT-IR, liquid chromatographic, UV-spectroscopy	[13, 21, 64]
Foreign fat/oil	DSC, MALDI-TOF-MS, FT-IR, TLC, GLC, electronic nose	[9, 10, 32, 43, 67, 83]
Low valued milk	PCR, ELISA	[25, 36, 70]
Melamine	Near-infrared spectroscopy, Raman spectroscopy, HPLC	[20, 23, 27, 68]
Nondairy protein	H NMR spectroscopy, low injection mass spectrometric, optical biosensor	[2, 14]
Synthetic milk	Paper-based impedance sensor	[24]
Urea	Voltamperometric, impedance sensor,	[24, 37]
Water	Impedance sensor, freezing point osmometry	[18, 22]
Whey	Capillary zone electrophoresis, laser-induced breakdown spectroscopy, fluorimetry SDS-PAGE	[16, 58]

6.9 MILK ADULTERATION IN INDIA

India is the highest milk producing country with almost 43 indigenous cattle breed and 18% progressive growth rate of cow population in 2019–2020 [54]. The continuous growth of the dairy sector has put the first-hand challenges to maintain the overall safety and quality of milk and dairy products. Adulteration of milk in India is a primary and serious challenge for the growth of the Indian dairy industry.

6.10 SOME RECENT STUDIES ON ADULTERATION OF MILK IN INDIA

FSSAI (in 2011) conducted a survey throughout different states of India (Figure 6.2). They collected and analyzed total of 1,791 samples of milk from 33 different states. It reported that 565 (31.5%) samples were found to meet the standards laid by FSSAI; however, 1226 (68.5%) samples of milk were adulterated [19].

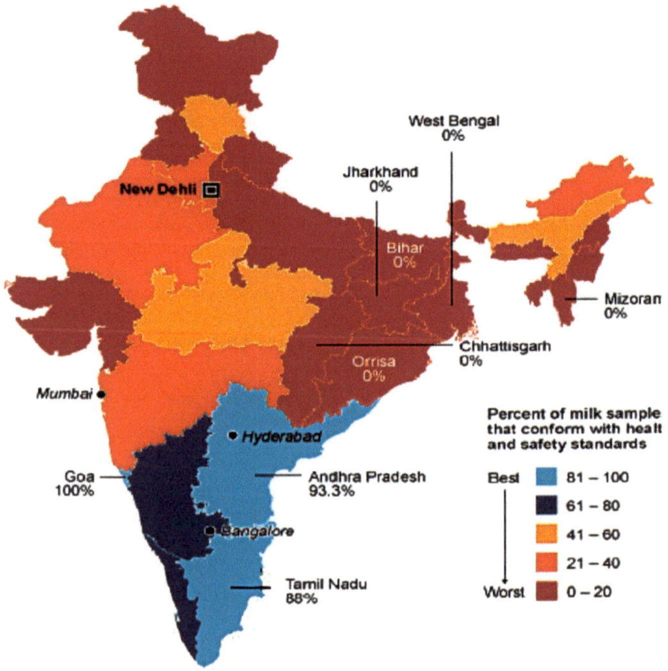

FIGURE 6.2 State-wise adulterants found in different states of India during FSSAI (in 2011) survey on milk.

The adulteration of milk is getting grim problem for public health concern; however, demand-supply gap and financial gain has been considered as one of main reasons for milk adulteration throughout India. Recent survey studies confirmed that adulteration in milk is one of the main hurdles of Indian Dairy Industry:

- Sahu and Chappalwar [71] reported that quality of milk sold in Parbani district was very poor.
- Palsra et al. [62] identified hazardous chemicals including alizarin, detergent, formalin, and neutralizers in milk.
- Swathi and Kauser [80] reported the presence of detergents in the milk sold in the local market of Hyderabad. They also reported 100% milk was adulterated with water.

- Kabariya and Ramani [41] reported that adulteration in milk samples were frequently practiced in the local market of Amreli (a district in Gujarat).
- Brindha et al. [17] found highly adulterated milk sold in Thuraiyur District in Tamil Nadu.
- Roy et al. [70] identified urea, detergents, and other adulterants in milk samples collected from the market of Faridabad, Gurgaon, and Noida.
- Singuluri and Sukumaran [77] reported that synthetic milk randomly sold in Greater Hyderabad market.

6.11 REASONS FOR MILK ADULTERATION IN INDIA [18, 44, 64]

- The Dairy plants mostly purchase milk through contractors and small vendors. These intermediaries sometimes resort to unlawful practices to get more profit by maximization of the quantity of milk.
- To extend the shelf life of milk.
- To earn more revenue by adopting illegal practices.
- The regulatory bodies fail to control the menace of adulteration due to lack of commitment coupled with inadequacy of basic infrastructure facilities for adulteration checks on milk.
- Empirical detection tests.
- The producer is unaware of the hazards produced to consumers due to adulteration.

6.12 RECOMMENDATIONS

- Stress should be given on clean milk production and hygienic practices.
- Effective implementation of the HACCP system.
- Milk should be purchased through co-operatives societies.
- Farmers should be educated to avoid indiscriminate use of antibiotics, synthetic fertilizers, and pesticides.
- To safe guard the public heath, continuous efforts are needed by the local, state, and national agencies to adopted strict measures to prevent adulteration.
- Rapid and strict execution of legal proceeding against adulterer of milk.

- Mass awareness programs among the consumers at urban and village level about quality culture.

Indian researchers are continuously searching to fight against the evil of adulteration. However, real-time sensors have been developed by scientists [22, 24, 41, 66] for the detection of very minimum quantity adulterants in milk at receiving docks. However, most of these techniques are patented technologies so far; hence the real-time uses of these techniques are not reaching to the quality control laboratories, especially rural-based areas. In India, most of the milk collection centers situated in rural areas. Therefore, GOI should take positive approaches for getting real time test protocols to the reach of the farmers or quality control officers.

6.13 SUMMARY

Adulteration in milk is a key problem faced by India and other developing countries around the world. Presently chemical analytical techniques are in practice to detect adulteration, but different sophisticated methods are being developed and employed for detection of different types of adulterants in milk. In the recent past, FSSAI, and other reports demonstrated that more than 68% of the milk was substandard. Therefore, a suitable analytical method for detection and controlling of this evil deed of adulteration in milk is required. Government agencies and food manufacturers (small or big) should take necessary steps to solve this problem. Government should also promote the growth of the organized dairy business.

KEYWORDS

- **adulteration**
- **carcinogenic**
- **contaminants**
- **enzyme thermistor**
- **hazards**
- **mass spectrometry**

REFERENCES

1. Abbas, O., Zadravec, M., & Baeten, V., (2017). Analytical methods used for the authentication of food of animal origin. *Food Chemistry, 246*, 6–17.
2. Abdulhalim, I., Zourob, M., & Lakhtakia, A., (2008). Overview of optical biosensing techniques. In: Marks, R., Cullen, D., Lowe, C., Weetall, H. H., & Karube, I., (eds.), *Handbook of Biosensors and Biochips* (pp. 413–446). New York: John Wiley & Sons.
3. Adam, B. J. D., (2009). Palette of our palates: Brief history of food coloring and its regulation. *Comprehensive Reviews in Food Science and Food Safety, 8*, 394–408.
4. Ai, K., Liu, Y., & Lu, L., (2009). Hydrogen-bonding recognition-induced color change of gold nanoparticles for visual detection of melamine in raw milk and infant formula. *Journal of the American Chemical Society, 127*(131), 9496, 9497.
5. Aishwarya, M. N. L., & Duza, M. B., (2017). A review on adulteration of milk. *Indo American Journal of Pharmaceutical Research, 7*(8), 630–637.
6. Akhtar, S., & Ahad, K., (2017). Pesticides residue in milk and milk products: Mini review. *Pakistan Journal of Analytical & Environmental Chemistry, 18*(1), 37–45.
7. Annasamy, G., Rajamani, A. S., Bandaru, R., & Sai, V. V. R., (2019). U-bent plastic optical fiber probes as refractive index based fat sensor for milk quality monitoring. *Optical Fiber Technology, 47*, 15–20.
8. Aparnathi, K. D., (2019). Milk adulteration detection-challenges in development of real-time test. e-souvenir. In: *National Conference on Adulteration of Dairy & Food Products: Accomplishment & Road Ahead* (pp. 1–5). Organized by college of dairy science and postgraduate institute of dairy education and research, Kamdhenu University, Amreli, Gujarat – 365601.
9. Aparnathi, K. D., Sharma, S., Antony, B., & Mehta, B. M., (2019). Development of method for detection and quantification of foreign oils and fats in ghee (heat clarified milk fat) using FT-NIR spectroscopy coupled with chemometric. *Indian Journal of Dairy Science, 72*(1), 12–22.
10. Ayari, F., Mirzaee-Ghaleh, E., Rabbani, H., & Heidarbeigi, K., (2018). Detection of the adulteration in pure cow ghee by electronic nose method (Case study: Sunflower oil and cow body fat). *International Journal of Food Properties, 21*, 1670–1679.
11. Azad, T., & Ahmed, S., (2017). Common milk adulteration and their detection techniques. *International Journal of Food Contamination, 3*, 1–9.
12. Bansal, S., Singh, A., Mangal, M., Mangal, A. K., & Kumar, S., (2017) Food adulteration: Sources, health risks and detection methods. *Critical Reviews in Food Science and Nutrition, 57*(6), 1174–1189.
13. Barui, A. K., Sharma, R., Rajput, Y. S., & Singh, S. A., (2013). rapid chromatographic method for detection of anionic detergent in milk. *Journal of Food Science and Technology, 50*(4), 826–829.
14. Berganaa, M. M., Adamsa, K. M., Harnlyb, J., Moorea, J. C., & Xiea, Z., (2019). Non-targeted detection of milk powder adulteration by 1H NMR spectroscopy and conformity index analysis. *Journal of Food Composition and Analysis, 78*, 49–58.
15. Barham, G. S., Khaskheli, M., Soomro, A. H., & Nizamani, Z. A., (2014). Extent of extraneous water and detection of various adulterants in market milk at Mirpurkhas, Pakistan. *Journal of Agriculture and Veterinary Science, 7*(3), 83–89.

16. Bilge, G., Sezer, B., Eseller, K. E., Berberoglu, H., Toccu, A., & Boyaci, I. H., (2016). Determination of whey adulteration in milk powder by using laser induced breakdown spectroscopy. *Food Chemistry, 212*, 183–188.
17. Brindha, N., Chitra, P., Janarthanan, R., & Murali, A. A., (2017). Study on detection of adulteration in milk samples from different regions of Thuraiyur District in Tamil Nadu, India. *International Journal of Current Microbiology and Applied Sciences, 6*(12), 3303–3310.
18. Buttel, B., Fuchs, M., & Holz, B., (2008). Freezing point osmometry of milk to determine the additional water content-an issue in general quality control and German food regulation. *Chemistry Central Journal, 2*(6), 1–7.
19. Chauhan, S. L., Priyanka, Mandal, K. D., Paul, B. R., & Maji, C., (2019). Adulteration of milk: A review. *International Journal of Chemical Studies, 7*(1), 2055–2057.
20. Cheng, Y., Dong, Y., Wu, J., Yang, X., Bai, H., Zheng, Z., Ren, D., et al., (2010). Screening melamine adulterant in milk powder with laser Raman spectrometry. *Journal of Food Composition and Analysis, 23*(2), 199–202.
21. Coitinho, B. T., & Cassoli, L. D., (2018). Adulteration identification in raw milk using Fourier transform infrared spectroscopy. *Journal of Food Science and Technology, 54*(8), 2394–2402.
22. Das, S., Sivaramakrishna, M., Biswas, K., & Goswami, B., (2014). A low-cost instrumentation system to analyze different types of milk adulteration. *ISA Transactions, 56*, 268–275.
23. Das, S., Goswami, B., & Biswass, K., (2015). Milk adulteration and detection: A review. *Sensor Letters, 14*, 4–18.
24. Das, S., Sivaramakrishna, M., Biswas, K., & Goswami, B., (2011). Performance study of a 'constant phase angle based' impedance sensor to detect milk adulteration. *Sensors and Actuators A, 167*, 273–278.
25. De, S., Brahma, B., Polley, S., & Mukherjee, A., (2011). Simplex and duplex PCR assays for species-specific identification of cattle and buffalo milk and cheese. *Food Control, 22*(5), 690–696.
26. FAO (Food and Agriculture Organization of the United Nations), (2013). p. 404. http://www.fao.org/3/i3396e/i3396e.pdf (accessed on 21 September 2021).
27. Filazi, A., Sireli, U. T., Ekici, H., Can, H. Y., & Karagoz, A., (2012). Determination of melamine in milk and dairy products by high performance liquid chromatography. *Journal of Dairy Science, 95*(2), 602–608.
28. FSSAI (Food Safety and Standards Authority of India), (2010). https://old.fssai.gov.in/Portals/0/Final_Regulations_2010.pdf (accessed on 21 September 2021).
29. FSSAI, (Food Safety and Standards Authority of India), (2015). https://old.fssai.gov.in/Portals/0/Pdf/Draft_Manuals/MILK_AND_MILK_PRODUCTS.pdf (accessed on 21 September 2021).
30. Fu, J., & Nicoll, G., (2011). The milk scandal and corporate governance in China. *Canberra Law Review, 10*(3), 103–124.
31. Fu, X., Kim, M. S., Chao, K., Qin, J., Lim, J., Lee, H., & Ying, Y., (2019). Detection of melamine in milk powders based on Nir hyperspectral imaging and spectral similarity analyses. *Journal of Food Engineering, 124*, 97–104.

32. Garcia, J. S., Sanvido, G. B., & Saraiva, S. A., (2012). Bovine milk powder adulteration with vegetable oils or fats revealed by MALDI-QTOF MS. *Food Chemistry, 131*, 722–726.
33. Ghasemi-Varnamkhasti, M., & Ghatreh-Samani, N., (2017). Development of two dielectric sensors coupled with computational techniques for detecting milk adulteration. *Computers and Electronics in Agriculture, 140*, 266–278.
34. Hattersley, J. G., (2000). The negative health effects of chlorine. *Journal of Orthomolecular Medicine, 15*(2), 89–95.
35. Hazra, T., (2017). *Triacylglycerol Profiling and Genomic DNA Analysis to Ascertain the Purity of Ghee* (p. 198). PhD Thesis. National Dairy Research Institute Karnal, India.
36. Hazra, T., Sharma, V., Sharma, R., & Arora, S., (2016). Simplex PCR assay for detection of cow milk presence in goat milk. *Indian Journal of Dairy Science, 69*(5), 621–625.
37. Hilding-Ohlsson, A., Fauerbach, J. A., Sacco, N. J., Bonetto, M. C., & Corton, E., (2012). Volt-amperometric discrimination of urea and melamine adulterated skimmed milk powder. *Sensors, 12*(9), 12220–12234.
38. Hong, E., Lee, S. Y., Jeong, J. Y., Park, J. M., Kim, B. H., Kwon, K., & Chun, H. S., (2017). Modern analytical methods for the detection of food fraud and adulteration by food category. *Journal of the Science of Food and Agriculture, 97*(12), 3877–3896.
39. Jaiswal, P., Jha, S. N., Kaur, J., & Borah, A., (2017). Detection and quantification of anionic detergent in milk using attenuated total reflectance-Fourier transform infrared spectroscopy. *Food Chemistry, 221*, 815–821.
40. Jawaid, S., Talpur, F. N., Sherazi, S. T. H., Nizamani, S. M., & Khaskheli, A. A., (2013). Rapid detection of melamine adulteration in dairy milk by SB-ATR-Fourier transform infrared spectroscopy. *Food Chemistry, 141*(3), 3066–3071.
41. Kabariya, R. B., & Ramani, V. M., (2018). Quality assessment of milk samples of Amreli region with nanotechnology-based dipstick and other known technology: A comparative study. *Bioscience Biotechnology Research Communications, 11*(2), 313–317.
42. Kalyankar, S. D., Khedkar, C. D., Patil, A. M., & Deosarkar, S. S., (2016). Milk: Sources and composition. In: Caballero B, Finglas P. M., & Toldra, F., (eds.), *Encyclopedia of Food and Health* (pp. 741–747). New York: Academic Press.
43. Kala, A. L., (2013). Detection of possible adulteration in commercial ghee samples using low-resolution gas chromatography triglyceride profiles. *International Journal of Dairy Technology, 66*, 346–351.
44. Kamthania, M., Saxena, J., Saxena, K., & Sharma, D. K., (2014). Milk adulteration: Methods of detection and remedial measures. *International Journal of Engineering and Technical Research, 1*, 15–20.
45. Kandpal, S. D., Srivastava, A. K., & Negi, K. S., (2012). Estimation of quality of raw milk (open and branded) by milk adulteration testing kit. *Indian Journal of Community Health, 24*(3), 188–192.
46. Khan, K. M., Krishna, H., Majumder, S. K., & Gupta, P. K., (2015). Detection of urea adulteration in milk using near-infrared Raman spectroscopy. *Food Analytical Methods, 8*(1), 93–102.

47. Kim, H. J., Park, J. M., Lee, J. H., & Kim, J. M., (2016). Detection for non-milk fat in dairy product by gas chromatography. *Korean Journal for Food Science of Animal Resources, 36*(2), 206–214.
48. Kumar, P., Manhas, S., & Navani, N. K., (2016). Simple method for detection of anionic detergents in milk using unmodified gold nanoparticles. *Sensors and Actuators B: Chemical, 233*, 157–161.
49. Kumar, N., Seth, R., & Kumar, H., (2014). Colorimetric detection of melamine in milk by citrate-stabilized gold nanoparticles. *Analytical Biochemistry, 456*, 43–49.
50. Luther, J. L., De Frahana, V. H., & Lieberman, M., (2017). Paper test card for detection of adulterated milk. *Analytical Methods, 9*(38), 5674–5683.
51. Madhuvilakku, R., Alagar, S., Mariappan, R., & Piraman, S., (2019). Glassy carbon electrodes modified with reduced graphene Oxide-MoS2-Poly (3, 4-Ethylene Dioxythiophene) nanocomposites for the non-enzymatic detection of nitrite in water and milk. *Analytica Chimica Acta, 1093*, 93–105.
52. Makhoul, S., Yener, S., & Khomenko, I., (2016). Rapid non-invasive quality control of semi-finished products for the food industry by direct injection mass spectrometry headspace analysis: The case of milk powder, whey powder and anhydrous milk fat. *Journal of Mass Spectrometry, 51*(9), 782–791.
53. Mehta, B. M., Antony, B., Sharma, S., Ratnam, K., & Aparnathi, K. D., (2018). Comparative appraisal of FT MIR spectra of ghee (Heat Clarified Milk Fat) with that of the mutton tallow and vegetable fat. *Indian Journal of Dairy Science, 71*, 240–245.
54. MFAHD (Ministry of Fisheries, Animal Husbandry & Dairying-Government of India), (2019). http://www.dahd.nic.in/about-us/divisions/cattle-and-dairy-development (accessed on 21 September 2021).
55. Mudgil, D., & Barak, S., (2013). Synthetic milk: Threat to Indian dairy industry. *Carpathian Journal of Food Science and Technology, 5*(1), 64–68.
56. Naik, P. P., Mishra, G. K., Danielsson, B., & Bhand, S., (2015). Android integrated urea biosensor for public health awareness. *Sensing and Bio-Sensing Research, 3*, 12–17.
57. Nedeljkovic, A., Tomasevic, I., Miocinovic, J., & Pudja, P., (2017). Feasibility of discrimination of dairy creams and cream-like analogues using Raman spectroscopy and chemometric analysis. *Food Chemistry, 232*, 487–492.
58. Neelima, R. S., Rajput, Y. S., & Mann, B., (2013). Chemical and functional properties of glycomacro peptide (GMP) and its role in the detection of cheese whey adulteration in milk: A review. *Dairy Science & Technology, 93*(1), 21–43.
59. Nieuwoudt, M. K., & Holroyd, S. E., (2016). Raman spectroscopy as an effective screening method for detecting adulteration of milk with small nitrogen-rich molecules and sucrose. *Journal of Dairy Science, 99*(4), 2520–2536.
60. Nisa, A., Zahra, N., & Butt, Y. N., (2016). Sudan dyes and their potential health effects. *Pakistan Journal of Biochemistry, 49*(1), 29–35.
61. Ntakatsane, M. P., Liu, X. M., & Zhou, P., (2013). Short communication: Rapid detection of milk fat adulteration with vegetable oil by fluorescence spectroscopy. *Journal of Dairy Science, 91*(4), 2130–2136.
62. Palsra, T., Khurana, S. K., Manel, B. G., & Sharma, A., (2019). Assessment of the quality and adulteration level of raw milk collected from consumer table in Himachal Pradesh. *International Journal of Livestock Research, 9*(9), 80–93.

63. Pathania, P., Sharma, V., Arora, S., & Rao, P. S., (2020). Novel Approach to detect highly manipulated fat adulterant as Reichert-Meissl value-adjuster in ghee (Clarified Butter) through signature peaks by gas chromatography of triglycerides. *Journal of Food Science and Technology, 57*(1), 191–199.
64. Poonia, A., Jha, A., Sharma, R., Singh, H. B., Rai, A. K., & Sharma, N., (2016). Detection of adulteration in milk: Review. *International Journal of Dairy Technology, 70*(1), 23–42.
65. Pradeep, S., Lakshminarayana, P., Varsha, R., & Shriya, K. K., (2016). Screening of adulterants in milk. *International Journal of Current Research and Review, 8*, 25–29.
66. Ramani, A., Hazra, T., Parmar, M. P., Sindhav, R. G., & Ramani, V. M., (2019). Simple rapid technique for detection of palm oil in ghee. *Indian Journal of Dairy Science, 72*(4), 441–444.
67. Rani, A., Sharma, V., Arora, S., Lal, D., & Kumar, A., (2013). Rapid reverse phase thin layer chromatographic protocol for detection of adulteration in ghee (clarified milk fat) with vegetable oils. *Journal of Food Science Technology, 10*, 2–6.
68. Reddy, D., Venkatesh, K., & Reddy, C., (2017). Adulteration of milk and its detection: A review. *International Journal of Chemical Studies, 5*(4), 613–617.
69. Ren, Q. R., Zhang, H., Guo, H. Y., Jiang, L., Tian, M., & Ren, F. Z., (2014). Detection of cow milk adulteration in yak milk By ELISA. *Journal of Dairy Science, 97*(10), 6000–6006.
70. Roy, B., Singh, J., Sunsunwal, S., Dayal, G., Yadav, B., Bhardwaj, C., & Teotia, A., (2017). Detection of harmful adulterants in milk supplied to Delhi, India. *Current Science, 112*(11), 2316–2320.
71. Sahu, R. K., & Chappalwar, A. M., (2019). Bacteriological quality assessment of market milk sold in Parbhani City - Maharashtra. *International Journal of Livestock Research, 9*(4), 68–75.
72. Salih, M. A. M., & Yang, S., (2017). Common milk adulteration in developing countries cases study in China and Sudan: Review. *Advances in Dairy Research, 5*(4), 192–195.
73. Sen, D., (2019). Be cautious from synthetic milk. E-souvenir. In: *National Conference on Adulteration of Dairy & Food Products: Accomplishment & Road Ahead* (pp. 1–5) Organized by College of Dairy Science and Postgraduate Institute of Dairy Education and Research, Kamdhenu University, Amreli, Gujarat – 365601.
74. Sharma, V., Hazra, T., Kandhol, R., Sharma, R., & Arora, S., (2018). Confirmation of buffalo tallow in anhydrous cow milk fat using gas-liquid chromatography in tandem with species-specific polymerase chain reaction. *International Journal of Dairy Technology, 71*(1), 158–163.
75. Sharma, R., Rajput, Y., Barui, A. K., & Naik, N. L., (2011). *Detection of Adulterants in Milk: A Laboratory Manual* (p.105). Karnal: National Dairy Research Institute.
76. Singh, P., & Gandhi, N., (2015). Milk Preservatives and adulterants: Processing, regulatory and safety issues. *Food Reviews International, 31*(3), 236–261.
77. Singuluri, H., & Sukumaran, M. K., (2014). Milk adulteration in Hyderabad - a comparative study on the levels of different adulterants present in milk. *Journal of Chromatography & Separation Techniques, 5*(1), 212–216.
78. Sukthija, Sharma, V., Arora, S., Singh, R., Raghu, H. V., & Raju, P. N., (2019). Evaluation of reversed-phase thin layer chromatographic protocol to detect vegetable oil/fat in milk. *Indian Journal of Dairy Science, 72*(4), 370–374.

79. Sun, F., Ma, W., Xu, L., Zhu, Y., Liu, L., Peng, C., & Xu, C., (2010). Analytical methods and recent developments in the detection of melamine. *Trends in Analytical Chemistry, 29*(11), 1239–1249.
80. Swathi, J. K., & Kauser, N., (2015). Study on adulteration of milk and milk products from local vendors. *International Journal of Biomedical and Advance Research, 6*(9), 678–681.
81. Tlustos, C., (2009). The dioxin crisis in Ireland 2008: Challenges in risk management and risk communication. *Organohalogen Compounds, 71*, 1152–1154.
82. Trivedi, U. B., & Lakshminarayana, D., (2009). Potentiometric biosensor for urea determination in milk. *Sensors and Actuators B: Chemical, 140*(1), 260–266.
83. Upadhyay, N., Goyal, A., Kumar, A., & Lal, D., (2017). Detection of adulteration by caprine body fat and mixtures of caprine body fat and groundnut oil in bovine and buffalo ghee using differential scanning calorimetry. *International Journal of Dairy Technology, 70*(2), 297–303.
84. Valand, R., Tnna, S., Lawson, G., & Bengtström, L., (2019). Review of Fourier transform infrared (FTIR) spectroscopy used in food adulteration and authenticity investigations. *Food Additives & Contaminants: Part A, 37*(1), 19–38.
85. Venkateswaran, H. N., Sharma, P. S., Dubey, A., Agarwal, A., & Goel, S., (2016). Rapid and automated measurement of milk adulteration using a 3-D printed optofluidic micro viscometer (OMV). *IEEE Sensors Journal, 16*(9), 3000–3007.
86. Walpole, S. C., Prieto-Merino, D., Edwards, P., Cleland, J., Stevens, G., & Roberts, I., (2012). The weight of nations: Estimation of adult human biomass. *BMC Public Health, 12*(439), 1–6.

CHAPTER 7

ANTIBIOTIC RESIDUES IN MILK: CONTROL STRATEGIES IN DAIRY ANIMALS

SUSHIL K. MOHAPATRA and TAPAS K. PATBANDHA

ABSTRACT

The antibiotic resistance (AR) leads to the development of new resistant bacteria or superbugs, resulting in failure of antibiotic therapy under clinical situations in both humans and animals. Situations, wherein antimicrobials are ineffective against certain strains, have also been observed in different studies carried out by researchers. Spread of resistance can happen between different bacterial species through cross-resistance without depending on pathogenic and nonpathogenic strains. AR development particularly in the pathogens has become the greatest threat to antibiotic therapy. Awareness and upgradation of knowledge regarding AR among all stakeholders, including farmers, veterinary, and para-veterinary staffs will be helpful to tackle AR issues. There is a need for successful implementation of strict regulation and surveillance monitoring of antibiotics residues in milk. Availability of reliable techniques (immunoassay, liquid chromatography (LC) and mass spectrometry) is quite essential for implementing national regulation, screening, and identifying antimicrobial residues in milk. In addition to judicial use of antibiotics, good hygiene and animal husbandry practices at dairy farms could reduce the antibiotic residues in milk.

7.1 INTRODUCTION

Milk production in India during 2018–2019 was 187.7 million tons with per capita availability of milk was 394 gm per day per head. India ranks top in the world in milk production since 1997 [24]. The milk produced in India is mostly used for domestic consumption in terms of liquid milk or for the preparation of various milk products. However, the milk produced in India could not achieve the export standard fixed by the developed countries due to its poor quality. Among different factors affecting the quality of milk, antibiotic residue is one of the important factors. In general, antimicrobial agents (AMAs) particularly antibiotics are commonly prescribed for prophylactic as well as therapeutic measures in human medicine, veterinary medicine, and aquaculture. Antibiotics, as growth promoters are also used in animals in a sub-therapeutic dose. However, in raw milk, the presence of antibiotic residues reflects unregulated and extensive overuse of antibiotics in dairy animals.

Though there are certain guidelines for discarding milk during treatment period based on their withdrawal time or at least seven days period [30], yet such guidelines are not being strictly followed by the farmers under field conditions in developed countries like India as it causes huge economic losses and also there is lack of strict monitoring protocol to detect residues of antibiotics in milk. Antibiotic residues in dairy products possess a huge public health importance, but all the stakeholders including farmers are unaware of this. Milk and milk-based products with antibiotic residues are allergic to human beings and alter intestinal microbial populations [22].

Further, consumption of dairy products containing antibiotic residue leads to antibiotic resistance (AR) in human beings and compels the medicos to use higher antibiotics for treatment which is detrimental to human health. However, a set of strategies (improved sanitation and hygiene to ensure lower overall infection rates) has been developed by the World Health Organization (WHO). This could prevent the overuse of antibiotics in dairy animals for therapeutic use [29]. There is need to understand the judicial use of antibiotics for treatment and prophylactic measures in dairy animals, their presence in milk and public health importance.

This chapter discusses control strategies and guidelines for antibiotic residues in milk in dairy animals.

7.2 ANTIBIOTICS USE IN DAIRY ANIMALS

7.2.1 HISTORY OF ANTIBIOTICS USE

The intensified livestock farming system, under the modern era owing to lower cost of production [3], may aggravate the spread of infections among the farm animals within a short period. This needs implementation of different strategies to reduce spread of the infection among the farm animals and antibiotics therapy is considered as one of the important measures [18]. With discovery, the antibiotics have been used in human beings as well as in animals and agriculture. Antibiotics are used for both better health and productivity of dairy animals. Antibiotics are used for prophylactic measures and growth promoters in lower doses (sub-therapeutic dose) in growing animals. The first use of antibiotics in animals started from 1938 onwards and sulphonamide was the first antibiotic used for farm animal treatment.

Later in the year 1940, gramicidin was used experimentally to treat udder infection in dairy cattle for the first time. Though penicillin was discovered by Alexander Fleming, a Scottish physician-scientist in the year 1928, its experimental use in farm animals was conducted 14 years later (in the year 1942). The therapeutic use of penicillin in mastitic animals was initiated during the year 1943 in Britain and Denmark [14]. Initially, penicillin was infused through the intramammary route and had a significant effect on mastitis cure in dairy cattle. The intramammary infusion of antibiotics is considered as an effective and most common route for treatment of mastitis which results in unexpected residues in the milk [17].

This could be minimized by discarding milk from treated animals depending on milk withdrawal period of antibiotics [11]. The withdrawal time of antibiotics also vary depending on their physical and chemical properties as well as route of administration [2]. Nowadays, antibiotics are not only used for treatment of mastitis but also for the prophylactic measure for mastitis.

In developed countries like the United States and European Unions (EUs), the intramammary infusion of antibiotics as a prophylactic measure under dry cow therapy is a common practice. This practice is followed to prevent the occurrence of new intramammary infection in dairy cows in the ensuing lactation [18]. Reports on the use of different AMAs for treatment of mastitis indicates that most commonly used AMAs are enrofloxacin (22%), ampicillin, and cloxacillin combination (18.57%), gentamicin (18.29) and ceftriaxone (12.57%). Among different classes of AMAs, penicillin, and their combination were most commonly used, which accounts about 29.43%. The percentage use of fluoroquinolone, aminoglycosides, cephalosporins, tetracyclines, and chloramphenicol groups of antibiotics was 22%, 21.43%, 16.86%, 9.14% and 1.14% cases, respectively for treatment of mastitis in dairy animals in India [20].

7.2.2 ANTIMICROBIAL RESISTANCE (AMR)

AMAs include drugs, which act against infections caused by a wide range of microbes like bacteria, virus, fungus, and parasites. Among these AMAs, drugs that act against bacterial infections (i.e., antibiotics) are most frequently used in food and dairy animals. Antibiotic use in humans and animals has been increasing to such an extent that they are even used for the common infection. However, continuous application of antibiotics in dairy animals both for prophylactic and therapeutic measures results in antimicrobial resistance (AMR). The microbes or bacteria that cause infections when develop AMR, survive well under exposure to the same drugs that have strong bactericidal or bacteriostatic properties [27].

The first report of AMR was noted in the year 1940 in *Escherichia coli* for the antibiotic benzyl-penicillin. Then later, first penicillin resistant *Staphylococcus aureus* bacteria were detected in the United States (US) in the year 1944 [5]. In the subsequent years till date several bacteria have developed resistance to different AMAs. There is also the appearance of 'superbugs' like Methicillin-resistant *Staphylococcus aureus* (MRSA) and tuberculosis causing extremely drug-resistant bacteria [27]. Now it is very difficult to treat these resistant bacteria using the existing AMAs. Hence, it is need of the hour to restrict use of AMAs and if required, then use in a judicial manner as per the guidelines.

Over and inappropriate AMAs use leads to the development of resistance genes in bacteria. Such bacteria transfer from animals to humans through food, direct contact or environment which results in significant public health threat. Further, the bacteria with acquired resistance genes can transfer to the next generation or to the other bacteria present in that environment, which is a key public health concern [3]. There is also evidence regarding spread of AMRs across the globe. In Vietnam, antibiotic-resistant Gonorrhea emerged in 1967, but later spread worldwide. Similarly, New Delhi-Metallo-beta-lactamase (NDM) enzyme producing bacteria were reported for the first time in Sweden in 2008, but later was found in different parts of the globe [19].

Furthermore, the condition owing to AMR is more aggravated as it increases the duration of infection, mortality, and morbidity along with social and economic costs. It has been estimated that by 2050 due to AMR about 10 million deaths per year will occur across the globe, which would cost an economic loss of about 100 trillion US dollars [29]. In India, nearly 60,000 infants die every year due to the infections caused by antibiotic resistant bacteria. Additionally, about 0.2 million deaths occur globally owing to the multidrug-resistant and extremely drug-resistant tuberculosis [27]. There are several causes for AR as depicted in Figure 7.1.

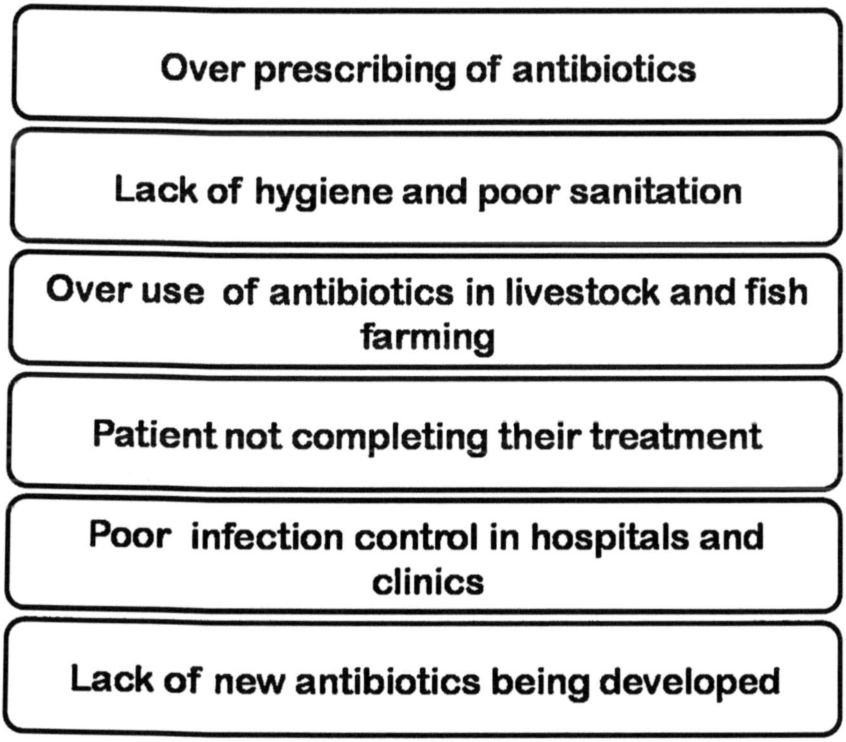

FIGURE 7.1 Causes of antibiotic resistance.

7.2.3 GLOBAL USE OF ANTIMICROBIAL AGENTS (AMAS)

Consumption of AMAs globally during 2010 was estimated to be 63,151 tons and projected to increase by 67% (105,596 tons) by 2030. China ranked top position during 2010 in terms of AMAs consumption for food animal productions and shared 23% of the total global consumption. The other countries under top five positions were: US (13%), Brazil (9%), India (3%) and Germany (3%) ranked 2nd, 3rd, 4th, and 5th position, respectively. This ranking is projected to be in the order of China (30%), followed by the US (10%), Brazil (8%), India (4%), and Mexico (2%) by the end of 2030. In Asia, the AMAs consumption by 2030 is estimated to be 82% of the total global consumption; whereas the AMAs consumption would rise by 99% in five BRICS (Brazil, Russia, India, China, and South Africa) countries alone [3].

However, it is projected that by 2030 global AMAs consumption may be higher than the consumption previously estimated by Boeckel et al. [3]; as in 2013 about 131,109 tons of AMAs were used globally for different purposes [4]. Boeckel et al. [4] also projected that global consumption of AMAs will reach to 200,235 tons by the end of 2030. Further, the maximum share of antibiotics consumed globally is used in farm animals producing food. In the US, about 80% of the annual antibiotics consumed were used in food animals, which was 4 times higher than the amount consumed in human beings [3] either for therapeutic use, prophylactic use or sub-therapeutic use [18]. Hence, there is need of the hour to develop policies, strict implementation of such policies to reduce the use of antimicrobials, specifically antibiotics in livestock and surveillance could minimize the risk of AMR.

Furthermore, the data on species-wise average consumption of AMAs per kg of animal weight revealed maximum for pig (172 mg) followed by chicken (148 mg) and cattle (45 mg) [3]. However, the AMAs consumption rate per kg of animal weight varies considerably among different countries, from 8 mg in Norway to 318 mg in China [4]. This indicated that China is the leading country in the globe not only in terms of total absolute quantity of antibiotic consumption but also per kg of animal products produced. Therefore, there must be more efforts from China as a leader to minimize the use of AMAs in animals producing food for human beings.

7.3 ANTIBIOTIC RESIDUES IN MILK

Antibiotic residue in milk and milk products is an important public health concern globally. Milk from the antibiotic treated cows is the main source for the presence of antibacterial drug residues depending on the withdrawal period and such milk must be discarded. Though there are several published reports regarding antibiotic residue in milk at global level but in India, yet only few reports are available. Kumar [15] observed antimicrobial residues in milk from cattle and buffaloes in Bihar. He screened 200 raw milk samples for the presence of tetracyclines and sulfonamides groups of antibiotics. Total 105 (53%) samples were tested positive for tetracyclines residues, whereas 137 (69%) samples were positive for sulfonamides residues. Out of this, 24 (12%) samples were positive for tetracyclines, 81 (41%) for oxytetracyclines, 64 (32%) for sulphamethoxazole and 73 (37%) for sulfadimethoxine residues. Furthermore, 4% (7), 8% (16), 6% (12) and 4% (8) positive samples had antibiotic residue level above the maximum residual limit (MRL) for tetracyclines, oxytetracyclines, sulphamethoxazole, and sulfadimethoxine, respectively.

Similarly, in Kerala about 8.48% (14/165) milk samples tested positive for the presence of antibiotic residues. The common antibiotics detected above permissible limit were: tetracycline, enrofloxacin, and ß-lactams in milk samples [16]. Moudgil et al. [22] found antibiotic residues in 11.3% milk samples out of total 168 samples. Further, 4.8, 3.0, 3.0 and 0.6% of samples had tested positive for enrofloxacin, penicillin-G, oxytetracycline, and sulphamethoxazole, respectively. Moreover, based on MRL cut off, about 1.7, 1.2 and 0.6% samples, respectively, had higher residual level for enrofloxacin, oxytetracycline, and penicillin-G than the cut-off.

In another study, Moudgil et al. [21] screened 524 milk samples in Punjab for the presence of antibiotic residues. They observed antibiotic residues in 16% and 12.5% samples, respectively collected from dairy farms and markets. Among the positive samples, 4% samples from commercial farms and 3.1% samples from the market had antibiotic residues higher than the MRL limit. Enrofloxacin residue was observed in a maximum number of milk samples followed by oxytetracycline, tetracycline, and sulphamethoxazole. These studies clearly indicate that human beings in India are consuming milk containing antibiotic residues higher than the maximum residual permissible limit (MRL) for antibiotics [7].

The tolerance limit (mg/Kg) of different antimicrobial drugs in milk as per Food Safety and Standard Authority (FSSAI) of India is depicted in Table 7.1.

TABLE 7.1 Residues of Antimicrobial Drugs in Milk with Tolerance Limit According to FSSAI

Antibiotic	Class of Antibiotics	Tolerance Limit (mg/Kg)
Ampicillin	Penicillin	0.01
Apramycin	Aminoglycoside	0.01
Cefquinome sulfate	Cephalosporin	0.01
Ceftiofur	Cephalosporin	0.01
Cephapirine	Cephalosporin	0.01
Chlortetracycline/oxytetracycline/tetracycline	Tetracycline	0.10
Cloxacillin	Penicillin	0.01
Colistin	Polymyxin	0.05
Dihydrostreptomycin/streptomycin	Aminoglycoside	0.02
Enrofloxacin	Fluoroquinolone	0.01
Lincomycin	Lincosamides	0.15
Monensin	Polyether ionophores	0.002
Neomycin	Aminoglycoside	1.50
Salinomycin	Aminoglycoside	0.01
Spectinomycin	Aminoglycoside	0.20
Sulfadiazine	Sulfonamide	0.01
Sulfadimidine	Sulfonamide	0.025
Sulfaguanidine	Sulfonamide	0.01
Sulfanilamide	Sulfonamide	0.01
Sulphadiazine	Sulfonamide	0.01
Sulphaquinoxaline	Sulfonamide	0.01
Sulphathiazole sodium	Sulfonamide	0.01
Trimethoprim	Sulfonamide	0.01
Tylosin tartrate	Macrolide	0.01
Virginiamycin	Streptogramin	0.01
Zinc bacitracin	Polypeptide	0.01

7.4 DETERMINATION OF ANTIBIOTIC RESIDUES

Chromatography is the most common method for detecting antibiotic residues and its levels in biological fluids, such as serum, milk, etc., and is able to detect and quantify the residues in minute quantities. GC, liquid chromatography (LC), thin-layer chromatography (TLC) and high-performance liquid chromatography (HPLC) techniques are most commonly used to detect the residues of antibiotics in the farm animal originated foods. Hence, chromatography technique is the most reliable quantitative test, and currently more than 80% of tests are being done using HPLC with mass spectrometry (MS) for detection of antibiotic residues. Requirement of expensive equipment and trained personnel are the main disadvantages associated with chromatography-based techniques. Moreover, nowadays some cow side screening tests are used for qualitative testing purposes. Such tests are only used to know the presence or absence of residue of a particular antibiotic or family of antibiotic in milk [24].

However, the quantitative tests are to be followed for confirmation of residue after preliminary screening of milk by qualitative test. There are also semi-quantitative indigenous commercial test kits available in the market to detect the presence of antibiotics residue in milk. A spore-based and a paper strip-based test kits have been developed at National Dairy Research Institute Karnal (NDRI) for the detection of antibiotics (such as: β-Lactam group, aminoglycosides, tetracycline, chloramphenicol, macrolides) and sulfa drugs at Codex MRLs within 2.3 hours and 1.3 hours, respectively in milk [26].

7.5 REDUCING ANTIBIOTICS USE IN ANIMALS AND HUMAN HEALTH

The global rise of AMAs consumption in animals is obvious because of the continuing and gradual increase in livestock products' demand by the consumers residing in the countries where income is low to moderate. Further, there is shifting of livestock farming towards more intensification where AMAs are used routinely for prophylactic purposes in sub-therapeutic doses. This increase of antibiotic use in animals results in AMR in animals and human beings. The main cause of AMR in animals and human beings is inappropriate and overuse of antibacterial agents in farm animals. This was recognized by the General Assembly of United Nations (UN) in September 2016 [4].

The AMR in animals and human beings could be minimized by restricting the use of AMAs in livestock. The reduction in AMAs used in the animals could have a beneficial effect on human health [6]. However, while formulating and implementing any policy for the reduction or restriction of the use of AMAs in animals, the food and nutritional security in low-income to lower-middle-income developed countries must be taken care [3]. Otherwise, there may be an increase in the gap between demand and supply of animal protein in the developing countries.

Following three-concept policy by Boeckel et al. [4] has been developed to minimize the antibiotic use in animals along with its assessment:

1. **Capping Excessive Use of Antimicrobials:** It is fixing the threshold value of antibiotics use per kg of animal product at 50 mg and with this cap, some European countries have been already maintaining higher livestock productivity.
2. **Restricting the Consumption of Animal Meat:** Reduce antimicrobial consumption by reducing animal protein intake particularly in those countries, where meat consumption is very high. Recently, China has revised its daily animal protein intake and has reduced it to half (40–70 g/day animal protein intake) of the current consumption level. The reduction of animal protein intake could also further counter the biodiversity losses and climate change.
3. **Introducing a Global User Fee on Use of Antimicrobials in Animals:** Antimicrobial use can be reduced by charging a user fee for the sale of antimicrobials for non-human use. This user fee may be used for raising funds for research and development of new antimicrobials while preserving existing drugs.

Implementation of these three concepts in an appropriate way could minimize the use of the AMAs globally in animals for food between 9% and 80% by the end of 2030, considering their consumption in the year 2013 as baseline data. Capping of antimicrobial use per kg of animal product at 50 mg per year could reduce the consumption of the total global antimicrobial drugs by 64%. Further, the total consumption of antimicrobial drugs globally in food producing farm animals could be curtailed by 66% in 2030 if worldwide meat intake is limited to 40 gm/day/head. This indicates a significant beneficial effect of reduced meat consumption on the use of AMAs in animals used for food globally, which could further not only reduce AMR in animals but also improve the environment and human health. Moreover, by introducing 50% user fees on AMAs used in animals could curtail antimicrobials consumption by 31% in 2030 globally. The implementation of the 50% user fees at current price could further generate revenue between 1.7 billion and 4.6 billion US dollars per annum [4].

In a meta-analysis study, it has been observed that minimizing or restricting the AMAs use in livestock could further reduce the occurrence of bacteria resistant to AMAs in both livestock and human beings. The reduction in the use of antibacterial agents in livestock along with interventions has revealed a decline in the occurrence of bacteria with antibiotic-resistant capacity by 10–15% in animals and the bacteria having resistance to multi-drugs by 24–32% as compared to those without any interventions. On the other hand, interventions of restricted use of antibacterial agents in animals had a 24% risk reduction on occurrence of bacteria with resistance to antibiotics in human beings.

Moreover, the human beings that were in close contact with animals (farm workers) had higher risk reduction on prevalence of antibiotic-resistant bacteria (29% versus 9%) than the persons without direct contact with animals [29]. This meta-analysis indicates that the antibiotic resistant bacteria could be reduced both in livestock and humans by restricting the use of AMAs in animals and the beneficial effect is comparatively more in farmworkers, who had direct contact with animals.

In another study, Bunnik and Woolhouse [6] used a simple mathematical model to quantify the beneficial effect of restricted use of antibiotics in food producing animals on human health. They observed little beneficial effect of curtailing the use of antibacterial agents in livestock on human beings. Moreover, they observed that minimization of resistant bacterial transmission rate from livestock to human beings was more effective as compared to the restricted use antibacterial agents use in livestock.

In addition to human and veterinary use, antibiotics are also very frequently used in agriculture, aquaculture, and horticulture purpose. This indicates that the resistant bacteria may be developed from the sources other than animals, and these bacteria may be transferred to human beings. Further, organic farming is a most important farming concept. Today, consumers also prefer more animal products from organic farming due to health issues. Moreover, organic farming concept may be implemented for rearing animals in developing countries like India without compromising the nutritional security of human beings.

7.6 REGULATION OF ANTIBIOTICS USE IN FARM OR DAIRY ANIMALS

Evidence of huge global consumption of AMAs and presence of antibiotic-resistant bacterial presence in farm animals and human beings compelled different world organizations and national governmental agencies to develop strategies to restrict overuse of AMAs in livestock. Worldwide, both developed and developing countries are taking keen interest in minimal application of AMAs in animal husbandry, agriculture, horticulture, and aquaculture.

The WHO along with Food and Agriculture Organization (FAO) and World Organization for Animal Health (OIE) have endorsed "One-Health approach" to address this issue on AMAs and have implemented global action plan (GAP) on AMRs during 2015 [30]. Under GAP, a set of strategies has been planned, which primarily focus on the improvement of sanitation and hygiene. The GAP strategies also include: creation of awareness among the farmers about AMR, strengthening surveillance of antibacterial use in animals, and promotion of research on the development of new antibacterial agents.

Reduction of prevalence of infection by proper implementation of improved interventions and optimization of antibiotics use in animals are also part of GAP strategy [31]. The OIE emphasizes strict execution of veterinary legislation and regulations for the use of AMAs of different categories, such as, veterinary critically important, veterinary highly important and veterinary important in a judicial manner for the farm animals [32, 33].

Many countries including India have implemented national targets to curtail the use of antibiotics in livestock. A task force has been constituted by the Ministry of Health and Family Welfare, Government of India (GOI) to frame anti-microbial policy at national level. The national policy for containment of AMR-India focusses on several aspects, such as: reduction in the application of antibacterial agents in livestock for non-therapeutic use (prophylactic use and growth promoter), strict surveillance for infections as well as for antibiotics resistance in hospitals, promotion of judicial use of antibiotics through education and awareness, close supervision and monitoring of antibiotics use, promotion of research for novel drug discovery and development and implementation of standard and strict prohibitory antibiotic policy [8]. In India, recently the National Action Plan (NAP) on AMRs (2017–2021) was implemented. Under this NAP-AMR, six strategic priorities have been identified. The primary focus area of this action plan is to create awareness on better understanding of bacterial resistance to antibiotics among the public through training and education, to minimize the prevalence of bacterial infection, to promote investments and to strengthen India's leadership on AMRs [30].

7.7 GUIDELINES FOR THE USE OF ANTIBIOTICS

The guidelines for proper antibiotics use by Jones [13] must be followed for an effective drug use program. These proposed guidelines are part of an efficient livestock management program and proper implementation of such essential steps may prevent antibiotics residue in milk for consumption. The different essential steps for effective drug use programs are mentioned below:

- **Step 1:** Properly read the "label" while purchasing any antibiotic. The person on the farm, who handles the drugs, must thoroughly understand their usage as well as the consequences of misuse.
- **Step 2:** Properly store the drugs as per instructions after purchasing or after reconstitution if needed. Otherwise, there is either loss of potency or growth of microbes which may have additional problems.
- **Step 3:** Properly administer the drug. In case of intra-mammary injection, the teat end and for parenteral route, the site of injection must be properly cleaned using alcohol before their administration.
- **Step 4:** Give proper attention to withdrawal time of antibiotics for milk, because of the wide variation of the withdrawal time for different antibiotics. The milk from the treated cows should not be mixed with milk from healthy cows. Such milk should always be discarded depending on the withdrawal period.
- **Step 5:** Special attention must be given to "extra-label" drug use, which is the permission given by the veterinarians to use an antibiotic in such a way that is usually not specified on the "label" or "instruction" inserted inside the package. This may be the higher

dose rate of drug or the use of drug to disease not mentioned on the label or instruction inserted inside the package.

- **Step 6:** Treated cows must be identified by using proper methods, so that the treated cows should be milked separately or at the end of milking and chances of mixing of contaminated milk to the healthy cows' milk could be avoided.
- **Step 7:** Maintain a written health record for all treated cows. The record must include all information such as date of diagnosis and treatment, the withholding time and the person treated the animals, etc.
- **Step 8:** Treated cows must be milked last with special precautions, which could minimize the chance of mixing with milk from a treated cow. This further reduces the chance of mixing milk from such cows with the healthy cows' milk.
- **Step 9:** The milk from all four quarters of treated cows should be discarded though she has been treated for one teat through intra-mammary route. This is because the drugs from one udder quarter reach to others through the circulatory system.
- **Step 10:** Always follow the recommended dose level specified on the label or instruction inserted inside the package or as prescribed by the veterinarians.
- **Step 11:** Several antibiotics should not be combined and administered together by the farmers. Always follow the instruction given by the veterinarians; otherwise, if any doubt arises, then the farmers should consult the veterinarians.
- **Step 12:** When offering medicated feed to the cows, then follow the instruction for feeding and withholding time as suggested by the veterinarians.
- **Step 13:** Always test the milk of the treated cow for drug residue after her recovery before the milk being sold for consumption.
- **Step 14:** The dairy producers should be trained and educated regularly on proper management/drug usage techniques every year.

7.8 ALTERNATIVES TO ANTIBIOTIC THERAPY

There are several alternatives to antibiotics therapy that are used by the veterinarians for the therapeutic and prophylactic measures in dairy animals. Administration of such non-antibiotic therapeutic agents may reduce the chance of AR in bacteria by restricting antibiotic use in animals. The common agents used as an alternative to antibiotics therapy are: (i) vaccines; (ii) probiotics, phages, and microbial derived products; and (iii) phytochemicals [1, 9]. These agents not only significantly minimize overall infection in animals but also minimize the mastitis-causing bacteria, such as: coliforms, streptococci, and staphylococci. The detailed effects of three categories of non-antibiotic therapeutic agents on animal health, particularly udder health is explained in subsections.

7.8.1 VACCINE

Vaccination practices are regularly carried out as per the schedule since long for herd immunity in dairy animals. Recently some new mono- or polyvalent vaccines have also been developed for the prevention of mastitis in dairy animals, particularly for the udder pathogens, such as: *Staphylococcus aureus, Streptococcus agalactiae and Escherichia coli*. However, these vaccines did not give consistent positive results. Hence, along with vaccination. The other health protocol and hygienic milking practices must be followed in a proper way [12].

7.8.2 PROBIOTICS, PHAGES, AND MICROBIAL DERIVED PRODUCTS

Probiotics are living microorganisms used for the treatment and prevention of mastitis, where it stimulates the host immune response and enhances clearance of pathogens causing mastitis. Thus, probiotics are also used for the prophylactic measures in dairy animals against mastitis. Recent reports cited the beneficial effects of probiotic based teat dip on dairy animals' udder health.

The effect of probiotics on the cure rate of udder infections is at par with antibiotics. There is evidence that teat washes and wipes containing antimicrobial peptides (i.e., bacteriocins) derived from certain bacteria have potency to prevent the spread of pathogens on the farm. Bacteriocins also help in diminishing of mastitis-causing bacterial pathogens from teats of cows and subsequently reduce udder infections. Now, the teat dips containing bacteriocins are also commercially available [1].

Bacteriophages or phages are the viruses, commonly known as self-replicating AMAs that infect and kill bacteria by the process of cell lysis. Scientists are able to conduct few studies to date to assess the efficacy of phage therapy for the control of mastitis and other diseases in bovines due to the narrow host specificity of phages. Though the *in-vitro* studies revealed inhibitory effect of phages on the growth of mastitis associated bacterial isolates from raw milk, yet the results from *in-vivo* studies in mastitic cattle are not conclusive [1]. However, the majority of studies in intensively reared pig and poultry have revealed significant positive effects of phage therapy on minimization of bacterial infections [10].

Moreover, endolysins, the phage derived specialized enzymes, have more success on mastitis treatment and prevention than the bacteriophage therapy. The endolysins kill bacteria without selecting their antibiotics resistance property [1, 9].

7.8.3 PHYTOCHEMICALS

Several scientific evidences are available regarding the efficacy of phytochemicals on mastitis treatment and prevention and other diseases. In a recent review, Mushtaq [23] cited the significance of medicinal plants on the prevention and treatment of diseases in bovines as an alternate therapy to antibiotics. The antibacterial agents present in plant extracts act as efflux pump inhibitors and reduce bacterial virulence properties. The plant extracts are also administered along with other existing drugs for the therapeutic use during mastitis infection. Thus, the application of alternate therapies to antibiotics could reduce the risk of antibiotic residue in milk and thereby the reduction in development of resistant bacteria in livestock and human beings.

In addition to alternate therapies, improved sanitation and hygienic practices could reduce the infections in dairy, which would further curtail the antimicrobial use. Proper cleaning and sanitation of udder and teats before milking reduces bacterial load. This subsequently reduces the chance of entry of microbes to the mammary glands through teat opening. It has been reported that good udder and teat hygiene reduce udder infection in dairy cows [28].

Teat dipping is an important managemental intervention, considered as the first line of defense for udder hygiene and mastitis prevention in dairy animals. Post-milking teat dipping blocks the teat opening and prevents bacterial influx into the udder. Nowadays companies like DeLaval are producing eco-friendly chemicals for teat dipping in organic dairy farming, which are also approved by the US and EU [9]. Thus, there must be proper implementation of sanitation and hygienic practices in livestock farms to reduce bacterial infections. The proper cleaning and hygienic practices just before, during, and after milking in dairy cattle is presented in Figure 7.2.

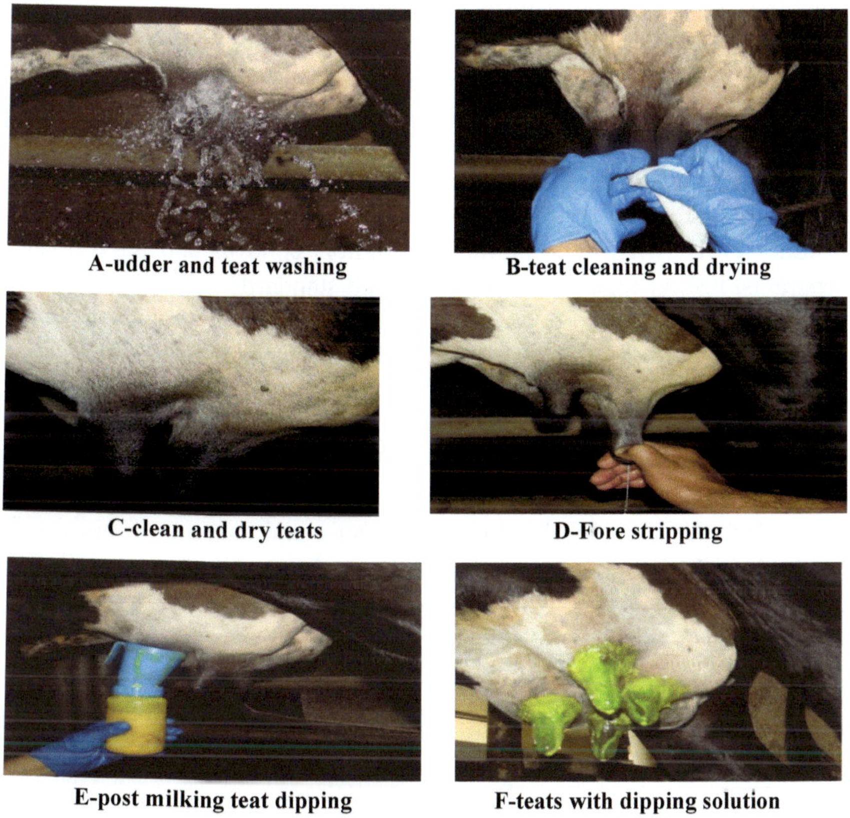

FIGURE 7.2 Proper cleaning and hygienic practices at milking: (A) udder and teat washing; (B) teat cleaning and drying; (C) clean and dry teats; (D) for stripping; (E) post milking teat dipping; and (F) teats with dipping solution.

To maintain good hygiene at milking, first properly wash the udder and teats with clean water, and then dry the teats with a clean towel or tissue paper. Then remove a few strips of foremilk (fore stripping) to check udder infection if any, followed by milking and finally after completion of milking apply dipping solution (post-milking teat dipping) to block the teat openings.

7.9 SUMMARY

Nowadays, antibiotics are emerging as micro-contaminants, due to their potential harmful effects on the health of animals and humans. Due to allergies, gut microflora damage, carcinogenesis, and many other risks, there is a need to regulate antibiotic residues in milk, which is the main food for babies and vegetarian peoples. The *Codex Alimentarius Commission* (CAC), European Economic Community (EEC), WHO, OIE, and FAO are actively involved in regulating the judicious use of antibiotics to control antimicrobial residues in milk of animals; These have suggested maximum residue levels (MRLs) for milk to monitor, control, and surveillance programs. FSSAI regulates food safety and standards in India. FSSAI in India also amended the Food Safety and Standards (Contaminants, toxins, and Residues) Regulations of 2011, to include new tolerance limits for 103 antibiotics and other veterinary pharmaceuticals in animal products on August 1, 2018. Judicial use of antibiotics, good hygiene and animal husbandry practices at dairy farms could reduce the antibiotic residues in milk. Practical alternatives to antibiotic therapy in farm animals could also reduce the overuse of antibiotics. Misuse and overuse of antibiotics are leading to the emergence of superbugs. Awareness and upgradation of knowledge regarding resistance to antibiotics of farmers, veterinary, para-veterinary staff will be able to tackle AR problems.

KEYWORDS

- **antibiotic residue**
- **antimicrobial resistance**
- **control strategies**
- **dairy animals**
- **gas chromatography**
- **non-antibiotic therapy**

REFERENCES

1. Angelopoulou, A., Warda, A. K., Hill, C., & Ross, R., (2019). Non-antibiotic microbial solutions for bovine mastitis-live biotherapeutics, bacteriophage and phage lysins. *Critical Reviews in Microbiology, 45*(5, 6), 564–580.
2. Beyene, T., (2016). Veterinary drug residues in food-animal products: Its risk factors and potential effects on public health. *Journal of Veterinary Science and Technology, 7*(1), 1–7.
3. Boeckel, T. P., Brower, C., & Gilbert, M., (2015). Global trends in antimicrobial use in food animals. *Proceedings of the National Academy of Sciences of the United States of America, 112*(18), 5649–5654.
4. Boeckel, T. P., Glennon, E. E., & Chen, D., (2017). Reducing antimicrobial use in food animals. *Science, 357*(6358), 1350–1352.
5. Box, R., (1998). Antibiotic resistance: A view from the pharmaceutical industry. *Pharmacochemistry Library, 29*, 237–241.
6. Bunnik, B. A. D., & Woolhouse, M. E. J., (2017). Modeling the impact of curtailing antibiotic usage in food animals on antibiotic resistance in humans. *Royal Society Open Science, 4*. Article ID: 161067.
7. FSSAI, (2011). *Food Safety and Standards (Contaminants, Toxins and Residues) Regulations*. https://www.fssai.gov.in/upload/uploadfiles/files/Compendium_Contaminants_Regulations_20_05_2019.pdf (accessed on 22 September 2021).
8. Ganguly, N. K., Arora, N. K., & Chandy, S., (2011). Rationalizing antibiotic use to limit antibiotic resistance in India. *Indian Journal of Medical Research, 134*(3), 281–294.
9. Garcia, S. N., Osburn, B. I., & Cullor, J., (2019). One health perspective on dairy production and dairy food safety. *One Health, 7*. Article ID: 100086.
10. Gigante, A., & Atterbury, R., (2019). Veterinary use of bacteriophage therapy in intensively-reared livestock. *Virology Journal, 16*(1), 155–160.
11. Gustafson, R. H., & Bowen, R., (1997). Antibiotic use in animal agriculture. *Journal of Applied Microbiology, 83*(5), 531–541.
12. Ismail, Z., (2017). Mastitis vaccines in dairy cows: Recent developments and recommendations. *Veterinary World, 10*(9), 1057 1062.
13. Jones, G. M., (2009). Preventing drug residues in milk and cull dairy cows. *Virginia Cooperative Extension Service, 403, 404*, 1–7.
14. Kirchhelle, K., (2018). Pharming animals: A global history of antibiotics in food production (1935–2017). *Palgrave Communications, 4(1)*, 96–96.
15. Kumar, P., (2010). *Studies on Antimicrobial Residues in Milk of Cattle and Buffaloes* (p. 119). Master's Thesis. Birsa Agricultural University, Kanke - Ranchi, Jharkhand.
16. Kumarswamy, N. P., Latha, C., Menon, V. K., Sethukekshmi, C., & Mercy, K., (2018). Detection of antibiotic residues in raw cow milk in Thrissur, India. *The Pharma Innovation Journal, 7*(8), 452–454.
17. Kurjogi, M., & Mohammad, Y., (2019). Detection and determination of stability of the antibiotic residues in cow's milk. *PLoS One, 14*(10). Article ID: e0223475.
18. Landers, T. F., Cohen, B., Wittum, T. E., & Larson, E., (2012). Review of antibiotic use in food animals: Perspective, policy, and potential. *Public Health Report, 127*(1), 4–22.
19. Laxminarayan, R., Duse, A., Wattal, C., & Zaidi, A. K. M., (2013). Antibiotic resistance-the need for global solutions. *The Lancet Infectious Diseases, 13*(12), 1057–1098.

20. Manimaran, A., Layek, S. S., & Kumaresan, A., (2014). Estimation of antimicrobial drug usage for treatment of clinical mastitis cases in organized dairy farm. *Indian Journal of Veterinary and Animal Sciences Research, 43*(2), 140–150.
21. Moudgil, P., Bedi, J. S., Aulakh, R. S., & Gill, J. P. S., (2019). Analysis of antibiotic residues in raw and commercial milk in Punjab - India vis-à-vis human health risk assessment. *Journal of Food Safety, 39*(4). Article ID: e12643.
22. Moudgil, P., Bedi, J. S., Aulakh, R. S., & Gill, J. P. S., (2019). Antibiotic residues and mycotoxins in raw milk in Punjab (India): A rising concern for food safety. *Journal of Food Science and Technology, 56*(11), 5146–5151.
23. Mushtaq, S., Shah, A. M., Shah, A., & Lone, S., (2018). Bovine mastitis: An appraisal of its alternative herbal cure. *Microbial Pathogenesis, 114*, 357–361.
24. Naik, L., Lata, K., Sharma, R., & Mann, B., (2016). Status of antibiotic residues in milk - a systematic review. *Indian Journal of Dairy and Biosciences, 27*, 25–33.
25. NDDB, (2020). *Milk Production in India*. https://www.nddb.coop/information/stats/milkprodindia (accessed on 22 September 2021).
26. NDRI, (2017). *Technologies Available at NDRI for Commercialization*. http://www.ndri.res.in/ndri/Design/documents/ITMC_Technologies_Available_2017.pdf (accessed on 22 September 2021).
27. O'Neill, J., (2016). *Review on Antimicrobial Resistance - Tackling Drug-Resistant Infections Globally: Final Report and Recommendations*. United Kingdom. https://amr-review.org/sites/default/files/160518_Final%20paper_with%20cover.pdf (accessed on 22 September 2021).
28. Schreiner, D. A., & Ruegg, P., (2003). Relationship between udder and leg hygiene scores and subclinical mastitis. *Journal of Dairy Science, 86*(11), 3460–3465.
29. Tang, K. L., Caffrey, N. P., & Nóbrega, D., (2017). Restricting the use of antibiotics in food-producing animals and its associations with antibiotic resistance in food-producing animals and human beings: Systematic review and meta-analysis. *Lancet Planet Health, 1*, E316–E327.
30. Walia, K., Sharma, M., Vijay, S., & Shome, B., (2019). Understanding policy dilemmas around antibiotic use in food animals and offering potential solutions. *Indian Journal of Medical Research, 149*(2), 107–118.
31. World Health Organization (WHO), (2015). *Global Action Plan on Antimicrobial Resistance*. www.wpro.who.int/entity/drug_resistance/resources/global_action_plan_eng.pdf (accessed on 22 September 2021).
32. World Organization for Animal Health (OIE), (2016). *OIE Annual Report on the Use of Antimicrobial Agents in Animals- Better Understanding of the Global Situation*. France: OIE. https://www.oie.int/fileadmin/Home/eng/Our_scientific_expertise/docs/pdf/AMR/Survey_on_monitoring_antimicrobial_agents_Dec2016.pdf (accessed on 22 September 2021).
33. World Organization for Animal Health (OIE), (2015). *OIE List of Antimicrobial Agents of Veterinary Importance*. France: OIE. https://www.oie.int/fileadmin/Home/eng/Our_scientific_expertise/docs/pdf/Eng_OIE_List_antimicrobials_May2015.pdf (accessed on 22 September 2021).

CHAPTER 8

HEAVY METAL CONTAMINATION IN MILK AND MILK PRODUCTS

APURBA GIRI, MONALISA ROY, SUCHETA SAHOO, and KHOKAN C. GAYEN

ABSTRACT

The contamination of milk and milk products can occur with heavy metals due to industrialization and human activities, etc. This review chapter discusses: heavy metals and their toxic levels, recommendation level, sources of contamination, contamination in different milk products, health effects and detection methods (such as: graphite furnace atomic absorption spectrometry, differential pulse anodic stripping voltammetric technique (DPASV), inductively coupled plasma atomic emission spectroscopy (ICP-AES), induced coupled plasma optical emission spectroscopy, ionic liquid, and bismuth oxide nanoparticles modified carbon paste electrode, inductivity coupled plasma mass spectrometry, flame atomic absorption spectroscopy, chronopotentiometric stripping analysis, and electrothermal atomic absorption spectroscopy).

8.1 INTRODUCTION

By different nature and activities, heavy metals (e.g., mercury, chromium, lead, arsenic, cadmium, zinc, iron) may enter into the food chain system via waste from industries, sewage, dust, polluted soils, etc. Also, in our body by replacement of essential metal ions of chelates, the toxic heavy metals are produced. Several researchers have reported that cadmium and lead are toxic for humans and are most sensitive in children. Although the essential heavy metals are zinc and copper with several biochemical roles, yet consumption of these metals in excess amount can be toxic.

Heavy metals may cause serious health issues, such as, nervous system disorders, genetic mutations, neurological disorders, different types of cancers, renal failure, infertility, cardiovascular disorders. Milk production and environment has significant relationships that depend largely on our activities. Pesticides as a contaminant can enter into animal feeds through medicines or injection. In contaminated milk, the amount of metal is really accurate, but the amount may vary noticeably through different processes, such as, production, and packaging. Still, heavy metals (such as: cadmium, lead, chromium, nickel, cobalt) can contaminate cows, camels, and other environments and can disturb their permissible limits in milk and milk products thus causing health problem.

This chapter deals with different types of heavy metals in milk and milk products, their estimation/identification and impact on human health.

8.2 HEAVY METALS

The metals with weight >5 g/cm^3 are known as heavy metals that have huge toxic effects on our body, plants, and foods. Heavy metals, such as, Mercury, Cadmium, Arsenic, Lead are poisonous. These metals have resistance power, get accumulated and not metabolized in other intermediate products and do not freely breakdown into the environment. From the first stage to the last stage (e.g., production to consumer intake), they can be introduced at any stage.

8.3 CLASSIFICATION OF HEAVY METALS BASED ON THEIR HEALTH IMPORTANCE

1. **Essential:** Zinc, copper, cobalt, chromium, iron, and manganese are essential for health [48] when taken within recommended limits (Table 8.1); and these show their toxic effect when taken in excess amount [12, 67].
2. **Non-Essential:** Aluminum, lithium, zirconium.
3. **Less Toxic:** Tin, aluminum (Table 8.1).
4. **Highly Toxic:** Mercury, lead, cadmium (Table 8.1).

The heavy metals are known as trace elements due to availability in trace quantities (10 mg/kg) in the environment [67].

TABLE 8.1 Limitations and Recommended Intake of Some Toxic Heavy Metals

Toxic Heavy Metals	Limitation	Recommended Intake/Safe Intake
Arsenic	3 mg/day (2–3 weeks)	15–25 µg/day
Cadmium	200 µg/kg weight	Adult: 15–50 µg/day; Children: 2–25 µg/day
Lead	500 µg/lit blood	Adult: 20–280 µg/day; Children: 10–275 µg/day
Zinc	150 µg/day	15 µg/day

8.4 SOURCE AND EFFECTS OF HEAVY METAL CONTAMINATION IN DAIRY PRODUCTS

Contamination can occur due to environmental pollution and human activities. From the polluted environment, polluted soil and water-heavy metals can be introduced into the living organisms (animals and humans) [8]. The heavy metals can originate in milk and milk products in several ways, such as contaminated environment, contaminated soil and water, application of veterinary medicine to the cow, etc. The non-biodegradable heavy metals are toxic and can cause serious health problems in the human body. The problem is quite common in urban areas due to rapid industrialization [60, 75].

The Cd and Pb are most toxic heavy metals [2]. Cd acts as a human carcinogenic, is responsible for lung cancer, kidney disease and problem of liver, disorder of cardiovascular systems [2]. Pb has a greater role in the damage of the peripheral and central nervous system, and children's intellectual capacity may be reduced [41]. Heavy metal changes the nutritional profile, when these are introduced in milk and milk products. Sometimes, the iron content may be decreased [46].

8.4.1 HEALTH IMPACTS OF SELECTED HEAVY METALS

8.4.1.1 CADMIUM

Cadmium is known as poisonous to animal body. Cadmium level in human body is nearly zero at birth, however with age it accumulates. At 50 years of age, 30 mg of cadmium may get accumulated. The well-known sources of cadmium are water pipes, coal burning, drinks (such as: tea, coffee) and refined/canned foods. For nonsmokers, generally the main cause of Cd accumulation is food. According to WHO, daily Cd intake through diet is 40–50 μg/day [69]. In the United States (US), a person may consume approximately 30 μg of Cd/day from foods.

The Cd can interact and bind with Fe, Zn, Se, and Cu for its structural matches. P or Ca concentration may be affected by Cd. Cadmium is responsible for expression of various genes, which are associated with carcinogenesis, and these are intermediated by early response genes (such as: c-jun, c-myc, and c-fos), heat-shock genes that are responsible for controlling glutathione and related proteins, and stress response gene (such as: metallothionein) and translation transcription factor [68] under various conditions. The Cadmium inhibits repair of DNA. Therefore, DNA repairment inhibition and oxidative stress increases and leads to DNA damage, arrests cell cycle, genomic instability and mutagenesis responsible for cancer and cell death [9].

8.4.1.2 LEAD

It is a most known industrial metal spread throughout the environment. In 2017, Health Metrics and Evaluation estimated that the longtime exposure to lead accounted for 1.06 million deaths. Low- and middle-income countries suffer more from this issue. It gets accumulated in bones, kidney, liver, brain, and other parts. Lead poisoning of animals is a serious issue and needs a special consideration by health personnel. Along with direct consumption, lead may increase its level in blood. The lead, which accumulates inside the body, behaves as an important source of blood lead burden [59]. The lead, which is accumulated, gets assembled from storage location with dropping of the level in blood. Lead may cause acute and chronic toxicity.

8.4.1.3 MERCURY

The toxic level of mercury largely depends on in which form it is present. If it is present in the form of methyl mercury, then it is most stable. The contaminant is a serious health hazard to all avian, mammalians, and predators. Mercury pollution is mainly caused by industrial waste.

Suggested urinary concentration for Hg is 1 to 2 μg/ml. Excessive Hg can cause Minamata disease with symptoms of loss of memory and vision [59].

8.4.2 CONTAMINATION IN COW MILK

Heavy metals are absorbed into the plants through roots, from the contaminated soil and groundwater. Consequently, the plant is infected and these metals enter into the human body and animals after ingestion. Heavy metals may be present in dairy products from different sources, such as contamination of raw milk from the cow due to the outside environment or intake of contaminated water or food. Toxic metals for humans are cadmium, lead, and mercury, and they can accumulate in food through various sources. In cow milk, the metal transfer is variable. Due to various industrial works, the contaminants are transferred to the ambient.

The existence of heavy metals residues (such as: cadmium and lead) in dairy products are a specific burden because these products are hugely consumed by infants, children, and adults. Carcinogenic elements in lead and cadmium are responsible for many disorders of the kidney, nervous system, cardiovascular system, skeletal system and blood. By different sources, lead may enter into the body through inhalation, animal feed and water.

8.4.3 CONTAMINATION IN CAMEL MILK

Camel milk has excellent dietary and therapeutic characteristics. This milk has been utilized for the treatment of autoimmune and metabolic diseases and can treat various other diseases. The water and roughage have large impact on the physical and chemical properties of camel milk [22]. From the health point of view, different milk products give out essential metals, such as, copper, iron, and zinc in low level; and beside this arsenic, lead, and chromium in low quantity are responsible for metabolic disorders [79]. The presence of these metals in excessive amounts can lead to genetic disorder, cancer, and neurological disorder [79]. Like different dairy products, camel milk can be exposed to different kinds of hazards that are mostly pollutants and toxic of heavy metals.

8.4.4 CONTAMINATION IN HUMAN MILK

Recently physiological and medical studies highlighted the appreciable prosperity of consumption of mother milk for infants. With increased concern in breast feeding, there is parallel development of concern about the contaminant of breast milk. Different kinds of environmental chemicals, contaminants, and drugs are found in breast milk. Heavy metals bind the milk protein, and thus the absorption rate of the metal is high. The lead, cadmium, and mercury are significantly poisonous to cause heamatotoxic, neurotoxic, nephrotoxic effects even at low concentrations. Dietary compounds containing lead show less risk than milk compounds containing lead, because the presence of lead in milk and dairy products is easily absorbed into the human body.

Different factors are responsible for the contamination of lead in breast milk. In industrial countries, the amount of lead in milk is probably 5–20 ppb, the risk is increased 20 times in polluted areas. Several studies reported that breast milk contains less amount of lead than the milk-based infant formulas. Infant blood samples contain high amounts of lead that are based on infant formulas than those dependent on breast feeding. The amount of mercury in breast milk is quite less than that of cadmium and lead. Mothers having fish in their regular diet can have a huge level of cadmium in their breast milk and approximately 20% of the entire mercury content converts into methyl mercury.

Although mother milk contains essential metals, such as: Cu, Zn, yet these can be toxic when consumed in excess amounts. The recommended value is essential to perform several kinds of biochemical functions in the human body.

8.4.5 CONTAMINATION IN YOGURT

The fermented products contain starter cultures, which are mainly *Lactobacillus bulgaricus* and *Streptococcus lactis, Streptococcus thermophilus*, etc. To maintain proper physiological processes, the role of calcium and magnesium is important. Yogurt can fulfill this recommendation for these nutrients. Zinc and phosphorus content in yogurt are also significant. Calcium and phosphorus have great impact on the maintenance of bone health [61]. Several studies reported that the yogurt blended with fruit can induce copper poisoning as a result of excessive Cu [13].

8.4.6 CONTAMINATION IN MILK POWDER AND INFANT FORMULA

These play a significant role in young children's health when breast milk is not supplied enough or the working mother who faced difficulties to feed their baby [24]. Infant formula is generally classified into two main types: (i) Products for 0–6 months baby; (ii) Products for >4–6 months baby.

Excessive amount of Mn in early age can initiate neurovascular disorder to impact the balance of dopamine activity. Lead is present in cereals, dairy products, milk powder, and these products have a significant role. Cereal products also contain Cadmium, especially in rice, which causes serious health hazards. From air through breathing, Pb, and Cd can be accumulated in the body.

8.4.7 CONTAMINATION IN CHEESE AND WHEY

Through different routes, heavy metals can enter into the cheese, e.g., environmental contamination, contamination of the life stock, contaminant salt, contamination from the process chain, contaminant valve, contamination from the equipment for cheese processing, pollution in the surrounding area, etc. During flocculation, the formulation of casein with lead and fat can boost the level of metal in cheese than in the milk. Sometimes, Cu contamination can also occur because of its addition copper to cheese during the processing.

8.5 METHODS TO IDENTIFY HEAVY METALS IN MILK AND MILK PRODUCTS

8.5.1 INDUCTIVELY COUPLED PLASMA ATOMIC EMISSION SPECTROSCOPY (ICP-AES)

These samples are homogenized and packaged in polyethene bags. Then, these samples are stored at < –20°C. In advance microwave digestion system, 0.5 mg samples are digested in solution containing 7 ml of HNO_3 (Concentration 65%) and 1 ml of H_2O_2. Then, analyzes of these samples are performed by plasma atomic emission spectroscopy [21].

8.5.2 ATOMIC ABSORPTION SPECTROPHOTOMETER WITH GRAPHITE FURNACE

The test is performed in cheese samples, which are collected at different ripening stages to detect Chromium, Cadmium, Lead, and Mercury. The collected samples are stored below –18°C. At first, the samples are properly dried, then 0.3 mg of finely crushed dried samples are digested in nitric acid and hydrogen peroxide mixture (6 ml of 65% HNO_3 and 30% of 1 ml H_2O_2). After that, samples are filtered and distilled water is added and then are stored in bottles. Then, analysis of Hg is detected by automatic mercury analyzer and other trace metals (such as: chromium, cadmium, lead) are analyzed by atomic absorption spectrometer attached with graphite furnace [6].

8.5.3 ELECTRO-THERMAL ATOMIC ABSORPTION SPECTROSCOPY (ETASS)

Standard solution of Cd, Al, Pb, and Ni is prepared by dilution with standard solution (1,000 ppm). Glass wares and plastic wares are washed and put into 10% HNO_3 and rinsed with ultrapure water to avoid metal contamination. Samples are kept in the microwave oven at separate heating power (20–80 W) for 1–5 min. From this microwave extraction, supernatant liquid evaporates and semi dried mass is transpired into a heating plate, and then dried sample is made up with 10 ml 0.2 mol L^{-1} and centrifuged for 5 min at 3,000 rpm. After that, analysis is done with supernatant using electrothermal atomic absorption spectroscopy (ETASS).

8.5.4 FLAME ATOMIC ABSORPTION SPECTROPHOTOMETER

Ten mL of concentrated milk sample is digested in HNO_3 and H_2O_2 solution (3:1) on a hot plate. The volume of solution is reduced to 2 ml, which is further diluted to 20 ml with distilled water and the solution is analyzed for presence of heavy metal with flame atomic absorption spectrophotometer (FAAS). Selected heavy metals with 1, 1.5 and 2 ppm are prepared for calibrating the heavy metals with FAAS [6].

8.5.5 DIFFERENTIAL PULSE ANODIC STRIPPING VOLTAMMETRIC METHOD

A known volume (25 ml) of sample is dried and is ashed and is dissolved in 10 ml HNO_3 (0.25% concentration). Then, samples are analyzed for heavy metals with differential pulse anodic stripping voltammetric technique (DPASV). Blank is also prepared by the same reagents and also analyzed with the same technique [66].

8.5.6 POTENTIOMETRIC AND CHRONOPOTENTIOMETRIC STRIPPING ANALYSIS (PSA)

1. **Sampling:** Different pasteurized fermented dairy products with 2.8% milkfat are collected for:
   i. **Dilution and Acidification:** 5 g sample is diluted with distilled water. Then, these samples are acidified with HCl (hydrochloric acid) and stirred for 5 min. For Pb and Cu, samples are acidified to 0.10 mol/L. When samples are more acidic, then metals extraction is done.
   ii. **Microwave Acid Digestion:** 1 g sample is extracted with 3 ml diluted HNO_3 in a Teflon cup though microwave treatment (160 W/40 s). Then, the solution is evaporated and the solution is almost dried. After that, the residue is dissolved and transferred into a volumetric flask (10 ml) having HCl (0.03 mol/L). Sometimes HNO_3 is also used for dissolving metals.
   iii. **Wet Acid Digestion:** It is done for the samples for destruction of organic components with a Kjeldahl flask with the addition of perchloric acid and nitric acid (1:1). In this method, 5 g of sample is treated with 30 ml of acid mix at 110°C/40 min. Then, 3 ml HNO_3 is added and the heat treatment (140°C) is continued for 30 min to complete the detraction. Under vacuum conditions, perchloric acid evaporates at 230°C. The resulting residue is dissolved in HCl solution (0.03 mol/L).
   iv. **Stripping Potentiometry:** Dissolved oxygen is used as an oxidant with stripping phase performing diffusive mass transfer. Sample contamination is prevented by inert gas

for deaeration. Also, chemicals must be added as oxidants. Mercury film electrode is developed due to constant current electrolysis. Same mercury film is used for 30 analyzes of the prepared sample.

v. In the 1st step, the prepared sample is electrolyzed under fixed potential at proper stirring (4,000/min). Higher concentration of HCl is for more effective metal extraction.

2. **Analytical Step:** After electrolysis of samples, inverse current is used in PSA mode. Metal deposit oxidation starts at the rest time period (15 S). Then, reduction time is carefully chosen and increasing time for metal oxidation (response time) is observed due to reduction of oxygen and partly reduction of freshly oxidized ions happens. The current in PSA should be in the range of 0.3–4.2 μA. Appropriate reduction time range should be 1.1–1.9 μA. Second-order polynomial expression ($\tau = 0.083$; $i_R^v - 0.082$; $i_R + 0.028$; and $R^2 = 0.09978$) describes the dependence of response time or oxidation time (τ).

8.5.7 IONIC LIQUID AND BISMUTH OXIDE NANOPARTICLES MODIFIED CARBON PASTE ELECTRODE

This voltammetric experiment is done in one-compartment cell. At 25 ± 1°C, the solution is stirred with a magnetic stirrer for pre-treatment and deposition step.

1. **Electrode Preparation:** For electrode preparation, bismuth oxide nanoparticle and ionic liquid modified carbon paste are used. Electrode conduction is done with copper wire.
2. **Milk Sample Extract Preparation:** Milk samples are collected from the supermarket. Then, are packed in the bottles and are stored under refrigerated temperature. After that, 6 ml sample is put in the centrifuge tube and is centrifuged at 4,000 rpm for 5 min. Then, lipid phase is precipitated and it is removed from the upper layer. The lower layer containing mineral is transferred to a separate tube; and 100 μL conc. HCl and 100 μL of glacial acetic acid are added. After that, it is carried out under centrifugal force at 6,000 rpm for 10 min. The supernatants including Cd and Pb

ions are collected by filtering with pore size 0.22. After that, the pH of these solutions is adjusted to 4.5 by adding NaOH.

3. **Method of Measurement:** Bismuth oxide is reduced to bismuth by KOH solution. To determine Cd, Pb square wave anodic stripping voltammetry is established. Before the measurement step of pre-condition, –0.3 V for 30s is applied [58].

8.5.8 INDUCTIVITY COUPLED PLASMA MASS SPECTROMETRY

Milk samples are collected from different farms and are stored in clean polyethylene bottles at –18°C. Then, samples are dried and are ashed at 95°C. Then, HCl (5 ml 6 M) is added and evaporated for drying. Next, the residue is dissolved in 0.1 M HNO_3 and test is performed using ICP-MS. In ICP-MS measurement, concentration of Pb, Zn, Cu, As, Cd, Ni, Cr, and Fe in solution extract is determined three times by Agilent 7500 ICP-MS. In this measurement for spectral purity, argon gas is utilized [46].

8.5.9 INDUCED COUPLED PLASMA OPTICAL EMISSION SPECTROSCOPY

Milk samples are collected from different locations and are kept in polypropylene flasks and are stored at –80°C in a refrigerator. Samples are lyophilized in the freeze-dry system and milk samples are also digested in the microwave oven. To 0.5 g of lyophilized sample, 10 ml of HNO_3 is added in tube that are placed inside the microwave oven at 1,600 W/15 min, 800 psi pressure and 200°C. After the microwave digestion, these samples are filtered by Whatman 42 paper, which is diluted to 50 ml by deionized water. Calibration standard for each metal is prepared with multi element standard. The Pb, Ni, Cr, Cu, and Zn are determined by this technique [64].

Heavy metals are contaminated in various ways and in polluted soil, water, sewage, processing equipment, etc., causing health issues in humans. These heavy metals are determined by various above methods in various countries [71]. Table 8.2 summarizes the detection method and sample used in these methods in different countries.

TABLE 8.2 Detection of Heavy Metal in Milk and Dairy Products

Researcher(s)	Sample	Method	Element	Country
Abdulkhaliq et al. [1]	Cow milk (dried and liquid), yogurt, hen's egg	GF-AA25S GF-AAS	Pb, Cd, Cu, Fe	Palestine
Ahmad et al. [3]	Camel, cattle, buffalo, goat milk	AAS	Cu, Ni, Zn, Cd, Cr, Fe, Mn	Pakistan (Khyber-Pakhtunkhwa)
Ahmadb et al. [3]	Camel milk, cattle, buffalo, goat	AAS	Mn, Fe Cd, Ni, Cr, Zn	Pakistan
Akhtar et al. [4]	Infant formula	AAS	Mn, Ni, Cd, Pb	Pakistan
Almayahi et al. [5]	Milk	FAAS	Zn, Fe, Cr, Cd, Cu	Iraq (Najaf city of Iraq)
Bakirciogl et al. [7]	Cheese	ICP-OES	Ni, Mn, Cu, Cr, Co, Cd, Zn, Se, Pb	Turkey
Bilandžić [10]	Cream cheese, and butter	ICP-OES	Mo, Al, Co, Cr, Li, Mn, Ni, Sr	Croatia
Birghila et al. [11]	Liquid milk	ICP-AES	Mn, Ni, Pb, Si, Sn, Zn, Ti	Romania (Constanata city)
Capcarova et al. [13]	(Strawberry, blueberry, and cherry) parts of commercially available yogurt	AAS	Hg, Zn, NI, Cr, Co, Mn, Pb, Cd, Cu	Slovak Republic
Castro et al. [14]	Cow milk, cheese	ICP-OES	Pb, Cr, Ni, Cu, Zn, As	Mexico (Pubela)
Cruz et al. [15]	Infant formula	AAS	Pb, Hg, As	Philippines
Dabeka et al. [16]		ICP-MS	Al, Cd, Pb	Canada
de Andrade et al. [17]	Yogurt	ETAAS	Cu, Cd, Pb, Cr	Brazil
de Castro et al. [18]	Infant formula	Voltametric; and spectroscopic	Pb, Cd	Brasilia
Derakhshesh et al. [19]	Goat milk and cow milk	GF-AAS	Pb	Iran

TABLE 8.2 *(Continued)*

Researcher(s)	Sample	Method	Element	Country
Dizaji et al. [20]	Milk	AAS	Pb, Cd	Iran
El Sayed et al. [21]	Raw cow milk	ICP-AES	Fe, Zn, Cr,	Egypt (Helwan, Tanash, Shubra, Menofia, Mansora, Awseen, Gharbia)
Enb et al. [23]	Buffalo milk, cow milk	AAS	Fe, Cu, Mn, Zn, Pb, Cd, Cr, Ni	Egypt
Isamov et al. [25]	Goat milk	FES	Co	Russia
Ismail et al. [26]	Milk	AAS	Cu, Ni, Cd, Co, Pb	Pakistan (Multan city)
Kanuspayeva et al. [29]	Camel milk, fermented camel milk	AAS	Fe, Mn, As, Zn, Cu, Pb	Kazakh-stan
Kazi et al. [32]	Infant formula	ETAAS	Al, Cd, Pb	Pakistan
Kazi et al. [27]	Raw and processed milk	AAS	Al, Cd, Ni, Pb	Pakistan (Hyderabad city)
Khan et al. [28]	Milk and yogurt	ICP-MSCd	As, Zn, Mn, Cu, Se, Cr	Korea
Konuspayeva et al. [30]	Camel milk, fermented camel milk	ICP	Pb, Cd, Zn, Cu	Kazakh-stan
Licata et al. [31]	Goat milk and ovine milk	ICP-AES	Cd, Se, As, Cu, Ni, Cr, Fe, Zn, Pb	Italy (Calabria)
Lutfullah et al. [32]	Infant food	AAS	Mn, Ca, Mg, Cu, Zn, Fe, Cr, Ni, Cd, Pb	Pakistan
Maas et al. [33]	Raw cow milk and comet cheese	AAS	Zn, Cu, Cd, Pb	France
Mania et al. [34]	Infant food	AAS	Hg, Pb, As	Poland
Mehrnia and Bashti [35]		AAS	Ni, Cd, Pb Mn	Iran

TABLE 8.2 *(Continued)*

Researcher(s)	Sample	Method	Element	Country
Miedico et al. [36]	Milk of goat and sheep milk	ICP-MS	Hg, As, Be, Ni, Sn, Fe, Zn, Se, Cu, Cd, Cr, Sr, Pb, Al, Mn, V, Co, Mo	Italy
Mostafidi et al. [37]	Camel milk	ICP	Ni, Cd, Pb	Iran
Muñoz et al. [38]	Milk	PSA	Cd, Pb, Cu	Spain
Najarnezhad et al. [40]	Buffalo milk, cow milk and ewe milk	FAAS	Cd, Pb	Iran
Najarnezhad et al. [40]	Raw cow milk	AAS	Pb, Cd, Hg	Iran
Najarneznad et al. [39]	Raw cow milk and ewe milk	AAS	Pb, Cd	North East Iran (Khorasan Razavi)
Nejatolahi et al. [42]	Raw cow milk	AAS	Pb	Iran
Nguta [43]	Camel milk	AAS	Pb, As	Kenya
Nnadozie et al. [44]	Milk and meat from: camel, goat, cow, sheep	AAS AES	Ni, Hg Mn, Zn, Cu, Ba, Cd, Pb, Co, Cr	Nigeria
Oliveira et al. [45]	Raw milk	GF-AAS	Pb	Brazil
Poti et al. [47]	Ewe milk	ICP-OES	Cd, Cr, Pb	Hungary
Rey-Crespo et al. [49]	Organic milk	ICP-MS	Zn, Se, Ni, Mo, Mn, I, Fe, Cu, Cr, Co	Spain
Rezaei et al. [50]	Pasteurized milk, yogurt, yogurt drink, cheese	ICP-SFMS	Pb, Al, As, Cd, Hg, Se	Iran
Sager et al. [51]	Formula and milk powder	ICP-MS	Pb	Tanzania

TABLE 8.2 *(Continued)*

Researcher(s)	Sample	Method	Element	Country
Salah et al. [52]	Cow and buffalo milk	FAAS	Pb, Cd, Al, Fe, Se, Mn	Egypt (Dakahlia Governorate)
Saleh et al. [53]		ICP-MS	Pb, Cd, Al, Cr, Ni	Egypt (Damanhour city)
Shahbazi et al. [54]	Milk, yogurt, cheese	Voltammetric	Pb, Zn Cu, Cd, Se,	Iran
Sipahi et al. [55]	Formulas	AAS	Al, Cr, Co, Mn, Pb, Cd	Turkey
Soltan et al. [56]	Camel milk, sheep milk	AAS	Fe, Cd Co, Zn, Mn, Pb, Ni,	Saudi Arabia
Starska et al. [57]	Milk and dairy products	FAAS	Hg, Pb, Cd, As	Poland
Suturović et al. [58]	Fermented milk	PSA	Cd, Pb, Cu	Leskovac (Serbia)
Suturović et al. [58]	Milk, fermented milk products	PSA	Cd, Pb, Cu	Serbia
Tassew et al. [62]	Cow milk	FAAS	Zn, Cd, Pb, Cr, Mn, Co, Ni, Cu	Ethiopia (Abaya district, Borenazone of oromiya region)
Temiz et al. [63]	Milk	ICP-MS	Pb, As, Cd, Ni, Cr, Zn, Fe, Cu	Turkey (Southeast of Samsun)
Tona et al. [64]	Cow milk, soft cheese goat milk, yogurt butter	AAS	Cd, Pb	Nigeria
Tripathi et al. [65]	Human milk	FAAS	Zn, Cu, Pb, Cd	India (Mumbai)
Yuzbasi et al. [70]	Kasser cheese	AAS	Fe, Cu, Zn, Pb, Cd	Turkey

8.6 RECOMMENDATIONS

Heavy metal pollution can be reduced by modification of the process of milk production. To decrease contamination in the processing phase of milk and dairy products, preventive measures should be taken at every step and only food material and food-grade materials should be used during processing. The dairy farm should be avoided from local polluted areas (like industrial areas) to prevent the unwanted contamination from metal. It is suggested that from production to ingestion of milk and milk products a special attention is required; otherwise, the reduction of these toxic heavy metals and heavy metal-related disorders can't be controlled.

8.7 SUMMARY

Contamination of milk and milk products is a very common problem throughout the world and it is growing rapidly. The milk and milk products get contaminated easily by several ways from unwanted pollution and manmade pollution. Contamination of heavy metals can occur from the production to consumption stages. Contamination by heavy metals can cause serious health problems. For example, lead can cause acute and chronic toxicity, mercury causes Minamata disease with symptoms of loss of memory, vision problem.

KEYWORDS

- **atomic absorption spectroscopy**
- **glutathione**
- **heavy metals**
- **inductively coupled plasma atomic emission spectroscopy**
- **inductively coupled plasma mass spectrometry**
- **inductively coupled plasma optical emission spectrometry**
- **mutagenesis**
- **pesticides**

REFERENCES

1. Abdulkhaliq, A., Swaileh, K., Hussein, R. M., & Matani, M., (2012). Levels of metals (Cd, Pb, Cu and Fe) in cow's milk, dairy products and hen's eggs from the west bank, Palestine. *International Food Research Journal, 19*(3), 1089–1094.
2. Abadin, H., Ashizawa, A., & Stevens, Y. W., (2007). *Toxicological Profile for Lead.* Atlanta - GA: Agency for toxic substances and disease registry (US), CDC. https://stacks.cdc.gov/view/cdc/37676/cdc_37676_DS1.pdf (accessed on 22 September 2021).
3. Ahmad, I., Zaman, A., Samad, N., Ayaz, M. M., Rukh, S., & Akbar, A., (2017). Atomic absorption spectrophotometry detection of heavy metals in milk of camel, cattle, buffalo and goat from various areas of Khyber-Pakhtunkhwa (KPK). *Pakistan. Journal of Analytical Bioanalytical Technique, 8*(367), 1–6.
4. Akhtar, S., Shahzad, M. A., Yoo, S. H., Ismail, A., Hameed, A., Ismail, T., & Riaz, M., (2017). Determination of aflatoxin m1 and heavy metals in infant formula milk brands available in Pakistani markets. *Korean Journal for Food Science of Animal Resources, 37*(1), 79–86.
5. Almayahi, B., Algburi, J., & Jarrah, N., (2019). Evaluation of the health impact of some heavy metals in milk from markets. *Prensa Med. Argent, 105*(5), 1, 2.
6. Anastasio, A., Caggiano, R., & Macchiato, M., (2006). Heavy metal concentrations in dairy products from sheep milk collected in two regions of Southern Italy. *Acta Veterinaria Scandinavica, 47*(1), 1–6.
7. Bakircioglu, D., Kurtulus, Y. B., & Ucar, G., (2011). Determination of some traces metal levels in cheese samples packaged in plastic and tin containers by ICP-OES after Dry, wet and microwave digestion. *Food and Chemical Toxicology, 49*(1), 202–207.
8. Beltrane, M. A., & Junior, M. M., (2005). *Principais riscos químicos no leite: Um problema de Saúde Pública* (Main Chemical Risks in Milk: A Public Health Problem). *Arquivos de Ciências da Saúde da UNIPAR, 9*(2), 140–145.
9. Bertin, G., & Averbeck, D., (2006). Cadmium: Cellular effects, modifications of biomolecules, modulation of DNA repair and genotoxic consequences. *Biochimie, 88*(11), 1549–1559.
10. Bilandžić, N., (2014). Trace elements content in cheese, cream and butter. *Mljekarstvo, 64*(3), 150–158.
11. Birghila, S., Dobrinas, S., Stanciu, G., & Soceanu, A., (2008). Determination of major and minor elements in milk through ICP-AES. *Environmental Engineering Management Journal, 7*(6), 805–808.
12. Blaylock, M. J., & Huang, J. W., (2000). Phytoextraction of metals. Chapter 7. In: Raskin I., & Ensley, B. D., (eds.), *Phytoremediation of Toxic Metal: Using Plants to Clean, Up the Environmental* (pp. 53–70). Toronto - Canada: John Wiley and Sons.
13. Capcarova, M., Harangozo, L., & Toth, T., (2017). Detection of selected trace elements in yogurt components. *Journal of Environmental Science and Health, Part B, 52*(12), 858–863.
14. Castro-González, N. P., & Calderón-Sánchez, F., (2018). Heavy metals in cow's milk and cheese produced in areas irrigated with waste water in Puebla, Mexico. *Food Additives & Contaminants: Part B, 11*(1), 33–36.

15. Cruz, G. C., Din, Z., & Feri, C. D., (2009). Analysis of toxic heavy metals (Arsenic, Lead and Mercury) in selected infant formula milk commercially available in the Philippines by AAS. *E-International Scientific Research Journal, 1*(1), 40–51.
16. Dabeka, R., Fouquet, A., Belisle, S., & Turcotte, S., (2011). Lead, cadmium and aluminum in Canadian infant formulae, oral electrolytes and glucose solutions. *Food Additives and Contaminants, 28*(6), 744–753.
17. De Andrade, C. K., De Brito, P. M. K., Dos, A. V. E., & Quináia, S. P., (2018). Determination of Cu, Cd, Pb and Cr in yogurt by slurry sampling electrothermal atomic absorption spectrometry: Case study for Brazilian yogurt. *Food Chemistry, 240*, 268–274.
18. De Castro, C. S., Arruda, A. F., & Cunha, L. R. D., (2010). Toxic metals (Pb and Cd) and their respective antagonists (Ca and Zn) in infant formulas and milk marketed in Brasilia, Brazil. *International Journal of Environmental Research and Public Health, 7*(11), 4062–4077.
19. Derakhshesh, A. M., & Rahimi, E., (2012). Determination of lead residue in raw cow milk from different regions of Iran by flameless atomic absorption spectrometry. *American-Eurasian Journal of Toxicological Sciences, 4*(1), 16–19.
20. Dizaji, A. A., Eshaghi, A., Golshani, A. A., & Nazeradl, K., (2012). Evaluation and determination of toxic metals (Lead and Cadmium) in cow milk collected from East Azerbaijan, Iran. *European Journal of Experimental Biology, 2*(1), 261–265.
21. El Sayed, E. M., Hamed, A. M., Badran, S. M., & Mostafa, A. A., (2011). Survey of selected essential and heavy metals in milk from different regions of Egypt using ICP-AES. *Food Additives and Contaminants: Part B, 4*(4), 294–298.
22. El-Agamy, E. I., (2006). Camel milk. In: Park, Y. W., & Haenlein, G. F., (eds.), *Handbook of Milk of Non-bovine Mammals* (pp. 297–344). Ames, IA: Blackwell.
23. Enb, A., Abou, D. M. A., & Abd-Rabou, N. S., (2009). Chemical composition of raw milk and heavy metals behavior during processing of milk products. *Global Veterinaria, 3*(3), 268–275.
24. Ikem, A., Nwankwoala, A., Odueyungbo, S., Nyavor, K., & Egiebor, N., (2002). Levels of 26 elements in infant formula from USA, UK, and Nigeria by Microwave Digestion and ICP–OES. *Food Chemistry, 77*(4), 439–447.
25. Isamov, N. N., Anisimov, V. S., & Grudina, N. V., (2011). Content of biogenic and toxic elements in goat feed and milk. *Russian Agricultural Sciences, 37*(5), 413–415.
26. Ismail, A., Riaz, M., Akhtar, S., Ismail, T., Ahmad, Z., & Hashmi, M. S., (2015). Estimated daily intake and health risk of heavy metals by consumption of milk. *Food Additives Contaminants - Part B, 8*(4), 260–265.
27. Kazi, T. G., Jalbani, N., & Baig, J., (2009). Assessment of toxic metals in raw and processed milk samples using electrothermal atomic absorption spectrophotometer. *Food and Chemical Toxicology, 47*(9), 2163–2169.
28. Khan, N., Jeong, I. S., & Hwang, I. M., (2014). Analysis of minor and trace elements in milk and yogurts by inductively coupled plasma-mass spectrometry (ICP-MS). *Food Chemistry, 147*, 220–224.
29. Konuspayeva, G., Faye, B., Loiseau, G., Diacono, E., & Akhmetsadykova, S., (2009). Pollution of camel milk by heavy metals in Kazakhstan. *Open Environmental Pollution and Toxicology Journal, 1*, 112–118.

30. Konuspayeva, G., & Jurjanz, S., (2011). Contamination of camel milk (heavy metals, organic pollutants and radionuclides) in Kazakhstan. *Journal of Environmental Protection, 2*(1), 90–96.
31. Licata, P., Di Bella, G., & Potortì, A. G., (2012). Determination of trace elements in goat and ovine milk from Calabria (Italy) by ICP-AES. *Food Additives and Contaminants: Part B, 5*(4), 268–271.
32. Lutfullah, G., Khan, A. A., Amjad, A. Y., & Perveen, S., (2014). comparative study of heavy metals in dried and fluid milk in Peshawar by atomic absorption spectrophotometry. *The Scientific World Journal, 2014*, 1–5.
33. Maas, S., Lucot, E., Gimbert, F., Crini, N., & Badot, P., (2011). Trace metals in raw cows' milk and assessment of transfer to comté cheese. *Food Chemistry, 129*, 7–12.
34. Mania, M., & Wojciechowska-Mazurek, M., (2015). Toxic elements in commercial infant food, estimated dietary intake, and risk assessment in Poland. *Polish Journal of Environmental Studies, 24*(6), 2525–2528.
35. Mehrnia, M. A., & Bashti, A., (2014). Evaluation of toxic element contents in infant foods commercially available in Iran. *Bulletin of Environment, Pharmacology and Life Sciences, 3*(6), 249–253.
36. Miedico, O., Tarallo, M., Pompa, C., & Chiaravalle, A. E., (2016). Trace elements in sheep and goat milk samples from Apulia and Basilicata regions (Italy): Evaluation by multivariate data analysis. *Small Ruminant Research, 135*, 60–65.
37. Mostafidi, M., Moslehishad, M., Piravivanak, Z., & Pouretedal, Z., (2016). Evaluation of mineral content and heavy metals of dromedary camel milk in Iran. *Food Science and Technology, 36*(4), 717–723.
38. Muñoz, E., & Palmero, S., (2004). Determination of heavy metals in milk by potentiometric stripping analysis using a home-made flow cell. *Food Control, 15*(8), 635–641.
39. Najarnezhad, V., & Akbarabadi, M., (2013). Heavy metals in raw cow and ewe milk from north-east Iran. *Food Additives Contaminants: Part B, 6*(3), 158–162.
40. Najarnezhad, V., Jalilzadeh-Amin, G., Anassori, E., & Zeinali, V., (2015). Lead and cadmium in raw buffalo, cow and ewe milk from West Azerbaijan, Iran. *Food Additives Contaminants: Part B, 8*(2), 123–127.
41. Neal, A. P., & Guilarte, T. R., (2013). Mechanisms of lead and manganese neurotoxicity. *Toxicology Research, 2*(2), 99–114.
42. Nejatolahi, M., Mehrjo, F., Sheykhi, A., & Bineshpor, M., (2014). Lead concentrations in raw cows' milk from Fars province of Iran. *American Journal of Food and Nutrition, 2*(5), 92–94.
43. Nguta, J. M., (2012). Heavy metal residues in camel milk from Kenya: Health implications. *Journal of Clinical Toxicology, 2*(9), 1–8.
44. Nnadozie, C. U., Birnin-Yauri, U. A., Muhammad, C., & Umar, A., (2014). Assessment of some dairy products sold in Sokoto metropolis, Nigeria. *International Journal of Advanced Research in Chemical Science, 1*(10), 31–37.
45. Oliveira, T. M., Augusto-Peres, J., Lourdes-Felsner, M., & Cristiane-Justi, K., (2017). Direct determination of Pb in raw milk by graphite furnace atomic absorption spectrometry (GF AAS) with electrothermal atomization sampling from slurries. *Food Chemistry, 229*, 721–725.

46. Patra, R. C., Swarup, D., Kumar, P., Nandi, D., Naresh, R., & Ali, S. L., (2008). Milk trace elements in lactating cows are environmentally exposed to higher level of lead and cadmium around different industrial units. *Science of the Total Environment, 404*(1), 36–43.
47. Póti, P., Pajor, F., Bodnár, Á., & Bárdos, L., (2012). Accumulation of some heavy metals (Pd, Cd and Cr) in milk of grazing sheep in North-East Hungary. *The Journal of Microbiology, Biotechnology and Food Sciences, 2*(1), 389–394.
48. Reeves, R. D., & Baker, A. J. M., (2000). Metal accumulating plants. Chapter 15. In: Raskin I., & Ensely, B. D., (eds.), *Phytoremediation of Toxic Metals: Using Plant to Clean up the Environment* (pp. 193–229). Toronto - Canada: John Wiley and Sons.
49. Rey, C. F., Miranda, M., & López, A. M., (2013). Essential trace and toxic element concentrations in organic and conventional milk in NW Spain. *Food and Chemical Toxicology, 55*, 513–518.
50. Rezaei, M., Dastjerdi, H. A., & Jafari, H., (2014). Assessment of dairy products consumed on the arak market as determined by heavy metal residues. *Health, 6*(5), 1–5.
51. Sager, M., McCulloch, C. R., & Schoder, D., (2018). Heavy metal content and element analysis of infant formula and milk powder samples purchased on the Tanzanian market: International branded versus black market products. *Food Chemistry, 255*, 365–371.
52. Salah, F. A. A. E., Esmat, I. A., & Mohamed, A. B., (2013). Heavy metals residues and trace elements in milk powder marketed in Dakahlia Governorate. *International Food Research Journal, 20*(4), 1807–1812.
53. Saleh, E., Elleboudy, A., & Elsakhawy, A., (2019). Heavy metals in raw milk and some dairy products at local markets. *Damanhour Journal of Veterinary Sciences, 1*(1), 27–30.
54. Shahbazi, Y., Ahmadi, F., & Fakhari, F., (2016). Voltammetric determination of Pb, Cd, Zn, Cu and Se in milk and dairy products collected from Iran: An emphasis on permissible limits and risk assessment of exposure to heavy metals. *Food Chemistry, 192*, 1060–1067.
55. Sipahi, H., Eken, A., Aydın, A., Şahin, G., & Baydar, T., (2014). Safety assessment of essential and toxic metals in infant formulas. *Turkish Journal of Pediatrics, 56*(4), 385–391.
56. Soltan, M. E., Al-ayed, A. S., Ismail, M. A., & Shadeed, M. A., (2017). Effect of the environmental factors on some element, on some element contents in camel and sheep milks: A comparative study between Qassim and Riyadh regions, KSA. *International Research Journal of Public and Environmental Health, 4*(8), 184–192.
57. Starska, K., & Wojciechowska, M. M., (2011). Noxious elements in milk and milk products in Poland. *Polish Journal of Environmental Studies, 20*(4), 1043–1051.
58. Suturović, Z., Kravić, S., Milanović, S., Đurović, A., & Brezo, T., (2014). Determination of heavy metals in milk and fermented milk products by potentiometric stripping analysis with constant inverse current in the analytical step. *Food Chemistry, 155*, 120–125.
59. Swarup, D., Patra, R. C., Naresh, R., Kumar, P., & Shekhar, P., (2005). Blood lead levels in lactating cows reared around polluted localities; transfer of lead into milk. *Science of the Total Environment, 15*(1–3), 67–71.

60. Tang, Z., Chai, M., Cheng, J., & Jin, J., (2017). Contamination and health risks of heavy metals in street dust from a coal-mining city in Eastern China. *Ecotoxicology and Environmental Safety, 138*, 83–91.
61. Tarakci, Z., & Dag, B., (2013). Mineral and heavy metal by inductively coupled plasma optical emission spectrometer in traditional Turkish yogurts. *International Journal of Physical Science, 8*(19), 963–966.
62. Tassew, A. H., & Rao, V. M., (2014). Determination of concentrations of selected heavy metals in cow's milk: Borena zone, Ethiopia. *Journal of Health Science, 4*(5), 105–112.
63. Temiz, H., & Soylu, A., (2012). Heavy metal concentrations in raw milk collected from different regions of Samsun, Turkey. *International Journal of Dairy Technology, 65*(4), 516–522.
64. Tona, G. O., Adetunji, V. O., Ameen, S. A., & Ibikunle, A. O., (2013). Evaluation of lead and cadmium heavy metal residues in milk and milk products sold in Ogbomoso, Southwestern Nigeria. *Pakistan Journal of Nutrition, 12*(2), 168–171.
65. Tripathi, R. M., Raghunath, R., Sastry, V. N., & Krishnamoorthy, T. M., (1999). Daily intake of heavy metals by infants through milk and milk products. *Science of the Total Environment, 227*(2, 3), 229–235.
66. Raikwar, M. K., Kumar, P., Singh, M., & Singh, A., (2008). Toxic effect of heavy metals in livestock health. *Veterinary World, 1*(1), 28–30.
67. Waisberg, M., Joseph, P., Hale, B., & Beyersmann, D., (2003). Molecular and cellular mechanisms of cadmium carcinogenesis. *Toxicology, 192*(2, 3), 95–117.
68. Wei, X., Gao, B., Wang, P., Zhou, H., & Lu, J., (2015). Pollution characteristics and health risk assessment of heavy metals in street dusts from different functional areas in Beijing, China. *Ecotoxicology and Environmental Safety, 112*, 186–192.
69. WHO, (1987). *Air Quality Guidelines for Europe*. WHO Regional Publication - European Series 23; World Health Organization (WHO), Copenhagen, Denmark. http://www.euro.who.int/document/e71922.pdf (accessed on 22 September 2021).
70. Yüzbaşi, N., Sezgin, E., Yildirim, Z., & Yildirim, M., (2009). Changes in Pb, Cd, Fe, Cu and Zn levels during the production of kaşar cheese. *Journal of Food Quality, 32*(1), 73–83.
71. Ziarati, P., Shirkhan, F., Mostafidi, M., & Zahedi, M. T., (2018). Overview of the heavy metal contamination in milk and dairy products. *Acta Scientific Pharmaceutical Sciences, 2*(7), 1–14.

CHAPTER 9

DETECTION OF PESTICIDES RESIDUES IN MILK AND MILK PRODUCTS

DIPESH AGGARWAL, SANTOSH ANAND, and DIWAKAR MISHRA

ABSTRACT

Extensive use of pesticides in agriculture, industries, public area, around the home has led to the accumulation of pesticides in the environment. Organochlorine (OCP) and organophosphate pesticides (OPP) can contaminate milk (3–5% fat) and its products. Pesticide residue is a cause of various acute and chronic health problems. The sensitive, reliable, widely applicable, waste-free, highly accurate, and rapid extraction and detection methods are required to confirm effective monitoring of pesticide residues. Novel extraction methods, such as, QuEChERS, MSPD, SPME, and PLE are giving good performance results and determination of pesticide residues is mainly done by GC and HPLC in combination with mass spectrometry (MS).

9.1 INTRODUCTION

Milk is an essential part of overall nutrition of infants and adults, and accounts for a serious risk of being a vector of pesticides [101]. A major concern to public health is the increasing pesticide contamination in milk and milk products. This has triggered scientists to come up with various analytical methods to identify pesticides residues in dairy products.

Pesticides presence is one of the prominent hazards to food safety having potential concern worldwide. Pesticides can be defined as a chemical substance that is used to repel, destroy or control pests, such as: weeds, rodents, insects, fungi, bacteria, or other organisms. These chemicals can find their way into metabolic systems including mammals especially via foods they consume [37]. Its uncontrolled application, volatile nature, chemical degradation, etc., are principal cause resulting in widespread contamination of environment.

Pesticide residue refers to the remains of pesticide present in foods after their application on them [89]. Consumption of pesticide-treated food or indirect contact of treated areas leads to its entry in the food chain of the general population. Chung and Chen [26] reported that persistent chemicals have the ability to bio-magnify and bio-accumulate in the food chain. These pesticides have been spotted in all major food items, such as, dairy, cereals, meat products, poultry, edible oils, nuts, and many fruits and vegetables, even in human milk [79].

In light of pesticide toxicity, some countries have banned its use, but still, a number of countries have been using these pesticides in large amounts. Several studies have reported pesticide residues in butter, cheese, yogurt, milk powder, etc., in different countries [9, 44, 59, 102, 106, 117]. There are several different techniques to detect pesticides residues in milk products. [35, 52, 75, 134]. With the advent of state-of-the-art techniques like GC-MS and HPLC-MS, the analytical approaches suitable for research and detection have now touched new heights.

This chapter discusses the occurrence of pesticides and viability of existing pesticide residue analytical methods.

9.2 PESTICIDES CONTAMINATION IN MILK AND MILK PRODUCTS

According to Battu et al. [8], majorly four groups of pesticides have been analyzed in products from animal origin, namely: organophosphorus pesticides (OPPs), carbamates (CBs), organochlorine pesticides (OCPs) and pyrethroids (PYRs). The MRLs for OCPs, OPPs, CBs, and PYRs in fluid milk and other milk-based products have been established by European Union. Table 9.1 presents maximum residue level of pesticides in milk products.

TABLE 9.1 Pesticides Maximum Residue Level (MRL) in Milk

Pesticide	Maximum Residue Level (mg/kg)	Type of Pesticide	Year of Adoption
2,4-D	0.01	Herbicide	2003
Acephate	0.02	Insecticide	2005
Acetamiprid	0.02	Insecticide	2012
Aldicarb	0.01	Insecticide	NA
Aldrin and Dieldrin	0.006	Insecticide	NA
Bifenthrin	0.2	Insecticide	2011
Carbaryl	0.05	Insecticide	2004
Carbendazim	0.05	Fungicide	2006
Chlordane	0.002	Insecticide	NA
Chlorothalonil	0.07	Fungicide	2011
Chlorpropham	0.01	Herbicide	2009
Chlorpyrifos-Methyl	0.01	Insecticide	2011
Clothianidin	0.02	Insecticide	2012
Cypermethrins (alpha- and zeta-)	0.05	Insecticide	2009
Cyprodinil	0.0004	Fungicide	2005
DDT	0.02	Insecticide	1997
Deltamethrin	0.05	Insecticide	2004
Diazinon	0.02	Insecticide	1995
Dicofol	0.1	Acaricides	1999
Difenoconazole	0.005	Fungicide	2011
Diflubenzuron	0.02	Insecticide	2004
Dimethipin	0.01	Plant growth regulator	2003
Dimethomorph	0.01	Fungicide	2008
Diphenylamine	0.006	Fungicide	NA
Diquat	0.01	Herbicide	1999
Endosulfan	0.01	Insecticide	2007
Ethoprophos	0.01	Insecticide	2005
Fenbuconazole	0.01	Fungicide	2010
Fenitrothion	0.01	Insecticide	2008
Fenpropimorph	0.01	Fungicide	2001
Fenvalerate	0.1	Insecticide	NA

TABLE 9.1 *(Continued)*

Pesticide	Maximum Residue Level (mg/kg)	Type of Pesticide	Year of Adoption
Flutolanil	0.05	Fungicide	2004
Glyphosate	0.05	Herbicide	2006
Heptachlor	0.006	Insecticide	NA
Lindane	0.01	Insecticide	2004
Methidathion	0.001	Insecticide	1995
Methoxyfenozide	0.05	Insecticide	2010
Paraquat	0.005	Herbicide	2006
Phorate	0.01	Insecticide	2006
Pirimicarb	0.01	Insecticide	2007
Pirimiphos-Methyl	0.01	Insecticide	2005
Profenofos	0.01	Insecticide	2009
Propiconazole	0.01	Fungicide	2008
Pyrimethanil	0.01	Fungicide	2008
Tebufenozide	0.01	Insecticide	2004

Source: CODEX Online Commodity Details for Milks; http://www.codexalimentarius.net/pestres/data/commodities/details.htm.

9.2.1 ORGANOCHLORINE (OCPS) PESTICIDES

OCPs are organic compounds with single covalently bonded chlorine atoms. Among the major classes of chlorinated hydrocarbon pesticides, aldrin or dieldrin, lindane, toxaphene are potential stereospecific inhibitors of binding of t-butylbicyclophosphorothionate (TBPs) to specific sites of the brain [67]. OCPs are majorly responsible for environmental pollution and health risks due to their lipophilic nature, resistant to degradation and prolonged shelf-lives [82]. OCPs have been used in agriculture to control pests and have high lipid solubility that tends to bio-accumulate in fatty food matrices [1, 100], such as dairy products and consumption of these products leads to exposure to residues [134]. Because of their high stability, less volatility, toxicity, non-polarity, and adverse harmful effects, these have been banned worldwide [119] by the Stockholm Convention on Persistent Organic Pollutants [31]. However, usage of these chemicals to control diseases like malaria, cholera, and typhoid still persists in developing countries [25]. The toxic effects of OCPs are endocrine disruption [56] and carcinogenic substances [113]. Human exposure is mainly from the contamination of food.

9.2.2 ORGANO-PHOSPHORUS (OPPS) PESTICIDES

OPPs are commonly used as insecticides and they are esters of phosphoric acid [90], for example: disulfoton, ethion, malathion, chlorfenvinphos, chlorpyrifos, dichlorvos, and phorate, etc. Salas et al. [99] investigated the presence of OPPs in milk due to pesticide (OPPs) contamination and their use in crops as animal feed. These substances directly act by inhibiting the expression of the acetylcholinesterase enzyme at cholinergic junctions of the nervous system [125].

9.2.3 CARBAMATE (CBS) PESTICIDES

Carbamate pesticides (CBs) are extensively used broad-spectrum insecticide or herbicide in the world [71]. CBs are similar to OPPs in the capacity to inhibit cholinesterase enzymes but have different affinity for cholinesterase unlike OPPs. Exposure to these pesticides can cause reversible neurological ailments [125] and few of them are known carcinogens [13].

9.2.4 SYNTHETIC PYRETHROIDS (PYRS) PESTICIDES

The pyrethroids (PYRs) have an excellent insecticidal property with minimal mammalian toxicity, shorter period of environmental persistence and non-accumulation in the environment [112]. Today, PYRs are used worldwide for the control of household, agricultural, and veterinary pests, industrial, and stored products under integrated pest management (IPM) program. PYRs are low-polarity lipophilic compounds [6]. PYRs are exposed to the food chain because of their metabolism and tendency to bio-accumulate in animal adipose tissues. These eventually lead to death due to dysfunction of the entire nervous system. It induces a continuous series of nerve impulse by interacting with sodium channels in nerve membranes.

9.3 SOURCES OF CONTAMINATION AND ENTRY OF PESTICIDES INTO MILK

Most of the chlorinated pesticides have been found in the environment due to their lipophilicity and their resistance to metabolic processes [37]. Their presence in milk can be a result of industrial contamination. These chemicals travel to animals from plants via biological food chain. Exposure to different organisms leads to its accumulation in tissues [10]. The infiltration of these toxic chemicals into the mammalian glands of human or non-human organisms has been observed. Occurrence of pesticides in feeds and dairy products can be due to direct treatment of animals and their dwelling places against parasites (shed-fumigation), control of flies/ insects in the milk processing area and use of contaminated water and forage of animals, and during processing of milk [46].

Even in the certain areas (where farming and industrial activities do not exist in pesticides), OCPs, and polychlorinated biphenyls (PCBs) can enter the environment from the water-soil-grass route, subjecting to exchanges between air-water and air-soil and atmospheric precipitation [91]. Similarly, the OPPs (such as: chlorfenvinphos and dichlorvos) were found in the food chain when they were used as ecto-parasitides for cattle. Milk may show the presence, if animal feeds (such as: alfalfa, soya, maize, and sorghum) were treated with these pesticides. Chlorpyrifos have also been found to bind with human and bovine serum albumin, thus causing a potential health hazard to the consumers [78].

9.4 PROCEDURES FOR SAMPLING AND ANALYSIS

In this section, traditional and modern methods are described for qualitative and quantitative monitoring of pesticides residues in milk products [12, 29, 66, 97, 131]. LeDoux [68] reviewed different methods for analysis in animal-based products. According to the Pesticide Residue Analysis Manual of ICAR (Indian Council for Agricultural Research), the five important steps in pesticide analysis are [108]: (i) sampling (includes collection, transport, and storage); (ii) sample preparation; (iii) extraction; (iv) clean-up or isolation; and (v) identification, quantification, and confirmation.

9.4.1 PREPARATION OF SAMPLE

The initial step in processing of sample is to decrease the sample size to an analyzable sample size that represents the whole sample, i.e., subsampling. In case of milk blending, it is an important step to get a homogeneous portion of the sample. Samples must be transported to testing sites using clean vials and further stored at –10°C or below [66].

9.4.2 EXTRACTION TECHNIQUE

Extraction can be defined as the process of obtaining constituents by separating it from the crude materials by use of solvents. During the process, interfering molecules in complex food matrix may hamper the examination of pesticides residues in food samples [108]. The accuracy of pesticide residue analysis is directly influenced by the selection of specific sample preparation methods [58]. Detection of pesticide residue from a food sample requires extraction of analytes of interest from the food matrix and then clean-up of analytes from extracted matrix [11].

Traditional methods are losing interest as they are time-consuming, labor-intensive, complex, and expensive in approach [127]. The perfect way for sample preparation should be quick with greater precision and accuracy, ease of application and requires inexpensive materials [92]. Different approaches are available for efficient extraction of pesticides in milk matrix, such as:

- Accelerated solvent extraction (ASE) [21];
- Automated size-exclusion solid-phase extraction (ASESPE) [123];
- Bio-optical method [3];
- Combination method of freezing-lipid filtration, liquid-liquid extraction (LLE) dispersive primary-secondary amine clean-up and solid-phase extraction [23];Flow injection analysis (FIA) method [35];
- Liquid-liquid extraction (LLE) [17];
- Magnetic solid-phase extraction (MSPE) [114];
- Matrix solid-phase dispersion (MSPD) [103, 107];
- Microwave assisted extraction (MAE) [34, 41];
- Pressurized liquid extraction (PLE) [85];
- QuEChERS sample preparation method [55, 111];
- Solid-phase extraction (SPE) [19, 28, 132];

- Solid-phase micro extraction (SPME) [37];
- Ultrasonic extraction [74].

9.4.3 LIQUID-LIQUID EXTRACTION (LLE)

Liquid-liquid extraction (LLE) is a common method for extraction of pesticide residues from milk samples [26]. This method involves shaking of fluid milk with selected organic solvents due to which compounds separate themselves into different immiscible liquids. However, separation efficiency relies on the affinity of compounds to the solvent, extraction volume and extraction number [54]. The selection of solvent completely depends on the solvent polarity and analyte solubility. Ahmed [2] reported the use of a combination of water-immiscible non-polar solvents (such as: dichloromethane and hexane) and those solvents having different polarity (such as: acetonitrile and ethyl acetate) to obtain the desired extraction. The limitations of LLE method are: higher duration of analysis and effort, complexity; and requires large amounts of solvents. In case of multi-residue analysis, analytes with significant difference in polarity are difficult to obtain good extraction without any additional solid phase extraction (SPE) step [54].

9.4.4 SOLID PHASE EXTRACTION (SPE)

Solid-phase extraction (SPE) is a rapid technique that is frequently applied as a substitute to LLE. SPE is the separation or removal of an analyte or analytes from a mixture of compounds between a solid phase (sorbent) and liquid phase (solvent). Majorly of SPE sorbents (such as: bonded silica (C8 and C18), porous graphite, cation exchangers and reversed-phase (RP) column) are used for the extraction of residues from milk-based matrices. Electrostatic interaction and Vander Waal forces are the two main forces due to which analytes are retained on sorbents [2].

The extraction process consists of four steps [82]: (i) conditioning of the sorbent with solvent to be used; (ii) loading the sample aids the retention of analytes to sorbents; (iii) selective washing of the packing to remove other impurities; and (iv) ultimately elution of the chemical compound of interest [82].

SPE has been proposed for the extraction of phenyl urea herbicides and triazine herbicides [5, 65], endocrine disruptors (ED) [114], OCPs [132], and OPPs [90] from milk samples. Gravimetric removal of fat through ice-cooling followed by SPE method is found to be a suitable method for the OPPs extraction [63].

SPE has proved to be a good substitute against gel-permeation chromatography clean-up, commonly in use for withdrawal of OPPs from bovine tissues [88]. For isolation of pesticide residues from different dairy products, SPE is often combined with other purification techniques. Recently, Gao, and Sun [41] developed a multi-walled carbon nanotubes (MWCNTs) dispersive solid-phase extraction (d-SPE) method with GC coupled electron capture detector (ECD), for detection of pyrethroid in fluid milk. A miniaturized SPE coupled with ultra HPLC and quadrupole time-of-flight MS was also used for tetracycline detection in milk [128].

9.4.5 MATRIX SOLID-PHASE DISPERSION (MSPD)

Matrix solid-phase dispersion (MSPD) was commonly used due to its feasibility and versatility; hence it can be applied to various analytes whether these are biological or environmental matrices (various solid, semi-solid as well as viscous matrices like milk). MSPD uses bonded-phase solid support and abound solvent, which aids in dispersion. When the sample is dispersed over the bonded phase due to affinity of the analytes (specific to solid supports), it separates from the milk matrices. The difference between SPE and MSPD is that liquid analytes are used in SPE and in case of MSPD, solid or viscous samples can be used for separation [66]. Normal phase dispersions (Florisil, alumina, and silica) and reversed-phase (most commonly used C18) can be used in MSPD procedure [82].

MSPD technique has been used by Martha et al. [81] for the quantitative extraction from 5 bovine samples; and the analysis of OPPs was done by using diode array diode array UV detector in HPLC. The MSPD extraction with octadecylsilyl sorbent combined with a silica-gel clean-up and acetonitrile elution with individual detection wavelengths has been optimized for: chlorpyrifos at 287 nm, chlorfenvinphos at 244 nm, diazinon at 244 nm, fenitrothion at 268 nm, and parathion-methyl at 273 nm. The benefits of MSPD are: less extraction time, use of small quantities of solvent and sorbent, simultaneous extraction and clean-up method and well-suited to multi-residue analysis.

9.4.6 SOLID-PHASE MICROEXTRACTION (SPME)

In this analytical procedure, solvent is not used that reduces the laboratory waste [37]. SPME is based on a single simultaneous process for sampling, extraction, concentration, and sample introduction, which results in higher sample thorough output [36]. SPME system consists of SPME syringe, which contains 1 cm of fiber located inside the syringe needle, made up of polymer deposited on the holder of fused silica. This fiber is drawn into a metal tube. When this syringe needle is inserted into a milk sample, the fiber is exposed to the bulk milk and fiber absorbs analytes after specific time.

A perspective view of the SPME technique used for milk samples is given in Table 9.2, which showcases the protocols in a nutshell. The clean-up is not important to be employed during sample extraction using SPME.

9.4.7 PRESSURIZED LIQUID EXTRACTION (PLE)

Pressurized liquid extraction (PLE) is an ASE method, wherein low volume of organic solvents and extraction are at high temperature (up to 200°C) and pressure (20 MPa) for a short period of time. This process results in better extraction efficiency due to increase in diffusion rates, solubility of the analytes and mass transfer [26, 42]. Application of elevated temperature increases the solvent strength and rate of diffusion of solvent into the food matrix. This results in higher rates of extraction in conjunction with decrease in solvent viscosity and surface tension [82]. Mezcua et al. [85] described the extraction of 12 OPPs and OCPs residues in milk-based infant formulae by using PLE analyzed by GC-MS.

9.4.8 MICROWAVE-ASSISTED EXTRACTION (MAE)

MAE method employs microwave technique to heat up the solvent and partitioning of analytes from milk matrix to solvent [33] in a closed

TABLE 9.2 Use of SPME for Recognition of Pesticide Residue in Dairy Products

Pesticide	Extraction	Clean-up	Detection	Detection limits	References
Pyrethroids; OCC	SPME procedure	–	GC-ECD analysis	Sub-ng/ ml level	[37]
Dimpylate; disulfoton, parathion, and its methyl ester, ethion, sulfotep, chlorpyrifos	SPME procedure	ECD	GC/ MS with selected ion monitoring (SIM); selective ECD	–	[131]
OCC	Headspace-SPME procedure	–	GC-ECD	–	[84]

Legend: OCC: Organochlorine compounds; GC: gas chromatography; MS: mass spectrometry; ECD: electron capture detector; μECD: micro-Electron capture detector; SPME: solid-phase microextraction.

pressurized vessel, and a 90% reduction can be achieved [38, 63]. The principle behind the microwave extraction is non-ionizing radiation, which causes molecular motion and rotation of dipoles due to which extraction takes place. In this method, the sample to be extracted is suspended in solvents and heated for 30s, without boiling the sample and this step is repeated to obtain the maximum extraction [2]. Fang et al. [34] developed an extraction method coupled with microwave and SPE to extract pesticides from different infant milk formulae and obtained the recoveries ranging from 72% to 111%. Wang et al. [122] demonstrated that MAE has excellent capability to extract pesticides from complex matrices.

9.4.9 ULTRASONIC EXTRACTION

Ultrasonication provides an efficient contact between solid and solvent and a higher rate of solvent penetration into the sample matrix thereby improving the mass transfer and accelerating the extraction process [18, 47]. Ultrasonic extraction is simple, rapid, and inexpensive technique. Table 9.3 lists ASE and external field assisted microwave and ultrasonication extraction for detection of pesticide in dairy products.

9.4.10 QUECHERS (QUICK, EASY, CHEAP, EFFECTIVE, RUGGED, AND SAFE)

It is a quick, easy, cheap, effective, rugged, and safe (QuEChERs) method for analysis of residues in milk stuffs. This method simultaneously extracts polar and non-polar compounds making it suitable for extraction of various compounds [121]. During its initial phase of development, QuEChERS is used for non-fatty and fatty foods. This method employs micro-scale extraction using acetonitrile with its purification by dispersive solid-phase extraction method [127]. Quantification and detection of pesticide residues in a complex milk matrix is a formidable challenge. Modified versions of QuEChERS have been employed successfully by different researchers confirming the versatility and compatibility of the method for the detection of residues [60].

The standardization of different chemicals employed in QuEChERS using RSM (response surface methodology) has come to the foray enabling

TABLE 9.3 Accelerated Solvent Extraction and External Field Assisted, Microwaves, and Ultra-Sonication Extraction for Detection of Pesticide Residues in Milk and Milk Products

Pesticide	Extraction	Clean-Up	Detection	Detection Limits	References
12 OCPs	ASE+ Na_2SO_4 ethyl acetate: Cyclohexane (1:1)	GPC SX3; Hexane: Ethyl acetate (1:1) SPE silica n-Hexane	GC-ECD, CPSil-2 Acetonitrile	–	[124]
12 OCPs	Microwaves + Na_2SO_4 Ethyl acetate: n-Hexane (1:1)	GPC SX3; Hexane: Ethyl acetate (1:1)	–	–	[120]
Quinolones	0.05 M phosphate buffer (pH-7.4)	SPE-Discovery C18, upon passing the extracts, rinsed with water and elution with methanol-30% ammonium hydroxide (75:25 v/v)	LC symmetry C18 (150 × 4.6 mM, 5 µM); two different runs: Acetonitrile-0.02 molar phosphate buffer pH 3.0 (18:82) at flow rate of 1 mL/minute at maximum wavelength of 280 and 450 nm. Acetonitrile-0.02 molar phosphate buffer pH 3.0 (34:36) at 1 mL/minute of flow rate (wavelength of 312 and 366 nm).	5–10 µg/kg	[95]

Legend: OCPs: Organochloro pesticides; GC: gas chromatography; MS: mass spectrometry; ECD: electron capture detector; ASE: automated solvent extraction; SPE: solid-phase extraction; LC: liquid chromatography.

faster detection of pesticide residues in milk [52, 53]. As milk being a fatty matrix, it necessitates the removal of fat, which requires a thorough extraction and clean-up protocol. QuEChERS method is based on two steps: extraction and clean-up. The QuEChERS approach involves extraction of a well-mixed sample by shaking it in a centrifuge tube containing magnesium sulfate and sodium acetate, with an equal amount of solvent (commonly acetonitrile). The clean-up is followed using solid-phase extraction in dispersive mode (dSPE).

Lehotay and Mastovska [69] developed two quick and simple methods for analysis of pesticide in fatty food matrix by using QuEChERs sample preparation technique and had obtained 100% recovery of residues. Selvi et al. [105] reported multi-residue analysis of OCPs in animal products by GC-ECD with recoveries of 72–105%. Another quantitative method was successfully developed based on the QuEChERS approach for liquid chromatography-tandem mass spectrometry (LC-MS/MS) for the determination of chloramphenicol in milk [111]. Recently this method has been commonly used by various researchers to analyze pesticide residues in milk and milk products [35, 129].

Liu et al. [72] devised a successful technique to determine amphenicol quantity in milk and honey samples using QuCHhERS-capillary LC-MS/MS method. Presently several companies have commercialized QuEChERS sample preparation kits, viz., Agilent, Restek, Supelco, Thermo-Fisher, Biotage, and Waters, etc.

9.4.11 CLEAN-UP

During the extraction of target pesticide residue from the food sample, several interfering compounds (such as: fat, sugar, proteins, etc.), may also be co-extracted thus hampering the determination of the target analyte [43, 135]. Clean-up or isolation refers to the separation of these interfering compounds during the analysis for pesticide residues. It is usually a combination of partitioning and purification. Purification is usually accomplished by preparative chromatography. The efficiency with which the partitioning solvent can remove pesticides from the sample extract helps in determination of the degree of the clean-up requirement. Various modification techniques may improve the efficiency of clean-up as well as the efficiency of detection.

In a recent study, Rejczak, and Tuzimski [96] proposed a method that involved liquid-liquid partitioning using acetonitrile. This is followed by dSPE clean-up using PSA along with zirconia coated silica particles to purify the extract. Though the clean-up is less important in modern pesticide residue extraction and analytical approaches as they are minimally affected by co-extracted matrix components. However, sample clean-up is an important step in the analysis of pesticides in fatty foods [2].

9.4.12 FREEZE CENTRIFUGATION

This method is important, where lipids are major interferences. It can be easily removed by freezing centrifugation. Not only lipids, the fatty compounds having a lower melting point in comparison to solvent can also be removed under frozen conditions by centrifugation or filtering. Freezing centrifugation significantly removes lipids from many fatty matrices but not from all, for example OCPs which remain dissolved in the solvent [26].

Chen et al. [23] used the freezing-lipid filtration along with a dispersive primary-secondary amine clean-up to effectively eliminate any matrix interferences in the quantification of OCPs, OPPs, carbamate, and pyrethroid residues in infant formulae and cow or human milk.

9.4.13 PARTITIONING WITH LIPID EXTRACTION

Usually the liquid-liquid partitioning is used before SPE to remove the co-extracts. However, solvent partitioning could result in some analyte (pesticide residues) losses resulting in a lower percentage of recoveries [68]. For example, OCPs are slightly soluble in polar solvent so that lower recovery rate may be expected for them [26].

9.4.14 ADSORPTION CHROMATOGRAPHY

In adsorption chromatography, analytes, and other compounds (which may interfere) are collected on a solid sorbent based on interaction of analytes dissolved in the solvent with the adsorptive surface. The target compounds are eluted from the substrate keeping the interfering substances arrested on substrate. It is one of the most commonly used clean-up procedures while applying the SPE technique [61, 64, 98].

Florisil, alumina, and silica gels are frequently used as adsorptive surface materials for adsorption chromatography. Major application involves purification of the extracts from lipid fractions [135]. Chung and Chen [26] reported its use to remove lipids during sample preparation with minimal effects on non-lipid compounds.

9.4.15 GEL PERMEATION CHROMATOGRAPHY (GPC)

It is a widely employed method to eliminate lipid components from fatty foods [82]. It separates lower molecular mass compounds (such as: OCPs) from lipids having high mass units ranging from 600–1,500 [26]. The basis for gel-permeation (or size exclusion) chromatography separation technique is the differences in molecular size of compounds. Samples are placed at the top of the preparative-sized columns and then eluted with solvent retaining major molecules in the deposit and thus eluting out minor ones. Copolymers of styrene and divinylbenzene are usually used in columns in gel chromatography. Commonly used mobile phases are cyclohexane or combinations of solvents having variable elution forces [135].

Many studies have reported using this clean-up technique to determine the OCPs in milk and its products preceded by different extraction techniques, such as: SPE [118], LL extraction [27, 88], Soxhlet extraction [133], SPE-GC-MS/MS [132], LC-MS/MS [130], and GC [50].

9.4.16 SULFURIC ACID TREATMENT

Sulphuric acid treatment is one of the purification procedures to elute out fatty co-extracted interfering substances [14]. Some OCPs (such as: endrin, endosulfan I and II, etc.), are sensitive to sulfuric acid. Hey get sulphonized and degraded with time [26]. This technique is advantageous for the determination of sulfuric acid resistant organochlorine pesticides (such as: endrin, dieldrin, and methoxychlor), which get destroyed on reaction with sulfuric acid. Many more samples of milk and its products have used sulfuric acid treatment as a clean-up process to remove any interference in the pesticide residue extraction [50, 62].

9.5 DETECTION METHODS FOR PESTICIDE RESIDUES

9.5.1 *GAS CHROMATOGRAPHY (GC)*

On the basis of the volatility of compounds, GC can separate them. Commonly used detectors for GC are: ECD, flame photometric detector (FPD), Nitrogen and Phosphorus (NPD), GC/MS/MS and MS [54]. Santos and Galceran [102] reported that most, of the pesticide residues and their degradation metabolites are analyzed by GC, whereas GC-ECD is most commonly used for determination of OCPs.

For separation and determination of pesticides using GC, various stationary phases (such as: phenyl-methyl-polysiloxane or polydimethyl-siloxane) have been used in capillary columns with Helium as a carrier gas [66]. More or less, every research group uses one or two columns (one for analytical purpose and the other for peak identification) for the same group of pesticides [30, 87]. Salem et al. [100] reported the analysis of pesticide residues in milk products in Jorden by using GC equipped with Ni-ECD detector, silica capillary column with helium carrier gas.

In Ghana, Darko, and Acquaah [29] analyzed levels of OCPs in milk products, using GC equipped with Ni-ECD detector and SPB-608 capillary column. Hernandes et al. [48] described a validated, simple, and economically feasible way to determine presence of deltamethrin and cypermethrin in bovine milk using GC-μECD. GC has also been used to analyze pesticide residues in dairy infant formulae [116].

9.5.2 *LIQUID CHROMATOGRAPHY (LC)*

Liquid chromatography (LC) is most commonly used for detection of polar non-volatile, and thermally labile groups of pesticides, such as triazines, and CBs with their degraded substances. LC may also be used when gas chromatography (GC) conditions are not appropriate for the detection of pesticide residues. RP chromatography on C-18 bonded silica column is a commonly used column for LC analysis [15, 70]. RP-HPLC (reversed-phase high-performance liquid chromatography) is the most widely used LC technique because of the varying variety of both stationary and mobile phases. HPLC pump forces the mobile phase and analyte is passed through a compact packed column, which provides analyte retention time due to analyte interaction with the stationary phase. Most commonly used detectors for pesticide residue analysis are: ultraviolet (UV), diode array detector (DAD), fluorescence, and chemiluminescence (CL) [54].

HPLC-UV detection (200–350 nm) techniques are usually used because of their excellent qualitative and quantitative versatility and simple protocol. DAD detectors are more sensitive than absorption detection. Barchanska et al. [7] reported the monitoring of atrazine in milk using ELISA and validation by HPLC coupled with DAD and C-18e (125 × 3, 5 μm) analytical column. In Iran, Ashnagar et al. [4] determined the OCPs in cow milk using RP-HPLC with UV detector. UV detector along with HPLC was also used to detect triazines in milk samples using cloud point extraction (CPE). CPE is a green pre-treatment technique used for efficient extractions and preconcentration of the analytes [75].

Seccia et al. [104] used HPLC with DAD to analyze and quantify pesticide residues in milk using Hydro-RPC18 column. An efficient analytical methodology has been proposed by Zhang et al. [131] for detecting seven sulphonamides from milk by using HPLC with ultraviolet detection (HPLC-UV), using analytical column Inertsil ODS-3. Lutter et al. [77] determined melamine in cow milk and infant formulae using a C-18 HPLC column with UV detector.

9.5.3 MASS SPECTROMETRY (MS) DETECTION

The performance capabilities of HPLC and GC in terms of identification and separation can be increased manifold together with mass spectrometry (MS). It is a tool that works as a qualitative detector and a powerful quantitative detector [2].

9.5.4 GAS CHROMATOGRAPHY-MASS SPECTROMETRY (GC-MS)

The important feature of using GC-MS method lies in its mode of ionization. The selection of ionization mode often depends on whether the analysis is target specific chemical class or multi-residue analysis method. The ionization methods are either electron capturing or electron impact, depending on the sample matrix and sample preparation procedures. In order to determine multi-residue by GC/MS, electron ionization is often applied. The popular GC-MS technique involves single quadrupole with electron impact ionization. Electron beam energy of 70 eV is commonly applied [24].

MS with selected ion monitoring (SIM) is currently used to detect number of targeted analytes in a given time. However, in the presence of interference matrix components, GC-MS analysis can be complicated and requires proper clean-up procedures. Hassine et al. [45] determined the OCPs and polychlorinated and polybrominated biphenyl ethers in human milk using GC-MS. Some pesticides have been quantified in bovine milk by GC-triple quadrupole MS [22]. Rodrigues et al. [97] developed head-space-SPME/ GC-MS equipped with capillary columns for determination of OPPs in cow milk. Many researchers have used MS [39] and tandem MS [20, 130, 132] along with GC to quantify pesticide residues in milk and milk products.

9.5.5 LC-MASS SPECTROMETRY (LC-MS)

LC coupled with MS is used for highly polar and thermolabile pesticide residues [51]. Silicon membrane separator, heated wire, introduction of direct liquid, electrospray ionization (ESI), etc., for positive ion and atomic pressure chemical ionization (APCI) for negative ion are different ionizing interferences in coupling HPLC with MS.

Tian [115] successfully determined pesticide residues in bovine milk by HPLC-MS/MS. MeCN-water is the most commonly used mobile phase employed in LC-MS but sometimes addition of acetate, formate, and formic acid to mobile phase may modify the sensitivity of LC-MS methods [66]. Power et al. [93] had used UHPLC MS/MS to analyze Tricladbendazole residues in some milk products. Wielogorska et al. [126] used UHPLC-MS/MS technique for extraction, detection, quantification followed by confirmation of 19 ED in milk.

9.5.6 CAPILLARY ELECTROPHORESIS (CE)

Capillary electrophoresis (CE) is a new detection technique used in pesticide determination [57]. Pico et al. [92] reviewed the off-column and on-column approaches for analysis and summarized the comparison between the conventional CE and CE with MS. CE involves a fused silica capillary filled with a running buffer, an electrolyte solution (sodium phosphate), two electrodes, a detector and a high voltage power supply. The distinctive characteristic of CE lies in the formation of an electro-osmotic flow inside the capillary. This causes all ions to migrate towards the detector for detection. LLE and SPE are commonly used sample preparation techniques with CE [32, 83]. Successful modes used in the analysis of pesticide by CE involve capillary zone electrophoresis (CZE), micellar electro-kinetic CE and capillary electromyography [92].

Hernandez-Borges et al. [49] studied the various sample preconcentration procedures viz. LLE, SPE, SPME, CPE, etc., and various detection modes like UV-Vis, laser-induced fluorescence (LIF) for pesticide residue analysis using CE. CPE was used as an efficient pre-treatment prior to CE to extract and pre-concentrate analytes and determine triazine herbicides in milk using HPLC [75]. The UV DAD is the most popular detector used in detection with MEKC mode. MEKC is the most commonly used mode to separate out neutral and compounds similar in structure through the formation of micelles that interact with compounds.

9.5.7 IMMUNOASSAY

Immunoassay is a technique based on interaction between antibodies and antigens. This method is rapid, sensitive, and specific to a particular analyte. The steps for analysis include: dilutions, additions, washings, and reading color intensity [94]. Most of the immunoassays are based on enzyme linked immune-sorbent assay (ELISA) procedure [40]. Quantification of analytes is based on the comparison between the signals generated by the unknown sample analytes with known standard curve.

Barchanska et al. [7] developed a simple rapid tube ELISA method for detection of atrazine residue in milk, the polyclonal antisera specific to atrazine was used and the plate coated with hapten-gel-conjugate was used and the color change due interaction was measured on a basis of spectrophotometer. An automated flow-based biosensor technique was developed by Mishra et al. [86] for the online detection of organophosphates in fluid milk by using three acetylcholinesterase enzymes immobilized on screen-printed electrodes to quantify OPPs.

Liu et al. [73]; and Marrazza [80] reviewed the advances in nanosensors and piezoelectric biosensors, respectively, for detection of OPPs. Campanella et al. [16] reported an immunosensor by using a transducer as an amperometric electrode for hydrogen peroxide and marker as peroxidase enzyme for atrazine residue detection.

9.5.8 THIN LAYER CHROMATOGRAPHY (TLC)

Thin-layer chromatography and HPTLC are used as complementary techniques for pesticide residue analysis. They have high sample throughput and low operating cost [109]. It is commonly used for the analysis of pyrethrins/ pyrethroids (PYRs), NMCs, and their metabolites [2]. HPLTC is made of smaller particle diameter due to which the particle size distribution is narrower and the layers are thinner that leads to higher separation efficiency, faster separation and improved detection limit [110].

Pesticide analysis by TLC and HPTLC are mostly performed on silica gel plates and the other less often used plates include cellulose, alumina, and kieselguhr. Normal-phase and RP (C-18) bonded silica gel layers have also been used for pesticide analysis. SPE and MSPD are the sample preparation methods for pesticide residue analysis through TLC [109].

9.6 SUMMARY

Due to the uncontrolled use of pesticides, necessary steps are needed for detection of these pesticide residues from farm to fork. Various novel mechanisms and technology for precise extraction and identification of residues from food samples is need of an hour and have seen tremendous development in the last decade and are in continuous research for making these techniques more sensitive and accurate. Clean and effective production and processing can be seen as a possible option for dissipating these pesticide residues from milk and other food products. While organic farming can be a better solution to kill this cause of contamination from the root.

KEYWORDS

- **contamination**
- **extraction**
- **maximum residue level**
- **organophosphorus pesticides**
- **pesticides**
- **pyrethroids**

REFERENCES

1. Ahmad, R., Salem, N. M., & Estaitieh, H., (2010). Occurrence of organochlorine pesticide residues in egg, chicken and meat in Jordan. *Chemosphere, 78*, 667–671.
2. Ahmed, F., (2001). Analyzes of pesticides and their metabolites in foods and drinks. *Trends in Analytical Chemistry, 20,* 649–661.
3. Appicciafuoco, B., Dragone, R., & Frazzoli, C., (2014). Microbial screening for quinolones residues in cow milk by bio-optical method. *Journal of Pharmaceutical and Biomedical Analysis, 106*, 179–185.
4. Ashnagar, A., Gharib, N. N., & Cheraghi, F., (2009). Determination of organochlorine pesticide residues in cow's milk marketed in Ahwaz city of Iran. *International Journal of Pharmatech Research, 1,* 247–251.
5. Balduini, L., Matoga, M., & Cavalli, E., (2003). Triazinic herbicide determination by gas chromatography-mass spectrometry in breast milk. *Journal of Chromatography B, 794,* 389–394.
6. Barbini, D. A., Vanni, F., Girolimetti, S., & Dommarco, R., (2007). Development of an analytical method for the determination of the residues of four pyrethroids in meat by GC-ECD and confirmation by GC-MS. *Analytical and Bioanalytical Chemistry, 389,* 1791–1798.
7. Barchanska, H., Jodo, E., Price, R. G., Baranowska, I., & Abuknesha, R., (2012). Monitoring of atrazine in milk using a rapid tube-based ELISA and validation with HPLC. *Chemosphere, 87,* 1330–1334.
8. Battu, R. S., Singh, B., & Kang, B., (2004). Contamination of liquid milk and butter with pesticide residues in the Ludhiana district of Punjab State, India. *Ecotoxicology and Environmental Safety, 59,* 324–331.
9. Bayat, S., Sari, A. E., Bahramifar, N., Younesi, H., & Behrooz, R., (2011). Survey of organochlorine pesticides and polychlorinated biphenyls in commercial pasteurized milk in Iran. *Environmental Monitoring and Assessment, 175,* 469–474.
10. Beek, B., & Bohling, S., (2000). The assessment of bioaccumulation. Chapter 4. In: Beek, B., (ed.)*Bioaccumulation-New Aspects and Developments* (pp. 235–276). Switzerland: Springer Nature AG.
11. Bennett, D. A., Chung, A. C., & Lee, S., (1997). Multiresidue method for analysis of pesticides in liquid whole milk. *Journal of AOAC International, 80,* 1065–1077.
12. Blasco, C., Pico, Y., & Andreu, V., (2009). Analytical method for simultaneous determination of pesticide and veterinary drug residues in milk by CE-MS. *Electrophoresis, 30,* 1698–1707.
13. Bogialli, S., Curini, R., Di Corcia, A., Lagana, A., Nazzari, M., & Tonci, M., (2004). Simple and rapid assay for analyzing residues of carbamate insecticides in bovine milk: Hot water extraction followed by liquid chromatography-mass spectrometry. *Journal of Chromatography A, 1054,* 351–357.
14. Bouwman, H., Sereda, B., & Meinhardt, H. M., (2006). Simultaneous presence of DDT and pyrethroid residues in human breast milk from a malaria-endemic area in South Africa. *Environmental Pollution, 144*(3), 902–917.
15. Buisson, S., Bouchart, V., Guerlet, E., Malas, J. P., & Costil, K., (2008). Level of contamination and impact of pesticides in cupped oyster, *Crassostrea gigas*, reared in

a shellfish production area in Normandy (France). *Journal of Environmental Science and Health B, 43,* 655–664.
16. Campanella, L., Eremin, S., Lelo, D., Martini, E., & Tomassetti, M., (2011). Reliable new immunosensor for atrazine pesticide analysis. *Sensors and Actuators B: Chemical, 156,* 50–62.
17. Campoy, C., Jiménez, M., & Olea-Serrano, M., (2001). Analysis of organochlorine pesticides in human milk: Preliminary results. *Early Human Development, 65,* 183–190.
18. Capelo, J. L., Maduro, C., & Vilhena, C., (2005). Discussion of parameters associated with the ultrasonic solid-liquid extraction for elemental analysis (Total Content) by electrothermal atomic absorption spectrometry. *Ultrasonics Sonochemistry, 12,* 225–232.
19. Celeiro, M., Facorro, R., Dagnac, T., Vilar, V. J., & Llompart, M., (2017). Photodegradation of multiclass fungicides in the aquatic environment and determination by liquid chromatography-tandem mass spectrometry. *Environmental Science and Pollution Research, 24*(23), 19181–19193.
20. Chen, M., Wen, K., Tao, X., & Ding, S., (2014). Novel multiplexed fluorescence polarization immunoassay based on a recombinant bi-specific single-chain diabody for simultaneous detection of fluoroquinolones and sulfonamides in milk. *Food Additives & Contaminants: Part A, 31*(2), 1959–1967.
21. Chen, S., Gfrerer, M., Lankmayr, E., Quan, X., & Yang, F., (2003). Optimization of accelerated solvent extraction for the determination of chlorinated pesticides from animal feed. *Chromatographia, 58,* 631–636.
22. Chen, X., Lin, Y., Dang, K., & Puschner, B., (2017). Quantification of polychlorinated biphenyls and polybrominated diphenyl ethers in commercial cows' milk from California by gas chromatography-triple quadruple mass spectrometry. *PloS One, 12*(1). Article ID: e0170129.
23. Chen, X., Panuwet, P., & Hunter, R., (2014). Method for the quantification of current use and persistent pesticides in cow milk, human milk and baby formula using gas chromatography-tandem mass spectrometry. *Journal of Chromatography B, 970,* 121–130.
24. Chen, Z. M., & Wang, Y., (1996). Chromatographic methods for the determination of pyrethrin and pyrethroid pesticide residues in crops, food and environmental samples. *Journal of Chromatography A, 754*(1, 2), 367–395.
25. Choi, M. P. K., Kang, Y. H., Peng, X. L., Ng, K. W., & Wong, M., (2009). Stockholm convention organochlorine pesticides and polycyclic aromatic hydrocarbons in Hong Kong air. *Chemosphere, 77,* 714–719.
26. Chung, S. W. C., & Chen, B. L. S., (2011). Determination of organochlorine pesticide residues in fatty foods: Critical review on the analytical methods and their testing capabilities. *Journal of Chromatography A, 1218,* 5555–5567.
27. Ciscato, C. H. P., Gebara, A. B., & Spinosa, H., (2002). Pesticide residues in cow milk consumed in São Paulo City (Brazil). *Journal of Environmental Science and Health, 37,* 323–330.
28. Dagnac, T., Garcia-Chao, M., & Pulleiro, P., (2009). Dispersive solid-phase extraction followed by liquid chromatography-tandem mass spectrometry for the multi-residue analysis of pesticides in raw bovine milk. *Journal of Chromatography A, 1216,* 3702–3709.

29. Darko, G., & Acquaah, S., (2008). Levels of organochlorine pesticides residues in dairy products in Kumasi, Ghana. *Chemosphere, 71,* 294–298.
30. Doong, R. A., & Lee, C., (1999). Determination of organochlorine pesticide residues in foods using solid-phase extraction clean-up cartridges. *Analyst, 124,* 1287–1289.
31. Earth Negotiations Bulletin, (2008). *Summary of the 4th Meeting of the Persistent Organic Pollutants Review Committee of the Stockholm Convention* (Vol. 15, No. 149, p. 1–17). International Institute for Sustainable Development (IISD).
32. Eash, D. T., & Bushway, R., (2000). Herbicide and plant growth regulator analysis by capillary electrophoresis. *Journal of Chromatography A, 880,* 281–294.
33. Eskilsson, C. S., & Bjorklund, E., (2000). Analytical-scale microwave-assisted extraction. *Journal of Chromatography A, 902,* 227–250.
34. Fang, G., Lau, H. F., Law, W. S., & Li, S. F. Y., (2012). Systematic optimization of coupled microwave-assisted extraction-solid phase extraction for the determination of pesticides in infant milk formula via LC-MS/MS. *Food Chemistry, 134,* 2473–2480.
35. Fernandes, F. C. B., Silva, A. S., Rufino, J. L., Pezza, H. R., & Pezza, L., (2015). Screening and determination of sulphonamide residues in bovine milk samples using a flow injection system. *Food Chemistry, 166,* 309–315.
36. Fernandez, A. M., Lamas, J. P., & Sanchez-Prado, L., (2010). Development of a solid-phase microextraction gas chromatography with microelectron-capture detection method for the determination of 5-Bromo-5-nitro-1, 3-dioxane in rinse-off cosmetics. *Journal of Chromatography A, 1217*(43), 6634–6639.
37. Fernandez-Alvarez, M., & Llompart, M., (2008). Development of a solid-phase microextraction gas chromatography with microelectron-capture detection method for a multiresidue analysis of pesticides in bovine milk. *Analytica Chimica Acta, 617,* 37–50.
38. Fish, J. R., & Revesz, R., (1996). Microwave solvent extraction of chlorinated pesticides from soil. *LC-GC, 14,* 230–234.
39. Frenich, A. G., Romero-Gonzalez, R., & Aguilera-Luiz, M., (2014). Comprehensive analysis of toxics (pesticides, veterinary drugs and mycotoxins) in food by UHPLC-MS. *Trends in Analytical Chemistry, 63,* 158–169.
40. Gabaldon, J. A., Maquieira, A., & Puchades, R., (1999). Current trends in immunoassay-based kits for pesticide analysis. *Critical Reviews in Food Science and Nutrition, 39,* 519–538.
41. Gao, Y. L., & Sun, P., (2017). Determination of five pyrethroid pesticides residue in liquid milk by gas chromatography using multi-walled carbon nanotubes as dispersion solid-phase extraction sorbent. *Acta Chromatographica, 30*(2), 1–6.
42. Giergielewicz, M. H., Dabrowski, L., & Namienik, J., (2001). Accelerated solvent extraction (ASE) in the analysis of environmental solid samples - some aspects of theory and practice. *Critical Reviews in Analytical Chemistry, 31,* 149–165.
43. Gilbert, L. B., García-Reyes, J. F., & Molina-Díaz, A., (2009). Sample treatment and determination of pesticide residues in fatty vegetable matrices: A review. *Talanta, 79,* 109–128.
44. Gutierrez, R., Ruiz, J. L., & Ortiz, R., (2012). Organochlorine pesticide residues in bovine milk from organic farms in Chiapas, Mexico. *Bulletin of Environmental Contamination and Toxicology, 89,* 882–887.

45. Hassine, S. B., Ameur, W. B., Gandoura, N., & Driss, M. R., (2012). Determination of chlorinated pesticides, polychlorinated biphenyls and polybrominated diphenyl ethers in human milk from Bizerte (Tunisia) in 2010. *Chemosphere, 89,* 369–377.
46. Heck, C., & Sifuentes, D. S. J., (2007). Estimation of children exposure to organochlorine compounds through milk in Rio Grande do Sul, Brazil. *Food Chemistry, 102,* 288–294.
47. Hemwimol, S., Pavasant, P., & Shotipruk, A., (2006). Ultrasound-assisted extraction of anthraquinones from roots of *Morinda citrifilia. Ultrasonics Sonochemistry, 13,* 543–548.
48. Hernandes, T., Dores, E. F. C. G., & Ribeiro, M., (2014). Simple method to determine residual cypermethrin and deltamethrin in bovine milk. *Journal of the Brazilian Chemical Society, 25*(9), 1656–1661.
49. Hernandez-Borges, J., Frias-Garcia, S., Cifuentes, A., & Rodriguez-Delgado, M., (2004). Pesticide analysis by capillary electrophoresis. *Journal of Separation Science, 24,* 947–963.
50. Hernik, A., Goralczyk, K., & Strucinski, P., (2014). Characterizing the individual health risk in infants exposed to organochlorine pesticides via breast milk by applying appropriate margins of safety derived from estimated daily intakes. *Chemosphere, 94,* 158–163.
51. Itoh, H., Kawasaki, S., & Tadano, J., (1996). Application of liquid chromatography atmospheric-pressure chemical-ionization mass spectrometry to pesticide analysis. *Journal of Chromatography A, 754,* 61–76.
52. Jeong, I., Kwak, B., Ahn, J., & Jeong, S., (2012). Determination of pesticide residues in milk using a QuEChERS-based method developed by response surface methodology. *Food Chemistry, 133,* 473–481.
53. Jia, W., Chu, X., Ling, Y., Huang, J., & Chang, J., (2014). Analysis of phthalates in milk and milk products by liquid chromatography coupled to quadrupole orbitrap high-resolution mass spectrometry. *Journal of Chromatography A, 1362,* 110–118.
54. Jin, B., Xie, L., Guo, Y., & Pang, G., (2012). Multi-residue detection of pesticides in juice and fruit wine: Review of extraction and detection methods. *Food Research International, 46,* 399–409.
55. Jones, K. C., & De Voogt, P., (1999). Persistent organic pollutants (POPs): State of the Science. *Environmental Pollution, 100*(1–3), 209–221.
56. Kalpana, B., (1999). Health risk assessment for exposures to pesticides: Case study of endocrine disrupters. In: *Proceedings of the Eighth National Symposium on Environment* (pp. 70–72). Kalpakkam, India.
57. Karcher, A., & El-Rassi, Z., (1999). Capillary electrophoresis and electrochromatography of pesticides and metabolites. *Electrophoresis, 20,* 3280–3296.
58. Kataoka, H., Lord, H. L., & Pawliszyn, J., (2000). Applications of solid - phase microextraction in food analysis. *Journal of Chromatography A, 880,* 35–62.
59. Kaushik, C. P., Kaushik, A., & Sharma, H. R., (2014). Seasonal trends in organochlorine pesticide residues in raw bovine milk from rural areas of Haryana, India. *Bulletin of Environmental Contamination and Toxicology, 92*(1), 15–22.
60. Keegan, J., Whelam, M., & Danaher, M., (2009). Benzimidazole carbamate residues in milk: Detection by surface plasmon resonance-biosensor, using a modified QuEChERS (quick, easy, cheap, effective, rugged and safe) method for extraction. *Analytica Chimica Acta, 654,* 111–119.

61. Klincic, D., Romanic, S. H., Saric, M. M., Grzunov, J., & Dukic, B., (2014). Polychlorinated biphenyls and organochlorine pesticides in human milk samples from two regions in Croatia. *Environmental Toxicology and Pharmacology, 37*(2), 543–552.
62. Kuba, J., & Tomza-Marciniak, A., (2015). Comparison of DDT and its metabolites concentrations in cow milk from agricultural and industrial areas. *Journal of Environmental Science and Health, Part B, 50*(1), 1–7.
63. Kuiniven, J., & Bengtsson, S., (2002). Solid-phase extraction and clean-up of organophosphorus pesticide residues in bovine muscle with gas chromatographic detection. *Journal of Chromatographic Science, 40,* 392–396.
64. Kumar, A., Dayal, P., Shukla, G., Singh, G., & Joseph, P. E., (2006). DDT and HCH residue load in mother's breast milk: A survey of lactating mother's from remote villages in agra region. *Environmental International, 32,* 248–251.
65. Lagana, A., Fago, G., & Marino, A., (1998). Determination of aryloxyphenoxy propionic acid herbicides in water using different solid-phase extraction procedures and liquid chromatography-diode array detection. *Journal of Chromatography A, 796,* 309–318.
66. Lambropoulou, D. A., & Albanis, T. A., (2007). Sample preparation for determination of pesticide residues in food matrices by chromatography-mass spectrometry-based techniques: Review. *Analytical and Bioanalytical Chemistry, 389,* 1663–1683.
67. Lawrence, L. J., Casida, J., (1984). Interactions of lindane, toxaphene and cyclodienes with brain-specific t-butylbicyclo-phosphorothioate receptor. *Life Sciences, 35,* 171–178.
68. LeDoux, M., (2011). Analytical methods applied to the determination of pesticide residues in foods of animal origin. A review of the past two decades. *Journal of Chromatography A, 1218*(8), 1021–1036.
69. Lehotay, S. J., & Mastovska, K., (2005). Evaluation of two fast and easy methods for pesticide residue analysis in fatty food matrixes. *Journal of AOAC International, 88,* 630–638.
70. Limsuwan, S., Prie, B., Tansakul, N., Nau, H., Kietzman, M., & Hamscher, G., (2007). Penetration studies of propoxur and phoxim from eggshell into whole egg after experimental exposure and application in henhouses. *Journal of Agricultural and Food Chemistry, 55,* 6401–6405.
71. Liska, I., & Slobodnik, J., (1996). Comparison of gas and liquid chromatography for analyzing polar pesticides in water samples. *Journal of Chromatography A, 733,* 235–258.
72. Liu, H. Y., Lin, S. L., & Fuh, M. R., (2016). Determination of chloramphenicol, thiamphenicol and florfenicol in milk and honey using modified QuEChERS extraction coupled with polymeric monolith-based capillary liquid chromatography-tandem mass spectrometry. *Talanta, 150,* 233–239.
73. Liu, S., Yuan, L., Yue, X., Zheng, Z., & Tang, Z., (2008). Recent advances in nanosensors for organophosphate pesticide detection. *Advanced Powder Technology, 19,* 419–441.
74. Liu, S., Zheng, Z., Wei, F., Ren, Y., Gui, W., Wu, H., & Zhu, G., (2010). Simultaneous determination of seven neonicotinoid pesticide residues in food by ultra-performance liquid chromatography-tandem mass spectrometry. *Journal of Agricultural and Food Chemistry, 58,* 3271–3278.
75. Liu, T., Cao, P., Geng, J., & Li, J., (2014). Determination of triazine herbicides in milk by cloud point extraction and high-performance liquid chromatography. *Food Chemistry, 142,* 358–364.

76. Lopez-Avila, L., Young, R., Benedicto, K., Ho, P., Kim, R., & Beckert, W., (1995). Extraction of organic pollutants from solid samples using microwave energy. *Analytical Chemistry, 67,* 2096–2102.
77. Lutter, P., Perroud, M. C. S., & Gimenez, E. C., (2011). Screening and confirmatory methods for the determination of melamine in cow's milk and milk-based powdered infant formula, validation and proficiency-tests of ELISA, HPLC-UV, GC-MS and LC-MS/MS. *Food Control, 22,* 903–913.
78. Lv, Y., Li, X., Wang, Z., Zheng, H., Zhang, Q., Huo, R., Chen, X., & Han, T., (2014). Short communication, interaction of bovine milk protein with chlorpyrifos. *Journal of Dairy Science, 97*(4), 2056–2060.
79. Man, Y. B., Chan, J. K. Y., Wang, H. S., Wu, S. C., & Wong, M. H., (2014). DDTs in mothers' milk, placenta and hair, and health risk assessment for infants at two coastal and inland cities in China. *Environmental International, 65,* 73–82.
80. Marrazza, G., (2014). Piezoelectric biosensors for organophosphate and carbamate pesticides: A review. *Biosensors, 4*(3), 301–317.
81. Martha, P., Garcia, D. L., Maria, L., & Reyes, R. A., (2009). Validated matrix solid-phase dispersion method for the extraction of organophosphorus pesticides from bovine samples. *Food Chemistry, 114,* 1510–1516.
82. Martins, J. G., Chavez, A. A., Waliszewski, S. M., Cruz, A. C., & Fabila, M. M. G., (2013). Extraction and clean-up methods for organochlorine pesticides determination in milk. *Chemosphere, 92,* 233–246.
83. Menzinger, F., Schmitt-Kopplin, P., Freitag, D., & Kettrup, A., (2000). Analysis of agrochemicals by capillary electrophoresis. *Journal of Chromatography A, 891,* 45–67.
84. Merib, J., Nardini, G., & Carasek, E., (2014). Use of doehlert design in the optimization of extraction conditions in the determination of organochlorine pesticides in bovine milk samples by HS-SPME. *Analytical Methods, 6,* 3254–3260.
85. Mezcua, M., Repetti, M. R., & Agüera, A., (2007). Determination of pesticides in milk-based infant formulas by pressurized liquid extraction followed by gas chromatography-tandem mass spectrometry. *Analytical and Bioanalytical Chemistry, 389,* 1833–1840.
86. Mishra, R. K., Dominguez, R. B., & Bhand, S., (2012). Novel automated flow-based biosensor for the determination of organophosphate pesticides in milk. *Biosensors and Bioelectronics, 32,* 56–61.
87. Muccio, A. D., Pelosi, P., & Camoni, I., (1996). Selective, solid-matrix dispersion extraction of organophosphate pesticide residues from milk. *Journal of Chromatography A, 754,* 497–506.
88. Mueller, J. F., Harden, F., Toms, L. M., Symons, R., & Furst, P., (2008). Persistent organochlorine pesticides in human milk samples from Australia. *Chemosphere, 70,* 712–720.
89. Nic, M., Jirat, J., & Kosata, B., (2006). Pesticide residue. In: *IUPAC Compendium of Chemical Terminology- The Gold Book* (2nd edn., p. 464). London: Blackwell Science.
90. Pagliuca, G., Gazzotti, T., Zironi, E., & Sticca, P., (2005). Residue analysis of organophosphorus pesticides in animal matrices by dual-column capillary gas chromatography with nitrogen-phosphorus detection. *Journal of Chromatography A, 1071,* 67–70.
91. Pan, J., Gai, N., Tang, H., Chen, S., Chen, D., Lu, G., & Yang, Y., (2014). Organochlorine pesticides and polychlorinated biphenyls in grass, yak muscle, liver,

and milk in ruoergai high altitude prairie, the eastern edge of Qinghai-Tibet plateau. *Science of the Total Environment, 491,* 131–137.
92. Pico, Y., Rodriguez, R., & Manes, J., (2003). Capillary electrophoresis for the determination of pesticide residues. *Trends in Analytical Chemistry, 22,* 133–151.
93. Power, C., Danaher, M., Sayers, R., & O'Brien, B., (2013). Investigation of the migration of triclabendazole residues to milk products manufactured from bovine milk, and stability therein, following lactating cow treatment. *Journal of Dairy Science, 96,* 6223–6232.
94. Pugazhenthi, T. R., & Ayyadurai, K., (2013). Immunoassay analysis: Tool for detection of pesticide residues in dairy products. *Indian Journal of Fundamental and Applied Life Science, 3,* 48–51.
95. Ramos, M., Aranda, A., Garcia, E., Reuvers, T., & Hooghuis, H., (2003). Simple and sensitive determination of five quinolones in food by liquid chromatography with fluorescence detection. *Journal of Chromatography B, 789,* 373–381.
96. Rejczak, T., & Tuzimski, T., (2017). QuEChERS-based extraction with dispersive solid-phase extraction clean-up using PSA and ZrO_2-based sorbents for determination of pesticides in bovine milk samples by HPLC-DAD. *Food Chemistry, 217,* 225–233.
97. Rodrigues, F. M., & Mesquita, P. R., (2011). Development of a headspace solid-phase microextraction/gas chromatography-mass spectrometry method for determination of organophosphorus pesticide residues in cow milk. *Microchemical Journal, 98,* 56–61.
98. Romero, M. L. L., & Dorea, J. G., (2000). Concentrations of organochlorine pesticides in milk of Nicaraguan mothers. *Archives of Environmental & Occupational Health, 55,* 274–278.
99. Salas, J. H., & González, M. M., (2003). Organo-phosphorus pesticide residues in Mexican commercial pasteurized milk. *Journal of Agricultural and Food Chemistry, 51,* 4468–4471.
100. Salem, N. M., Ahmed, R., & Estaitieh, H., (2009). Organochlorine pesticide residues in dairy products in Jordan. *Chemosphere, 77,* 673–678.
101. Santonicola, S., De Felice, A., & Cobellis, L., (2017). Comparative study on the occurrence of polycyclic aromatic hydrocarbons in breast milk and infant formula and risk assessment. *Chemosphere, 175,* 383–390.
102. Santos, F. J., & Galceran, M., (2004). The application of gas chromatography to environmental analysis. *Trends in Analytical Chemistry, 21,* 672–685.
103. Schenck, F. J., & Wagner, R., (1995). Screening procedure for organochlorine and organophosphorus pesticide residues in milk using matrix solid-phase dispersion (MSPD) extraction and gas chromatographic determination. *Food Additives & Contaminants, 12,* 535–541.
104. Seccia, S., Fidente, P., Montesano, D., & Morrica, P., (2008). Determination of neonicotinoid insecticides residues in bovine milk samples by solid-phase extraction clean-up and liquid chromatography with diode-array detection. *Journal of Chromatography A, 1214*(1, 2), 115–120.
105. Selvi, C., Paramasivam, M., Rajathi, D. S., & Chandrasekaran, (2012). Multiresidue analysis of organochlorine pesticides in milk, egg and meat by GC-ECD and confirmation by GC-MS. *Bulletin of Environmental Contamination and Toxicology, 89,* 1051–1056.
106. Shaker, E. M., & Elsharkawy, E., (2014). Organochlorine and organophosphorus pesticide residues in raw buffalo milk from agro-industrial areas in Assiut, Egypt. *Environmental Toxicology and Pharmacology, 39*(1), 433–440.

107. Shamsipur, M., Yazdanfar, N., & Ghambarian, M., (2016). Combination of solid-phase extraction with dispersive liquid-liquid microextraction followed by GC-MS for determination of pesticide residues from water, milk, honey and fruit juice. *Food Chemistry, 204*, 289–297.
108. Sharma, K., (2010). *Pesticide Residue Analysis Manual* (p. 90). Directorate of information and publications of agriculture, New Delhi, India.
109. Sherma, J., (2003). Recent advances in the thin layer chromatography of pesticides: A review. *Journal of AOAC International, 86,* 602–611.
110. Sherma, J., (2000). Thin-layer chromatography in food and agricultural analysis. *Journal of Chromatography A, 880,* 129–147.
111. Sniegocki, T., Posyniak, A., Gbylik-Sikorska, M., & Zmudzkia, J., (2014). Determination of chloramphenicol in milk using a QuEChERS-based on liquid chromatography-tandem mass spectrometry method. *Analytical Letters, 47*(4), 568–578.
112. Stefanelli, P., Santilio, A., Cataldi, L., & Dommarco, R., (2009). Multiresidue analysis of organochlorine and pyrethroid pesticides in ground beef meat by gas chromatography-mass spectrometry. *Journal of Environmental Science and Health B, 44,* 350–356.
113. Surendranath, B., & Unnikrishnan, V., (1998). Organochlorine pesticides in animal tissues and their excretion through milk. *Journal of Food Science and Technology, 35,* 547–548.
114. Synaridou, M. S., Sakkas, V. A., Stalikas, C. D., & Albanis, T., (2014). Evaluation of magnetic nanoparticles to serve as solid-phase extraction sorbents for the determination of endocrine disruptors in milk samples by gas chromatography-mass spectrometry. *Journal of Chromatography A, 1348,* 71–79.
115. Tian, H., (2011). Determination of chloramphenicol, enrofloxacin and 29 pesticides residues in bovine milk by liquid chromatography-tandem mass spectrometry. *Chemosphere, 83,* 349–355.
116. Tolentino, R. G., Leon, S. V., & Beatriz, S. B., (2014). Organochlorine pesticides in infant milk formulas marketed in the South of Mexico City. *Food Nutrition Sciences, 5*(13), 1–9.
117. Tsakiris, I. N., & Goumenou, M., (2015). Risk assessment for children exposed to DDT residues in various milk types from the Greek market. *Food Chemistry and Toxicology, 75,* 156–165.
118. Tsydenova, O. V., & Sudaryanto, A., (2007). Organohalogen compounds in human breast milk from the republic of Buryatia, Russia. *Environmental Pollution, 146,* 225–232.
119. United Nations Environmental Program, (2004). *Master List of Actions on the Reduction and/or Elimination of the Releases of Persistent Organic Pollutants* (5th edn., p. 195). Stockholm Convention. https://wedocs.unep.org/rest/bitstreams/13329/retrieve (accessed on 22 September 2021).
120. Vetter, W., Weichbrodt, M., Hummert, K., Glotz, D., & Luckas, B., (1998). Combined microwave-assisted extraction and gel permeation chromatography for the determination of chlorinated hydrocarbons in seal blubber and cod livers. *Chemosphere, 37,* 2439–2449.
121. Vidal, J. L. M., Plaza-Bolaños, P., Romero-González, R., & Frenich, A. G., (2009). Determination of pesticide transformation products: A review of extraction and detection methods. *Journal of Chromatography A, 1216,* 6767–6788.
122. Wang, W. T., Meng, B. J., Lu, X. X., Liu, Y., & Tao, S., (2007). Extraction of polycyclic aromatic hydrocarbons and organochlorine pesticides from soils: A comparison between Soxhlet extraction, microwave-assisted extraction and accelerated solvent extraction techniques. *Analytica Chimica Acta, 602,* 211–222.

123. Wang, X., & Li, P., (2015). Rapid screening of mycotoxins in liquid milk and milk powder by automated size-exclusion SPE-UPLC–MS/MS and quantification of matrix effects over the whole chromatographic run. *Food Chemistry, 173,* 897–904.
124. Weichbrodt, M., Vetter, W., & Luckas, B., (2000). Microwave-assisted extraction and accelerated solvent extraction with ethyl acetate cyclohexane before determination of organochlorines in fish tissue by gas chromatography with electron-capture detection. *Journal of AOAC International, 83,* 1334–1343.
125. Wesseling, C. M., & Keifer, M., (2002). Long-term neurobehavioral effects of mild poisonings with organophosphate and n-methyl carbamate pesticides among banana workers. *International Journal of Occupational Medicine and Environmental Health, 8,* 27–34.
126. Wielogorska, E., Elliott, C. T., Danaher, M., Chevallier, O., & Connolly, L., (2015). Validation of an ultra-high performance liquid chromatography-tandem mass spectrometry method for detection and quantitation of 19 endocrine disruptors in milk. *Food Control, 48,* 48–55.
127. Wilkowska, A., & Biziuk, M., (2011). Determination of pesticide residues in food matrices using the QuEChERS methodology. *Food Chemistry, 125*(3), 803–812.
128. Xu, J. J., An, M., Yang, R., Tan, Z., Hao, J., Cao, J., Peng, L. Q., & Cao, W., (2016). Determination of tetracycline antibiotic residues in honey and milk by miniaturized solid-phase extraction using chitosan-modified graphitized multiwalled carbon nanotubes. *Journal of Agricultural and Food Chemistry, 64*(12), 2647–2654.
129. Yague, C., Bayarri, S., Lazaro, R., Conchello, P., Arino, A., & Herrera, A., (2001). Degradation of organochlorine contaminants by yoghurt culture organisms and the effect on their growth. *Bulletin of Environmental Contamination and Toxicology, 66,* 29–35.
130. Yang, A., El-Aty, A. M. A., & Park, J., (2014). Analysis of 10 systemic pesticide residues in various baby foods using liquid chromatography-tandem mass spectrometry. *Biomedical Chromatography, 28*(6), 735–741.
131. Zhang, W., Duan, C., & Wang, M., (2011). Analysis of seven sulphonamides in milk by cloud point extraction and high-performance liquid chromatography. *Food Chemistry, 126,* 779–785.
132. Zheng, G., Han, C., Liu, Y., Wang, J., Zhu, M., Wang, C., & Shen, Y., (2014). Multiresidue analysis of 30 organochlorine pesticides in milk and milk powder by gel permeation chromatography-solid phase extraction-gas chromatography-tandem mass spectrometry. *Journal of Dairy Science, 97*(10), 6016–6026.
133. Zhou, P., Wu, Y., & Yin, S., (2011). National survey of the levels of persistent organochlorine pesticides in the breast milk of mothers in China. *Environmental Pollution, 159,* 524–531.
134. Zhou, X., & Zhao, X., (2015). Susceptibility of nine organophosphorus pesticides in skimmed milk towards inoculated lactic acid bacteria and yogurt starters. *Journal of the Science of Food and Agriculture, 95*(2), 260–266.
135. Zwir-Ferenc, A. Z., & Biziuk, M., (2004). An analysis of pesticides and polychlorinated biphenyls in biological samples and foods. *Critical Reviews in Analytical Chemistry, 34,* 95–103.

PART III

Food Safety Hazards: Practices and Management

CHAPTER 10

FOOD SAFETY MANAGEMENT SYSTEMS IN THE FOOD PROCESSING INDUSTRY

NANCY AWASTI, VENKATESWARLU SUNKESULA, and
AHMED R. A. HAMMAM

ABSTRACT

Food safety management systems (FSMS) and regulations in the food and dairy sectors have been established with several revisions with a purpose to reduce risks and contamination hazards and concerns on food safety. These past and present situations have raised many questions to the regulators on the efficiency and effectiveness of these mechanisms to prevent food safety hazards. This chapter focuses on the current FSMS and practices to ensure food safety. The chapter also provides an advanced understanding on efficient implementation of food safety systems.

10.1 INTRODUCTION

Food safety is an essential global public health issue for all countries. According to Havelaar et al. [12], about 8% of the entire world's population is affected annually by foodborne diseases resulting in 420,000 deaths, out of which 92,000 are children [44]. Unsafe food containing harmful viruses, bacteria, chemicals, parasites, or chemical compounds can cause >200 different health issues, namely, diarrhea as a simple symptom and different types of cancer as critical issues. The impact on global public health is not compatible with the maximum burden of illness episodes on the South East Asian and African regions [21]. Numerous factors resulting in foodborne outbreaks have been reported by Wu et al. [46], such as cross-contamination, improper storage of raw or prepared foods, contaminated additives or ingredients, and inadequate cooking.

Yan et al. [47] reported *Listeria monocytogenes* outbreak in several food products, and the existence of *L. monocytogenes* was identified in frozen foods made of rice and wheat products, cooked meat, raw meat, seafoods, and other non-fermented food products. Also, foodborne illness was caused by ingestion of botulinum toxin produced by *Clostridium botulinum* due to storage of contaminated sausages at room temperature for an extended period [49]. Kendal et al. [20] reported improper storage of raw and finished food products as one of the significant factors responsible for foodborne illness. Different factors can result in contamination of food from farm to table.

Wheeler et al. [43] reported contaminated raw green onions in salsa preparation due to foodborne illness (hepatitis A virus). In addition to this, bad hygienic practices and poor food handling were also determined to cause the Nevada outbreak in the United States of America, where infected individuals were found positive for *enterotoxigenic Escherichia coli* due to ingestion of contaminated sushi in 2004 [18]. Microbial evaluation of >50 restaurants (1994) in Seattle showed that foodborne disease outbreaks have resulted from poor hygienic practices and the use of raw fish and pre-prepared rice [24].

Wang et al. [42] reviewed 2447 published papers on occurrence of foodborne outbreaks between 1994 to 2005 in China. They reported that 1082 outbreaks were caused due to bacterial pathogens. Recently, the world has faced over 14 million globally confirmed cases of novel coronavirus (COVID-19) with approximately half a million deaths reported (on July 20th, 2020) [44], and this virus is known to be originated from the seafood market in Wuhan, China.

These present and past situations have raised numerous questions and concerns to food safety regulators on the ineffectiveness and inefficiency of food safety management systems (FSMS) tools. Therefore, these persisting foodborne illnesses and outbreaks have been the driving force to develop more stringent measures and policies to ensure food safety. Foodborne diseases and food safety hazards are preventable by the implementation of food safety systems and techniques, such as, ISO:22000 and ISO:9001 that were developed by the International Organization for Standardization (ISO), HACCP, risk-based analysis, and other FSMS. The number of outbreaks and foodborne illness cases increases worldwide, food safety issues and benefits of implementing FSMS have gained more importance. Overall, improving existing food safety systems is mandatory for regulators and health authorities to decrease the load of foodborne outbreaks of diseases.

The main aim of this chapter is to identify implementation methodology for FSMS systems and to review the challenges and benefits of these implementations to produce safe food.

10.2 FOOD SAFETY MANAGEMENT SYSTEMS (FSMS)

The aim of FSMS is to detect and eradicate chemical, physical, and microbiological contamination from farm to table, thus ensuring the safety of foods for consumption [16]. Implementing FSMS by any food organization is a planned decision that can support or help them to improve overall food safety performance. The potential benefits of FSMS implementation to any organization are: it increases their ability to continually provide safe food services and products and it keeps them in compliance with all required standards. Besides this, FSMS implementation also addresses food safety risks and non-conformities associated with each step of food processing.

Food safety is associated with the existence of food safety hazards from the time of raw food handling to its consumption. Thus to prevent the occurrence of food safety hazards, adequate controls are essential throughout the food chain. The current chapter elaborates an overview of FSMS and its requirements with recognized vital elements, such as HACCP, ISO 22000, good hygienic practices (GHP), and good manufacturing practices (GMPs), etc.

With over 150 years of advances in molecular biology and microbiological techniques, there is a rise in the number of reports on foodborne illnesses that indicate lack of safety of food [38]. Several reasons resulting in illnesses and outbreaks include [11]: the change in shopping and cooking practices, food consumption patterns, farming practices, increase in gaps between production and consumption of food, and most significant evolution of novel and exiting pathogens [11].

The US Food and Drug Administration (FDA) has endorsed that all regulatory agencies should focus on reducing the occurrence of factors (such as: temperature abuse, unsafe source of received raw materials, poor hygienic practices, sanitation programs, improper usage of chemicals, pest control, contaminated equipment and inadequate cooking) that are responsible for causing foodborne outbreaks [28]. Several studies have been carried out to understand the implementation and certification of FSMS, and most of them were based on principles of HACCP [26, 41].

10.3 THE ISO 22000: AN OVERVIEW OF REVISED VERSION

The ISO is a global federation of national standards bodies or ISO member bodies. ISO developed ISO-22000 in September of2015, "2005 an auditable standard that meets the FSMS requirements based on prerequisite programs (PRPs), HACCP, interactive communication, and system management." ISO-22000–2005 applies to all food processing organizations and includes the criteria explained in ISO-9001.

Recently, the revised version of ISO-22000 has been released and it is referred to as ISO 22000–2018. In addition to the previous version (ISO 22000:2005), the revised version includes internal and external issues concerning process and organization. The ISO 22000–2018 [5] provides a high-level structure (HLS) and harmonization with FSMS; contains the requirements explained by quality management-ISO 9001, environment-ISO 14001, food safety-ISO-22000, and occupational and health safety-ISI-18001.

According to the new process of ISO 22000–2018 (based on the CODEX HACCP), the risk (which was initially evaluated by means of the possibilities and severity) could be transformed into an opportunity after the risk evaluation. Based on organization requirements and choices, ISO-22000 can also be combined with other food safety systems, for example, ISO 14001, ISO 9001, and OHSAS 18001. ISO has developed guidelines for the text and structure to ensure that significant development in management system standards can be easily considered and integrated.

Additional guidance on ISO-22000 is available on ISO website (https://www.iso.org/standard.html) that includes guidance on the application of various food safety-related standards, such as, ISO/TS 22004–2014 (FSMS-Guidance on the implementation of ISO 22000–2005) is a guide on the implementation of ISO 22000-2005.

ISO/TS 22002–2009 is the guidance of PRPs on food safety for food manufacturing, whereas ISO 22005–2007 provides the guidelines on traceability in food and feed chains.

ISO-22000 supports the HACCP system in several ways; and therefore requires policies, planning, implementation, action, assessment of performance, analysis by management, and improvement. It restructures the traditional idea and concepts of distributing control measures into critical control points (CCPs) and GHP. Therefore, ISO covers all control measures and addresses them by dividing them into three categories: (1). PRPs: To manage all essential activities and conditions such as maintaining processing environment, hygienic production, handling environment; (2). OPRPs (operational prerequisite programs) control the control measures that are not controlled by CCPs and identified as significant hazards by hazard analysis; and (3). the HACCP plan to control and manage all identified critical hazards to an acceptable level by generating critical limits and applying CCPs.

ISO-22000 also describes the way to associate the HACCP design with OPRPs and PRPs into a single integrated hazard control and food safety system. CODEX-HACCP emphasizes an internal food safety system, whereas ISO-22000 encourages the acceptance of a food chain system to control food-associated hazards.

The overall objective is to improve the integration and alignment with other ISO management systems by using the risk-based thinking, process approach, and plan-do-check-act (PDCA) cycle. The process approach is based on designing and implementing FSMS and improving its effectiveness to enhance the production of food services by following appropriate requirements. It also helps to understand and to manage interrelated processes to accomplish projected results in accordance with policies of food safety.

The management of the process and the overall system could be achieved by using the PDCA cycle with significant attention on risk-based thinking (targeting at taking advantage of opportunities and avoiding unwanted outcomes). With few changes and addition of several clauses, ISO 22000–2018 also includes PDCA cycle and risk-based approach at organizational and at process level. The process approach using the concept of PDCA is the key change in ISO 22000–2018 covering risks and opportunities. It includes two separate PDCA cycles that work together with a communication system as an essential link between operational processes (HACCP principles) and FSMS.

10.3.1 PDCA CYCLE

PDCA is elaborated as a Plan-Do-Check-Act. The PDCA cycle (Figure 10.1) allows an organization to confirm that its processes are effectively managed and resourced, and prospects for improvement are determined and acted on. The concept of PDCA in ISO 22000–2018 has been introduced to combine it with HACCP and to support the FSMS. Explanation below covers the cycle for organizational planning and control, operational planning and control:

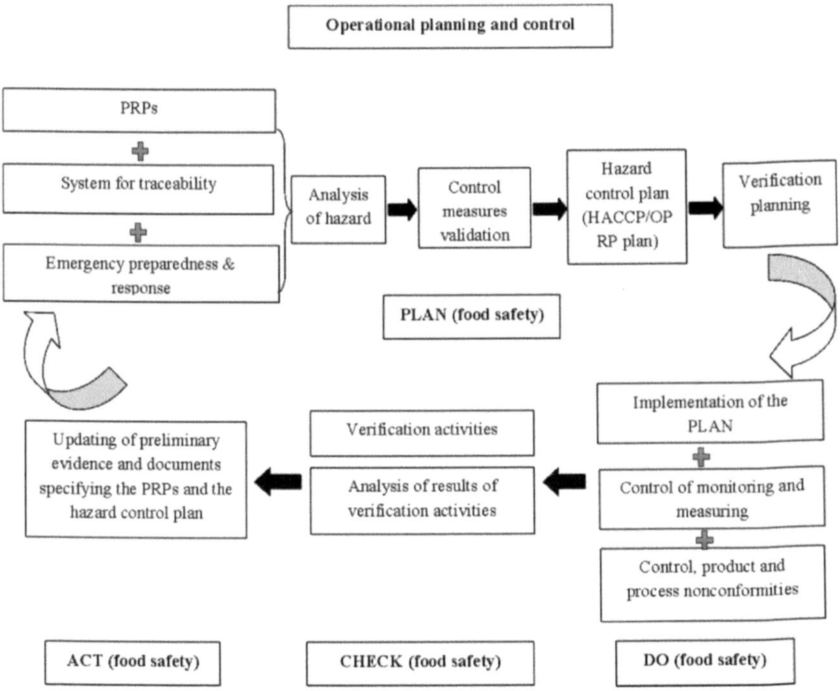

FIGURE 10.1 PDCA (Plan-do-check-act) cycle at the operational level.

1. **Plan:** This focuses on establishing the objectives of the system and providing the resources required to deliver the results and identifying the opportunities and risks associated with the process.
2. **Do:** It includes the implementation of the system planned.
3. **PDCA:** This part of the PDCA cycle integrates a second PDCA cycle implemented at the operational level, thus referred under

the category of operational planning and control. Explains and covers the requirement of PRPs, effective traceability systems, emergency preparedness and response, documentation, verification, and integration of HACCP within ISO 22000–2018. Flow chart representing a second PDCA cycle is shown in Figure 10.1.

4. **Check:** This step monitors each stage of the process, resulting in products and services. It also analyzes and evaluates the data and information from monitoring, measures, and verifies the process activities and preparation of results report.
5. **Act:** during any non-conformity, this step elaborates on necessary actions taken to improve the results.

10.3.2 RISK-BASED THINKING

This step helps in identifying the factors that can result in a deviation of processes and FSMS from the planned results and can help to minimize or to prevent adverse effects by implementing adequate controls. Like PDCA cycle, risk-based thinking is addressed at two levels: organizational and operational level. As an example, during any change in statutory, legal regulations and standards than the risk must be assessed at the corporate level. One can refer to ISO 22000–2018 for more details [5].

10.4 HAZARD ANALYSIS CRITICAL CONTROL POINT (HACCP)

The HACCP system was developed in 1960 as a preventive approach in controlling and monitoring food processing procedures. Globally, HACCP is referred to as the utmost efficient management system, in which food safety issues are processed and ensured by evaluation and control of physical, chemical, and biological hazards in all segments of the food processing chain from receiving to distribution. HACCP was first presented in 1971 to FDA, and was proposed in 1999 as a procedure to assure the food safety in restaurants [2].

In addition to health hazards, biological contamination caused by illness also results in financial losses to consumers. Therefore, these hazards should be controlled and prevented in food processing facilities. To ensure food safety, the execution of the HACCP plan is significant for all food organizations that deal with processing, production, and catering of food to consumers. HACCP plan (Figure 10.2) ensures to prevent and to control the entry of hazards in all stages of food handling process. Initially, the HACCP plan had three main principles [7].

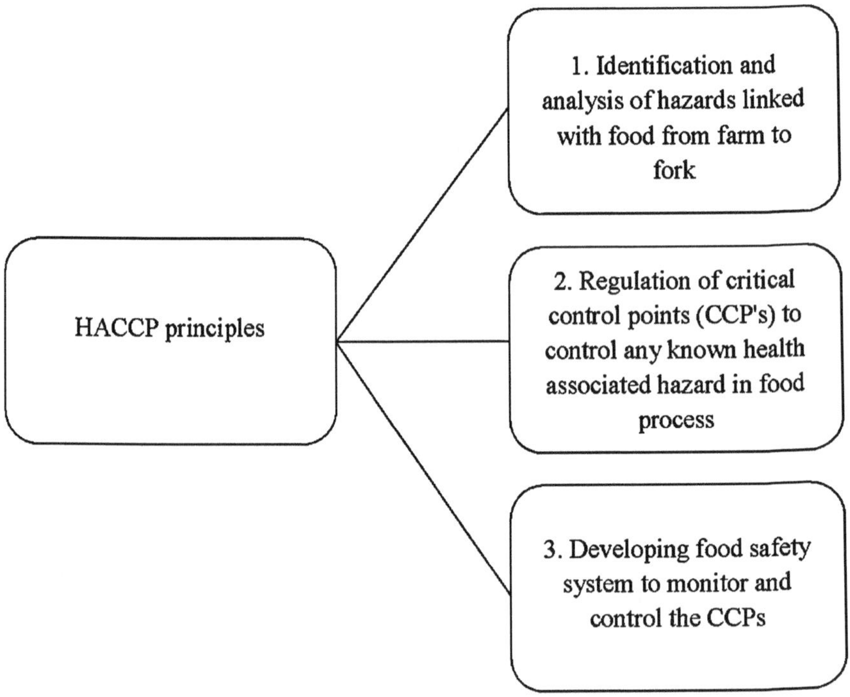

FIGURE 10.2 The initially developed HACCP system with three principles.
Source: Ref. [10].

By the end of 1980s, several food safety companies started to offer training for development and implementation of HACCP as a food safety management system and finally in 1998, HACCP certification was a mandatory for all food organizations with an exception to the smaller food business, due to their financial incapability to implement HACCP plan [38]. In the early 1990s, the HACCP guidance standards were revised, and five preliminary steps were added to previous principles of the HACCP system [31].

The revised HACCP guidance is an internationally recognized system for domestic and international food production for reducing hazards risk in foods. The current HACCP guidelines include seven principles that cover food safety at each step in the food handling process ranging from receiving, storage, processing, and distribution of raw food products to final products. HACCP is based on scientific investigation and control of physical, chemical, and biological hazards [10]. Figure 10.3 represents the current seven essential principles of HACCP systems that are implemented internationally.

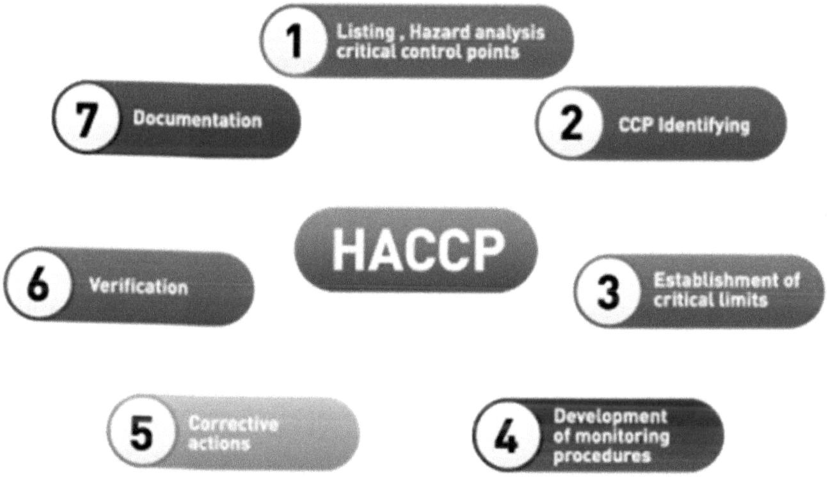

FIGURE 10.3 Seven principles of HACCP to ensure the food safety.
Source: Ref. [10].

In the HACCP system, the identification of potentially hazardous food (PHF) is essential as defined by Food Safety Organization (FSO). PHF is known to support the growth of microbes and their toxins if proper control measures are not implemented properly. FSO and FDA food code classify PHF as the food that involves time-temperature monitoring and control to prevent the risk of foodborne illness. According to the FDA, most of the food and food products that have water activity and pH of 0.85 and 4.6 or above falls under the category of PHF [9]. Therefore, the HACCP plan should be appropriately designed and implemented after a complete examination of PHF, CCPs, and critical limits in a specific food processing facility.

This section briefly describes the existing seven principles of the HACCP system that are established and implemented in most of the food organizations.

- **Principle 1:** This includes identifying hazards by performing hazard and risk analysis. Hazards are identified and analyzed by the HACCP team in each step of the food processing. After identification and analysis of hazards, methods are developed that can be integrated into the HACCP system to eliminate, control, or prevent these known hazards in the food processing facility.
- **Principle 2:** This includes the identification of CCPs and a control technique that can be used in each step of the food processing to either reduce the hazard to an acceptable level or eliminate it from the final product. In a food processing facility, CCPs are identified using a standard CCPs decision tree method, and the number of CCPs depends on food type, associated hazards, and controls needed to prevent these hazards to ensure the safety of food. Figure 10.4 provides an illustration of the CCPs decision tree for a single food processing step. In some cases, one or more steps in the food process or food safety hazard can be controlled by only one CCP whereas, in other cases, more than one CCP may require controlling a single hazard. In other words, the number of CCPs required for any food organization setup depends on numerous factors, for example, type, and nature of food being handled and controls required to guarantee the safety of food [10, 15]. CCP decision tree involves a sequence of questions with answers, yes or no. This process of questions and answers assist in deciding on setting up a CCP.
- **Principle 3:** This includes establishing a critical limit for a process. After identifying a CCP for a step-in food process, the critical limit is established where the physical, biological, and chemical hazards can be controlled. The critical limit can be time, pH, temperature, water activity, moisture, or any additional technique that is acceptable by food safety governing standards. As an example, for the cooking step, there is a minimum temperature of cooking that is required as a critical limit to reduce or eliminate microbial contamination if present.
- **Principle 4:** It is based on the implementation of a continuous monitoring system where each food process step is monitored by measuring already established critical limits. For this step, CCPs must describe the frequency, time, method of measurement, and person responsible for its measurement during the food production process.

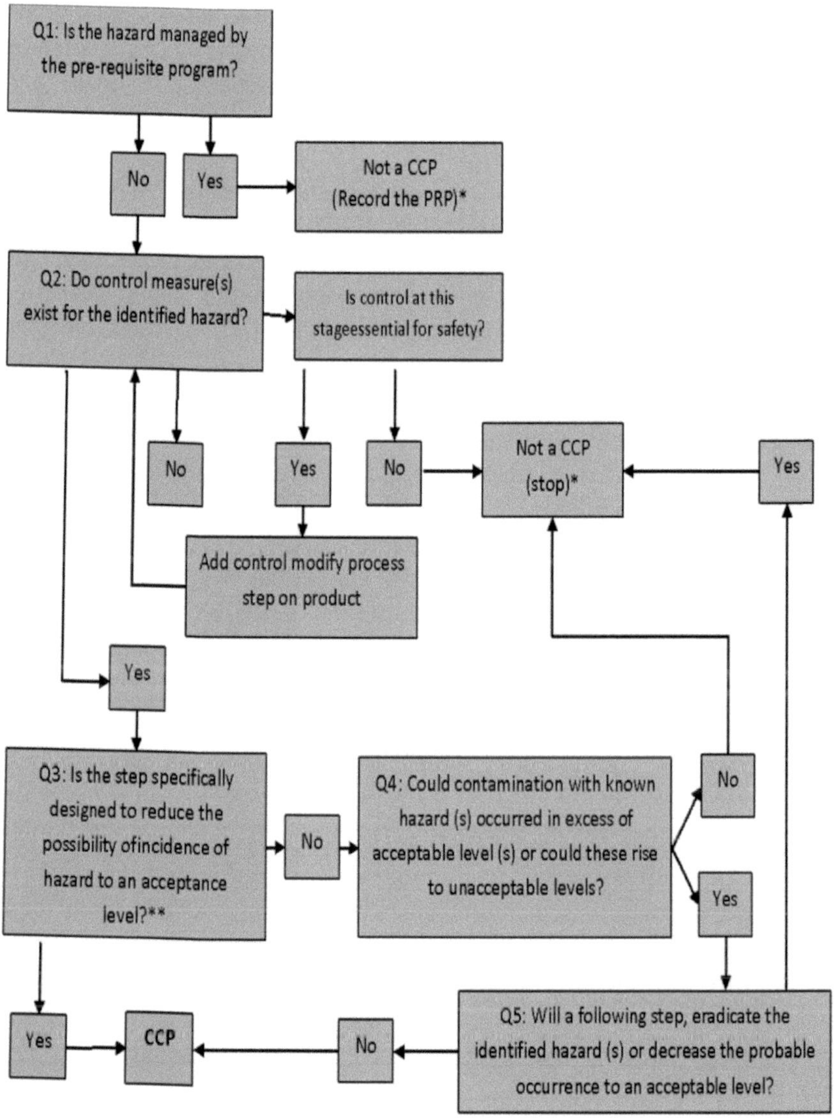

FIGURE 10.4 CCPs decision Tree for a one-step process [*continue to next identified hazard in the process; **ensure these levels are defined within the HACCP plan].

Source: Ref. [10].

- **Principle 5:** According to the 5th principle, during any process deviations from the already established critical limit, corrective action must be implemented. Therefore, to produce safe food, appropriate corrective action is important during an event of CCP deviation. A corrective action specifically includes identification of issues and taking a corrective step to ensure that the problem never happens again.

- **Principle 6:** The 6th principle explains the methods and procedures involved in the verification of the HACCP system. The verification step determines the validity of the HACCP system to understand if the process is functioning as described by the plan and verified by instrument calibration such as calibration of thermometers, machines, the temperature of coolers and freezers, CCPs auditing, microbial testing of food and record review. Therefore, individuals handling these verification steps must have specific technical expertise which is either performed by HACCP professionals from corporate, regulatory agencies, or a third party.
- **Principle 7:** The seventh principle of HACCP includes documentation and record-keeping. This step is an essential step of the HACCP plan that includes information on records and documentation practices to verify the food safety produced for consumption. Complete documentation should have information on product description, hazard analysis, identified CCPs, flow diagrams, monitoring systems involved, critical limits, management, and employee training on HACCP, verification, and record-keeping procedures.

For a successful implementation of HACCP, continuous monitoring by the HACCP team is required, which includes inspection and audit. Mostly both these terms are used interchangeably, but they both have different goals and procedures. The key objective of the inspection is to deliver answers to questions "is it clean?" [3] and these questions are mainly pointed on the hygiene of the workplace and environment where food is processed or handled. The inspection also involves temperature and time monitoring of food.

In addition to this, the audit process is focused on assuring that the HACCP system is being followed properly and making sure that the entire process is examined and recorded properly. HACCP audit is a time taking process and emphasizes more on the documentation related to preparation, storage, handling, service, packaging, and distribution, with a focus on specific probable danger points [3].

Finally, it should be noted that execution of HACCP as a standalone process cannot ensure the complete safety of food and must be built with other food safety interventions, such as: Sanitation Standard Operation Procedures (SSOPs) or GHP and GMP.

10.5 CURRENT GOOD MANUFACTURING PRACTICES (CGMPS)

The current good manufacturing practices (CGMPs) are regulated by the US FDA to examine the design, control, manufacturing maintenance, and facility operations in the United States (US). The current term is referred to as the status of companies that maintain the latest technology. The CGMPs are applied to food companies, pharmaceutical, and biotechnology companies to assure the quality of the products before being reached to the consumers. The companies will pay penalties if they do not follow the FDA regulations. The GMP includes a set of requirements at the operational level to report food quality and safety issues, required for successful development and implementation of the HACCP plan. CGMP guidelines provide regulation for sanitation protocols, quality assurance (QA), testing protocols, and food manufacturing processes to support in eliminating or reducing food associated risk causing illness or outbreaks. They also provide guidelines on secure production, handling, and distribution of safe food for human consumption with acceptable quality.

CGMP includes following standard principles [15]:

- Maintenance of sanitation and cleanliness of manufacturing workspace (tables, floor, equipment's, cutting boards, knives, manufacturing tools, etc.);
- Prevention of cross-contamination between different food products by controlling environmental conditions;
- Clearly define, develop, and control of manufacturing process;
- CCPs validation to confirm the safety of end food products, the detailed and clear written format of production instructions;
- Recordkeeping, good documentation practices;
- Developing protocols and procedures to minimize the contamination risk during shipping and distribution; and
- To establish procedure or system for quick food recall or market withdrawal in case of any illness or unsafe food distribution.

10.6 SANITATION STANDARD OPERATION PROCEDURES (SSOPS)

To ensure suitable cleaning and sanitation of both non-food contact and food contact areas, SSOPs should be followed continuously. It covers the standard sanitation procedure of raw materials, personal hygiene, sanitation of equipment, food product handler practices, and work surface sanitation [33]. The requirement of a food HACCP plan for SSOPs includes proper documentation and periodic review of records by responsible and professional management. As per SSOPs rules, raw food resources must be accepted from certified and approved suppliers or sources with adequate standard conditions. SSOPs provides guidelines for the various process as follows, guidelines for storage of sensitive and hazardous raw materials, hair restraints and gloves, certification, and permit requirements for inspection, employee health, and food safety, employee cleanliness including hand wash procedures and requirements, chemical selection and sanitation protocols, sanitation, and maintenance of equipment's, periodic sanitation of workspaces and utensils and verification protocols.

10.7 GOOD HYGIENIC PRACTICES (GHP)

The GHP usually applies to the live animals and their products, and it is crucial to several organizations, namely Food and Agriculture Organization (FAO) and the World Health Organization (WHO), as well as the World Organization for Animal Health (OIE). GHP is determining and renewing the food regulations by following those organizations. FAO, WHO, and OIE legalized new rules on foods (such as meat and poultry) because they are mainly consumed by many people around the globe. The objective of these new rules is to protect the consumer from many foodborne diseases or toxins by surveillance and to screen the products from different aspects, including biological risks. The role of GHP is not just examining the risks in the product but also determining the possibility of preventing the risk from reaching to the foods.

Recently, the hygiene and safety practices of small and large dairy farms were assessed and it was found that poor to basic hygienic practices were one of the major reasons for causing bacterial and chemical contamination of fresh milk at the farm level [23]. Based on the responses and evaluation in comparison to big farms, small scale dairy farmers were carrying out poor to the basic level of food safety practices and thus demonstrated the need for the improvement of training plans for employees.

10.8 FOOD SAFETY TRAINING

Food safety training plays a crucial role in the successful implementation of any food safety program, and it is one of the effective methods to control or avoid the foodborne outbreak. Several studies have supported the positive influence of food safety training to employees and management in control of food safety issues. Untrained or inadequately trained staff mainly contribute to bad hygienic and handling practices, food safety ignorance, bad manufacturing practices, and others. Previous studies [14, 45] reported that about 97% of cases of foodborne illness or poisoning were due to inadequate training of staff. Reduction in microbial counts of food was reported by [27] before and after food safety training. Therefore, these previous studies support the importance of training on food safety and education in improving the knowledge and awareness among staff.

10.9 RISK PERCEPTION, CHALLENGES, AND BENEFITS OF FSMS IMPLEMENTATION

Presently, all international standards for food safety have disadvantages and advantages. For first-rate food safety measures, the major focus of any food organization is to implement FSMS effectively. The effectiveness of any food safety system audit depends on several factors such as criteria used and the skills and ability of the auditor. Internal and external audits assist as a crucial means to identify food safety gaps and ensure that food safety procedures and systems are effective and compliant. When it is about the safety of food, the safety of the consumer is supreme. Therefore, it becomes important for the regulators to assure food safety by conducting food safety assessments and audits without compromising consumer safety. The implementation of any FSMS comes with challenges and benefits that are detailed below in this section. But before understanding the challenges and benefits associated with FSMS implementation, it is vital to know the meaning of risk perception by consumers.

10.9.1 PERCEPTION OF RISK BY THE CONSUMER

Risk perception is an essential factor that helps understand the potential influence it has on consumers. Risk perception includes two dimensions, first, the occurrence probability of an event during a specific period of duration (how often foodborne outbreak or illness occurs) and secondly, the extent or seriousness of outbreak consequences [4]. Yeung and Morris [46] have defined risk perception as "*the uncertainty that clients come across as soon as they cannot predict the consequences of their buying decisions.*" Perception of risk by consumers comprises time loss, financial, and physical loss which usually affects the willingness of consumers to buy any food product [29, 30].

10.9.2 CHALLENGES

In comparison to large food facilities, HACCP application is less common in small food organizations. A previous study [36] identified that most of the time, the HACCP structure is not implemented effectively in small food facilities. Taylor [40] determined the problems associated with HACCP system implementation in small food businesses. Especially for small food facilities, the adoption of the f FSMS such as HACCP is difficult and requires different expertise, mindsets, money, documentation handling, knowledge, skills, training, and time. A previous study [13] indicated that most of the foodborne outbreak transmission occurs through restaurants, which usually have either weak or no HACCP implementation.

A comprehensive review by Lee et al. [13] has covered the historical and safety perspective of popular ethnic foods in the US. They also focused on various food safety systems, implementations, challenges, training, and other useful techniques for the prevention of outbreaks associated with food. Simonne et al. [35] indicated that lack of food knowledge among employees, cultural issues, and unfamiliarity with ingredients are a few challenges that can steer towards the cause of the foodborne outbreak.

Moreover, several studies have pointed out that insufficient guidance and support [22], lack of knowledge, experience, human, and financial resources, information [34], employee, and management commitment [1], and personnel time restrictions sum up to discouragement of FSMS implementation. Therefore, continuous monitoring and regulation are required for successful implementation of all food safety systems for producing safe food.

10.9.3 BENEFITS

If the above-mentioned challenges and difficulties are conquered, then the HACCP implementation can benefit food facilities in different ways. Numerous studies have reported a positive impact of HACCP implementation on food safety from small food facilities to airline catering industries [19, 37]. A previous study [40] listed several benefits to food facilities after successful implementation of HACCP such as reduction of cost, reduced wastage, increased confidence and provides legal defense during any foodborne outbreak situation and finally, it also helps organizations to adopt new circumstances during any in regulations or process.

10.10 TECHNIQUES TO IMPROVE EXISTING ASSESSMENT TOOLS

Food processing operators and organizations are challenged at every step to use the combination of requirements from different stakeholders such as retails, government agencies, consumer demands, etc., to improve FSMS. This section of the chapter focuses on the selection and analysis of tools that can help in improving the performance of FSMS.

Based on European food safety standards, a previous study [17] presented tools to assess and improve the FSMS. The presented tools were validated in the European meat and dairy food processing industries. They also demonstrated the usefulness of online easily accessible FSMS support tools that can be utilized for the internal auditing process (www.pathogencombat.com). The study by Jacxsens et al. [17] has elaborated on the definition of three tools (Pathogen Combat as Diagnostic, selection, and improvement) that were applied in the European meat and dairy food chain to improve food safety.

Among the three tools used, the diagnostic tool enabled systematic assessment and analysis of a detailed list of indicators for analysis of controls and QA activities that are addressed in the company's food safety system. These control activities are aimed at preventing growth and contamination by microbes during the food production process and include various interventions to reduce the level of pathogens. Whereas, quality control activities aim at assuring food safety and include disinfection, cleaning, CCP analysis, and other intervention steps such as filtration, pasteurization, etc.

The selection tool involving the microbial assessment scheme and QA evaluation grid were evaluated and elaborated by Jacxsens et al. [17] to understand its purpose in supporting FSMS. It helped in assessing the microbiological performance of implemented FSMS by opting for globally standardized sampling and analysis methods such as ISO QA evaluation grid used in the above-mentioned study can assist food industries as a decision-making tool when selecting QA guidelines and standards. It mainly supports small and medium-sized food companies in balancing the efforts against the benefits of implementing QA guidelines and standards. Finally, improvement tools involved mapping the problem area, identifying the effects and causes of problems, and redesigning and implementing solutions. An example illustrating an improvement in sanitation practices at the plant level was mapped using road map technique. For more details on emerging FSMS tools, the reader is referred to the previous chapter [17].

Many studies pointed to insufficiencies in the performance of hygiene and safety control practices of food and dairy chains [25, 39]. A recent study [23] evaluated hygiene and safety control practices of small and large dairy farms by using a customized tool. The study aimed to develop a customized tool in emerging dairy chains, evaluate on-farm practices, and monitor the chemical and microbial safety of fresh milk. Based on the responses and evaluation, they reported that in comparison to big farms, small scale dairy farmers were carrying out poor to a basic level of food safety practices and thus demonstrated the need for the advance of training agendas for employees.

10.11 SUMMARY

Food safety remains a major challenge and a public health concern. Several factors negatively impacting the safety of food at organizational and operational levels must be evaluated thoroughly with continuous monitoring of FSMS for producing safe foods. Even though different forms of food safety systems are implemented globally to control such challenging situations, yet there is a need for upgraded and stringent food safety systems. FSMS of a specific company must be seen as a dynamic system, which needs to be upgraded and changed with respect to requirements and evolving food safety scenarios. Therefore, depending on a specific company's requirements and implemented system, identification, and analysis of new food safety tools can empower FSMS to control and prevent food-related issues or outbreaks.

KEYWORDS

- **food safety management systems**
- **food safety systems**
- **foodborne illness**
- **good manufacturing practices**
- **hazard analysis critical control point**
- **plan-do-check-act**

REFERENCES

1. Aggelogiannopoulos, D., Drosinos, E. H., & Athanasopoulos, P., (2007). Implementation of a quality management system (QMS) according to the ISO 9000 family in a Greek small-sized winery: A case study. *Food Control, 18*(9), 1077–1085.
2. Allwood, P. B., Lee, P., & Borden-Glass, P., (1999). The vital role of restaurant health inspections. *Journal of Environmental Health, 61*(9), 25–29.
3. Boehnke, R. H., & Graham, C., (2000). *International Survey on Public Posting of Restaurant Inspection Reports, and/or Grade Card Posting Schemes Based upon Health Inspections* (p. 68). Region of Ottawa-Carleton Health Department. http://citeseerx.ist.psu.edu/viewdoc/download?doi=10.1.1.410.3676&rep=rep1&type=pdf (accessed on 22 September 2021).
4. Breakwell, G. M., (2014). *The Psychology of Risk* (2nd edn., p. 395). UK: Cambridge University Press.
5. Chen, H., Liu, S., Chen, Y., Chen, C., Yang, H., & Chen, Y., (2020). Food safety management systems based on ISO 22000: 2018 methodology of hazard analysis compared to ISO 22000: 2005. *Accreditation and Quality Assurance, 25*(1), 23–37.
6. (2007). *Document - ISO 22000: Food Safety Management Systems: Requirements for Any Organization in the Food Chain* (p. 42). https://luk.staff.ugm.ac.id/phk/inherent/UGM-K1-2007/ftp/files/cea/content/ISO_22000_2005_E.pdf (accessed on 22 September 2021).
7. El-Mesery, H. S., Mao, H., & Abomohra, A. E. F., (2019). Applications of non-destructive technologies for agricultural and food products quality inspection. *Sensors, 19*(4), 846–851.
8. FDA (Food and Drug Administration), (2013). *Food Code-Annex 4 Management of Food Safety Practices - Achieving Active Managerial Control of Foodborne Illness*

Risk Factors (pp. 547–586). https://www.fda.gov/media/87140/download (accessed on 22 September 2021).
9. FDA (Food and Drug Administration), (1997). *HACCP Principles & Application Guidelines*. https://www.fda.gov/food/hazard-analysis-critical-control-point-haccp/haccp-principles-application-guidelines (accessed on 22 September 2021).
10. Gehring, K. B., & Kirkpatrick, R., (2020). Hazard analysis and critical control points (HACCP). Chapter 8. In: Demirci, A., Feng, H., & Krishnamurthy, K., (eds.), *Food Safety Engineering* (pp. 191–204). Switzerland: Springer.
11. Griffith, C. J., (2006). Food safety: Where from and where to? *British Food Journal, 108*(1), 6–15.
12. Havelaar, A. H., & Kirk, M. D., (2015). World health organization (WHO) global estimates and regional comparisons of the burden of foodborne disease in 2010. *PLoS Medicine, 12*(12). Article ID: e1001923.
13. Hedberg, C. W., Smith, S. J., Kirkland, E., Radke, V., Jones, T. F., & Selman, C. A., (2006). Systematic environmental evaluations to identify food safety differences between outbreak and non-outbreak restaurants. *Journal of Food Protection, 69*(11), 2697–2702.
14. Howes, M., McEwen, S., Griffiths, M., & Harris, L., (1996). Food handler certification by home study: Measuring changes in knowledge and behavior. *Dairy, Food and Environmental Sanitation, 16*, 737–744.
15. Ibrahim, O. O., (2020). Introduction to hazard analysis and critical control points (HACCP). *EC Microbiology, 16*, 1–7.
16. ISO, B., (2005). *9000: 2005 Quality Management Systems, Fundamentals and Vocabulary* (p. 58). London: British Standards Institution.
17. Jacxsens, L., & Luning, P. A., (2011). Tools for the performance assessment and improvement of food safety management systems. *Trends in Food Science & Technology, 22*, S80-S89.
18. Jain, S., Chen, L., Dechet, A., Hertz, A. T., Brus, D. L., Hanley, K., & Bopp, C. A., (2008). An outbreak of enterotoxigenic *Escherichia coli* associated with sushi restaurants in Nevada, 2004. *Clinical Infectious Diseases, 47*(1), 1–7.
19. Kang, Y. J., (2000). Safe food handling in airline catering. Chapter 6. In: Farber, J. M., & Todd, E. C. D., (eds.), *Safe Handling of Foods* (pp. 197–233). New York: CRC Press.
20. Kendall, M. E., & Mody, R. K., (2013). Emergence of salsa and guacamole as frequent vehicles of foodborne disease outbreaks in the United States, 1973–2008. *Foodborne Pathogens and Disease, 10*(4), 316–322.
21. Kirk, M. D., & Pires, S. M., (2015). World Health Organization estimates of the global and regional disease burden of 22 foodborne bacterial, protozoal, and viral diseases, 2010: A data synthesis. *PLoS Medicine, 12*(12). Article ID: e1001921.
22. Küpper, G., & Batt, P. J., (2009). Barriers to the adoption of quality assurance systems in the food and beverage sector. *Stewart Post-Harvest Review, 5*(3), 1–5.
23. Ledo, J., Hettinga, K. A., & Luning, P. A., (2020). Customized assessment tool to differentiate safety and hygiene control practices in emerging dairy chains. *Food Control, 111*. Article ID: 107072.
24. Lee, J. H., Hwang, J., & Mustapha, A., (2014). Popular ethnic foods in the United States: A historical and safety perspective. *Comprehensive Reviews in Food Science and Food Safety, 13*(1), 2–17.

25. Lindahl, J. F., Kagera, I. N., & Grace, D., (2018). Aflatoxin M1 levels in different marketed milk products in Nairobi, Kenya. *Mycotoxin Research, 34*(4), 289–295.
26. Macheka, L., & Manditsera, F. A., (2013). Barriers, benefits and motivation factors for the implementation of food safety management system in the food sector in Harare province, Zimbabwe. *Food Control, 34*(1), 126–131.
27. Martínez-Tomé, M., Vera, A. M., & Murcia, M. A., (2000). Improving the control of food production in catering establishments with particular reference to the safety of salads. *Food Control, 11*(6), 437–445.
28. Mauer, W. A., Kaneene, J. B., DeArman, V. T., & Roberts, C. A., (2006). Ethnic-food safety concerns: An online survey of food safety professionals. *Journal of Environmental Health, 68*(10), 32–38.
29. Mitchell, V. W., (1998). Role for consumer risk perceptions in grocery retailing. *British Food Journal, 100*(4), 171–183.
30. Mitra, K., Reiss, M. C., & Capella, L. M., (1999). Examination of perceived risk, information search and behavioral intentions in search, experience and credence services. *Journal of Services Marketing, 13*(3), 208–228.
31. NACMCF (National Advisory Committee on Microbiological Criteria for Foods), (1997). Hazard analysis and critical control point principles and application guidelines. *Journal of Food Protection, 61*, 762–775.
32. Redmond, E. C., & Griffith, C. J., (2003). Consumer food handling in the home: Review of food safety studies. *Journal of Food Protection, 66*(1), 130–161.
33. Sanitation Standard Operating Procedures (SSOPs), (2019). *USDA-FSOS*, p. 40, https://www.fsis.usda.gov/wps/wcm/connect/4cafe6fe-e1a3-4fcf-95ab-bd4846d0a968/13a_IM_SSOP.pdf?MOD=AJPERES (accessed on 22 September 2021).
34. Semos, A., & Kontogeorgos, A., (2007). HACCP implementation in Northern Greece: Food companies perception of costs and benefits. *British Food Journal, 109*(1), 5–19.
35. Simonne, A. H., Nille, A., Evans, K., & Marshall, Jr. M. R., (2004). Ethnic food safety trends in the united states based on CDC foodborne illness data. *Food Protection Trends, 24*(8), 590–604.
36. Soriano, J. M., Rico, H., Molto, J. C., & Manes, J., (2002). Effect of introduction of HACCP on the microbiological quality of some restaurant meals. *Food Control, 13*(4, 5), 253–261.
37. Sun, Y. M., & Ockerman, H. W., (2005). Review of the needs and current applications of hazard analysis and critical control point (HACCP) system in food-service areas. *Food Control, 16*(4), 325–332.
38. Surak, J. G., (2009). *The Evolution of HACCP* (p. 1). Food quality magazine. https://www.foodqualityandsafety.com/article/the-evolution-of-haccp/?singlepage=1 (accessed on 22 September 2021).
39. Swai, E. S., & Schoonman, L., (2011). Microbial quality and associated health risks of raw milk marketed in the Tanga region of Tanzania. *Asian Pacific Journal of Tropical Biomedicine, 1*(3), 217–222.
40. Taylor, E., (2001) HACCP in small companies: Benefit or burden? *Food Control, 12*(4), 217–222.
41. Teixeira, S., & Sampaio, P., (2013). Food safety management system implementation and certification: Survey results. *Total Quality Management & Business Excellence, 24*(3, 4), 275–293.

42. Wang, S., Duan, H., Zhang, W., & Li, J. W., (2007). Analysis of bacterial foodborne disease outbreaks in China between 1994 and 2005. *FEMS Immunology & Medical Microbiology, 51*(1), 8–13.
43. Wheeler, C., Vogt, T. M., & Armstrong, G. L., (2005). Outbreak of hepatitis a associated with green onions. *New England Journal of Medicine, 353*(9), 890–897.
44. World Health Organization, (2020). *Novel Coronavirus (2019-nCoV) Situation Reports.* https://www.who.int/emergencies/diseases/novel-coronavirus-2019/situation-reports (accessed on 22 September 2021).
45. Worsfold, D., & Griffith, C. J., (2003). Survey of food hygiene and safety training in the retail and catering industry. *Nutrition & Food Science, 33*(2), 68–79.
46. Wu, Y. N., Liu, X. M., Chen, Q., Liu, H., Dai, Y., Zhou, Y. J., & Chen, Y., (2018). Surveillance for foodborne disease outbreaks in China, 2003 to 2008. *Food Control, 84*, 382–388.
47. Yan, H., Neogi, S. B., & Mo, Z., (2010). Prevalence and characterization of antimicrobial resistance of foodborne *listeria monocytogenes* isolates in Hebei province of Northern China, 2005–2007. *International Journal of Food Microbiology, 144*(2), 310–316.
48. Yeung, R. M., & Morris, J., (2006). Empirical study of the impact of consumer perceived risk on purchase likelihood: A modeling approach. *International Journal of Consumer Studies, 30*(3), 294–305.
49. Zhang, S., Wang, Y., Qiu, S., & Dong, Y., (2010). Multilocus outbreak of foodborne botulism linked to contaminated sausage in Hebei province, China. *Clinical Infectious Diseases, 51*(3), 322–325.

CHAPTER 11

CURRENT TRENDS AND ADVANCES IN FOOD ALLERGY MANAGEMENT

ROHIT G. SINDHAV, TANMAY HAZRA, and ANKIT J. THESIYA

ABSTRACT

Abnormal immunological reactions have been characterized as allergies due to different foods or food components (such as: poultry products, milk, and milk products, marine products, nuts, and some cereals, etc.). Overall proper food leveling systems for different food composition, dietary consultation, or proper education are the basic preventive measures to restrict the exposure of food allergens to humans.

11.1 INTRODUCTION

The change of food consumption pattern is due to hectic work schedule that may be prone to food allergy to people throughout the world [48]. The immunological adverse reaction to a particular food can because of allergic reaction to the individual person not to whole population. However, food allergy and food intolerance must be distinguished from each other. Allergy, due to consumption of food, is classified into three groups: IgE intermediated, non-IgE intermediated and due to both. IgE causing allergy can occur because of the entrance of certain peptides or protein from food into skin, gut or lining that produce antibodies due to the immune defense mechanism of the body [21].

Scientific studies demonstrated that 90% of the food-induced allergy is mainly because of milk, soy, egg, crustacean/shellfish, tree nuts (cashew nut), fish, wheat, and peanut [21]. In adults, tree nuts, peanuts, fish or shellfish are common sources of allergens [54]. Research has proved that the occurrence of food allergies increases by 4–8% in kids and 1–5% in adults [47]. For protection against food allergy, some countries are mandating the labeling of allergenic compounds on the label of food packages. This labeling and amount of allergenic compound may differ from region to region due to change in dietary habits [47]. According to the site of sensitization of food for allergic reaction, allergens are categorized into following two classes:

- Class I type can be recognized as whenever food passes through the GI tract, causing sensitization. This type of allergens is resistant to acid degradation, heat, and proteolysis resistant. Generally, this type of allergy is seen in children and sometimes in adults also [54]; and
- Class II type of allergy can be from food due to inhalation, therefore creating sensitization in the respiratory system [9].

This chapter explores major food allergens in routine foods with their modern detection methods.

11.2 SYMPTOMS

The common symptoms of food allergies are: respiratory issues, cardiovascular problems, life threatening anaphylaxis and sometimes irritation in the gastrointestinal (GI) including vomiting, diarrhea, abdominal pain, etc. [21].

11.3 ALLERGENS IN FOODS

11.3.1 WHEAT

Wheat is an integral cereal, which is consumed highest through-out the world with a production of 734 MT in 2018 [24]. Around 66% of total production is directly consumed for food and remaining is used as feed or non-feed applications [55].

Whole wheat is considered as a major source of carbohydrate and also contains a good amount of fiber. As the consumption of wheat is high, the percentage of allergy from wheat is high. The allergy due to milk and egg protein decreases as age increases while allergy due to wheat protein increases from 2 to 9% with age [44]. The common symptoms of allergic reaction due to consumption of wheat are: vomiting, diarrhea, abdominal pain, exacerbation of atopic dermatitis, etc.; and these may be observed after 24 h of consumption of wheat [42, 68]. Apart from consumption, those workers, who are used to be exposed to wheat flour, may have a chance of wheat allergy through breathing causing asthma or rhinitis [16].

11.3.2 PEANUT

Worldwide peanut is consumed as peanut butter, peanut oil, roasted peanut, etc. Around 0.6 to 1% of the total population in the US and EU suffer from peanut allergy [69]. Peanut allergy is common among all other legumes [36]. Generally, peanut allergy occurs in the beginning of life; and infants may feel some allergic reaction on their first introduction to peanuts after consumption [28].

Scientific studies confirm that children, who are introduced to peanuts in earlier stages of life, have a low risk of allergy as compared to those avoiding it [28]. This type of allergy sometimes is non-curable; therefore, such patients having peanut allergy may suffer for a lifetime [69]. It is also noticed that a very nominal quantity of peanut is enough to produce the allergy to the individual [64]. The 7S globulin (which is also termed as Ara h1) has been recognized as the most collective allergenic protein peanut. 2S albumin (identified as Ara h 2 and Ara h3) has also been identified as a major cause of peanut mediated allergen [69].

Peanut mediated allergic reaction can be exploited by certain processing factors of peanuts that include ripening, cooking, roasting, concentration of allergen, solubility, etc. [69]. During roasting of peanut, IgE binding capability of protein increases and it decreases during boiling of peanut [69]. Methods applied for processing of peanut also impart in the development of allergenic compounds [58].

Roasting of peanuts at elevated temperature increases the allergenicity of protein as compared to steaming and frying of peanuts [8, 43]. Due to thermal processing of peanuts, allergic compounds are formed by modification of the structure of protein in the food matrix [69]. In general, around 95% of the patients having peanut allergy can tolerate soy and another legumes allergy [61].

11.3.3 FISH

Out of total fish production of 171 MT in the world, 88% is used as human consumption, while the remaining 12% is utilized as non-food uses [22]. China is the largest producer and consumer of seafood in the world [22, 73]. The consumption of fish increases worldwide due to the presence of essential and polyunsaturated fatty acids and fat-soluble vitamins with high nutritional components [1, 38].

Most common fish (codfish) allergen is Gad c 1 (parvalbumin) with molecular weight of 12.3 kD. This allergen is capable of resisting enzymatic and thermal degradation [74]. In fish and fish products, around 70–95% allergic reactions come by parvalbumin allergen [57]. In addition to this, gelatin (type I collagen) was recognized as a rare allergen in some fishes [74].

The allergic reactions can occur if a person comes in direct contact or through consumption or inhaled the cooking vapors. The person having fish and shellfish allergy, it is advisable to avoid the same throughout the life as it does not resolve with increase in age. Report suggested that in high consumption countries like Asia, Australia, and some parts of Europe, shellfish, and fish allergy have increased [49]. Some common symptoms due to fish mediated allergic reactions are: swallowing, vomiting, diarrhea, breathing problems, loss of consciousness, even death, etc. [51]. Figure 11.1 shows major allergenic foods and some allergic reactions.

11.3.4 MILK

In infants, allergy due to consumption of milk is very common and it increases at the rate of 2.5% [19, 33, 34, 50]. It is observed that those people, who are hypersensitive to milk, can also be affected [33].

It has been noticed that allergy due to consumption of cow milk is predominated in children or toddler, although breast milk may also cause an allergic if mother is consuming milk or milk products [2, 33, 67], because these milk proteins are absorbed through gut and are released in human milk [70]. It is advisable for mothers to eliminate the consumption of cow milk and milk products during this period. Around 90% is the cross-reactivity between bovine's milk and goat's milk, whereas it is believed that people suffering from cow's milk allergy also can be sensitive to goat's milk [5]. This shows that shifting of milk will not eliminate the allergenicity. On the other hand, only around 4% of people are sensitive to mare's milk; those also suffer from cow milk allergy [13].

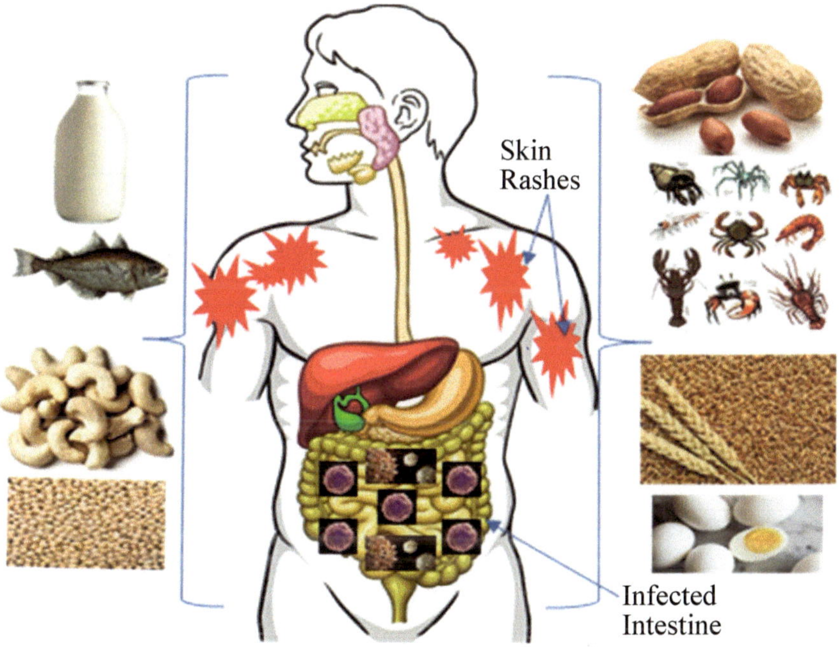

FIGURE 11.1 Common allergic foods and their symptoms on the human body.

Scientific literature has reported that almost 85% of total children having cow's milk allergy can endure soy-based formula. However, food formulation from soy-milk may be equally allergic [12]. Recently, it has been observed that the constituent A5B3 glycinin in soy protein may cross-react with casein [56].

When milk is heated at high temperature, then there could be some loss of tertiary structure of protein but cannot eliminate the chance of milk protein allergy. Production of hypoallergenic infant formula through processing for milk substitutes may decrease the chance of allergic reaction [27]. Therefore, it is advisable to provide such types of infant formula to minimize the allergic reaction due to cow milk. As earlier discussed, providing mare's milk to the infant can also be fruitful for the elimination of the chance of milk protein allergy in infants. Table 11.1 shows major food allergens.

TABLE 11.1 List of Common Food Allergens

Food	Name of Allergen	Common Name	References
Cashew nut	Ana o 1, Ana o 2, Ana o 3	Vicilin, legumin, 2S albumin	[45, 66]
Crustaceans	Tropomyosin	–	[74]
Egg	Gal d 1, Gal d 2, Gal d 3, Gal d 4	ovomucoid, ovalbumin, ovotransferrin, lysozyme	[6]
Fish	Gad c1	Parvalbumin	[62]
Milk	Bos d 8, Bos d 5, Bos d 4	Casein, β-lactoglobulin, α-lactalbumin	[71]
Peanut	Ara h 1, Ara h 2, Ara h 3	Vicillin, conglutin, glycinin	[60]
Soy	Gly mBd 30K	–	[69]
Wheat	Tri a 19, Tri a 20, Tri a 21	Alpha gliadin, omega 5-gliadin, gamma gliadin.	[15]

11.3.5 CRUSTACEANS

Seafood, such as, shellfish is a major source of dietary protein in coastal areas. According to FAO [23], the per capita consumption of seafood has increased to 20 kg in 2010 from 10 kg in 1960. As per the UN report, the current world population is near 8 billion that will increase to around 9 billion in 2030 and around 10 billion in 2050. This trend demands more food to satisfy the needs of individuals [31]. Therefore, with an increase in food consumption, the chance of food allergy also increases. In many countries with low intake of total protein, seafood protein plays an important role and provides essential nutritional elements to persons.

Crustaceans and mollusks come under a shellfish group that is invertebrates, unlike fish. Tropomyosin is one of the major allergens identified in shellfish that is a very vulnerable cause of allergy for toddler, young or aged people [39]. In invertebrates and invertebrates, tropomyosin is a very unique protein for muscle retrenchment [73]. The tropomyosin is a cross reactive allergen between mollusks and crustaceans [41]. As earlier discussed, shellfish allergy also continues till the end of life. In case of eggs from hen and allergies due to milk from cows as time passes, children develop natural tolerance; and in cash of shellfish allergy this does not happen [65].

11.3.6 EGG

Around 42% of eggs are produced by China, which is the biggest egg producer in the world. The US and India comes after China with 7 and 6% egg production, respectively. With more than 60%, Asia is the largest region in egg production. World egg production has grown by 150% in the last three decades. The world egg production increased from 15 in 1961 to 87 MT in 2017 [25].

Egg allergy is roughly around 1–9% in the world [63]. It is second in rank after milk allergy in children with symptoms of eczema [59]. Egg white has been proved to be the most prudent cause of allergy [75]. Gal d 1, 2, 3 and 4 have been characterized as major allergens in eggs. Among these, Gal d 1 is heat stable, while remaining are unstable proteins. Therefore, heating, and enzymatic degradation reduces the allergic reaction due to Gal d 2, 3, and 4 [10].

Boiled eggs have proved to be less prone to allergic reaction among 70% population affected by egg protein allergy [40]. At around teenage age, 60–75% infants develop tolerance towards egg allergy [63]. There is less chance of increase in egg allergy in children, who have a higher concentration of IgE antibodies focused against ovomuciod (Gal d 1) [6]. Egg allergy may resolve with age [49].

11.3.7 CASHEW NUT

The total cashew nut with shell production in the world was 5.9 MT in 2018 [24]. Worldwide cashew nuts are consumed as salted and are roasted as snacks. People consume around 60% as a snack, whereas remaining is converted in products like bakery and confectionery products, ice cream and chocolates items [66]. Cashew-nut has been protected with two shell layers. The shell is poisonous in nature; therefore, it requires some processing before consumption.

Different health benefits of cashew nuts have been acknowledged by different scientific studies [45]. Like tree nuts and peanuts, cashew nuts show similar allergic reactions (Figure 11.2) in all age groups [66]. Scientific studies proved that life-threatening allergic reactions are sometimes more vulnerable in cashew nuts rather than peanuts [66]. Allergic components in cashew nuts are Ana o 1, 2 and 3 with molecular weight of 50, 33 and 13 kDa, respectively [66]. People suffering from cashew nut allergy are sometimes susceptible to pistachio-nut allergy; therefore, it is advisable not to consume pistachio in cash of cashew nut allergy [3].

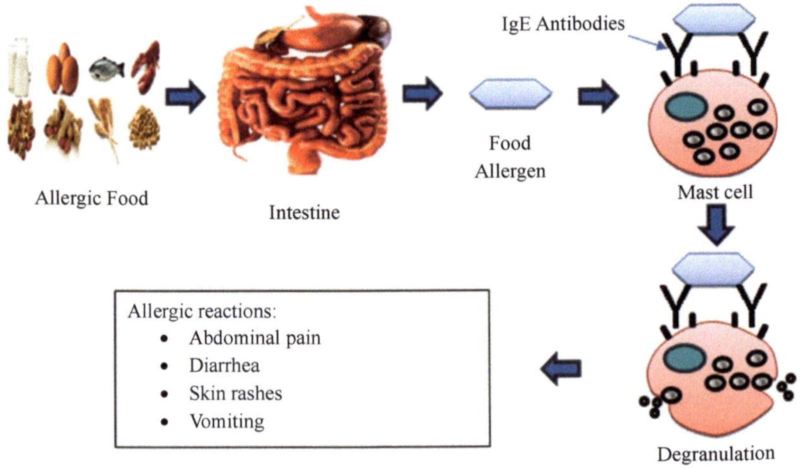

FIGURE 11.2 Mechanism of food allergy.

11.3.8 SOYBEAN

The soybean is an essential ingredient for dairy, food, and pharma applications [24, 69]. As the consumption of soybean has increased, the risk of allergy has also increased simultaneously. Around 6% of sensitive children are affected by soy allergy [69]. As per the report of researchers, the threshold levels for allergic reaction of soybean vary from 0.0013 to 500 mg [4].

The major allergen in soybean is Gly mBd 30K. Around 60% of allergy due to soybean is caused by this allergen only [69]. The decrease in IgE binding in soybean is not shown, even if the soybean protein extracts were heated at 100°C for 60 min [11]. Microwave application and boiling (120 min) can reduce the chances of allergenicity of soybean [69].

11.4 MECHANISM OF FOOD ALLERGY

Whenever a person has an allergy towards a particular food and one consumes that food; the allergenic food with allergens enters into the intestine that binds with IgE and mast cells. This mast cell is filled with basophil granules, and whenever it is degranuted or blasted, it releases histamine that is a cause of inflammation and other allergic symptoms (Figure 11.2).

11.5 DETECTION OF FOOD ALLERGENS

One of the major health threats for sensitive consumers is food allergy. Therefore, efforts have been customized by food processors to provide safe foods to allergy prone customers. Detection of level of allergens (in raw material, in semi-finished food or in the finished food) provides information to customers about purchasing decisions of a product. Several analytical methods have been used for detection and identification of various food allergens [77].

These detection techniques are reliable on the sample collection method and the procedure followed during the collection.

11.5.1 SKIN PRICK TEST (SPT)

Skin prick test (SPT) is the most common method that is applied to detect the food allergy [20]. This test indicates the sensitization of IgE towards a particular protein in a person. In this test, on mast cells in IgE is marked by swelling, whenever allergen extract is applied on skin [69]. In this test, pricking or scratching the surface of skin is done. After that for activation of skin mast cells, commercial extract or fresh food is added; and wheal and flare reaction can be calculated within a short time of addition [15]. The study suggested that the combination of both food-specific IgE and SPT can help in the detection of food allergens [15].

The advantage of skin prick test (SPT) is less time taken for interpretation. It takes 15–20 minutes after application of reagent on skin. In addition, this test gives a visible sign of the symptoms [32]. As per the report of Cetinkaya et al. [14], SPT wheel size was the best indicator to evaluate the reactivity for both cashew and pistachio nuts.

11.5.2 ORAL CONSUMPTION OF FOOD UNDER TEST

This test is both useful for IgE and non-IgE mediated allergic response [69]. Double-blind placebo-controlled food, challenge is another and important method for diagnosis of food allergens [69].

Generally, food allergy occurs in children or in infants. In this test, a small amount of allergenic food is given to the infants and the first allergic reaction is noticed. As infants are non-verbal, care must be taken while identification of symptoms while conducting this test. If symptoms were not seen in 15–20 minutes of feeding, then the next step is to feed larger amounts of food to patients. The amount of food given to patients depends on the age and past history of patients.

11.5.3 BIOSENSOR

Due to their sensitivity, low cost, specificity, and on-site detectability, nano-materials (such as: oxides of gold, titanium, silver, graphene nano-tubes, quantum dots, and carbon) are extensively applied for detection of food allergens. Among these materials, gold nano-particle and quantum dots are widely used in nano-technologies [51]. For detection of allergens from fish, a magnetic nano-particle probe can detect the allergens from 0.01 to 100 μg/mL [51]. To detect the specific allergenic epitopes from food protein, microarrays are used [7].

Generally, lab methods are commonly used in detection due to its accuracy and sensitivity. However, these methods require skilled persons with scientific knowledge and highly expensive instruments. Fu et al. [30] developed a smart phone-based microplate reader that can detect biomarkers at 340–680 nm visible range. They validated the same using ELISA method.

In 2016, Weng et al. [72] developed microfluidic ELISA biosensor to detect wheat and peanut allergens in samples. Total assay time was reduced from 3–4 h to 15–20 minutes. This technique also consumes lesser amount of reagent than the commercially available ELISA kits with high sensitivity. For instant detection of milk allergen casein, Jiang et al. [37] developed an electrochemical mast cell-based paper biosensor. This biosensor can detect 3.2×10^{-8} g/mL of allergen.

11.5.4 ELISA

ELISA method can capture an antigen using a specific enzyme-labeled antibody. It is easy to use as compared to other complicated test methods. It captures targeted antigen/antibody in a sample using specific antibody/

antigen and detects its quantity using enzyme reaction with its substrate. This method is used for verification for newly developed techniques. It is one of the important quantitative methods for detection of food allergens at low level [35].

11.5.5 ATOPY PATCH TEST

The main principle behind this test is to delay the hypertensive reaction during allergic immunological reaction [69]. It is one of the costlier and time-consuming method for detection of food allergy. As response in atopy patch test is cell-mediated, it is supportive for diagnosis of the overdue reactions to oral food challenges [69].

11.5.6 FOOD-SPECIFIC IGG TESTS

This method is useful for detection of the level of specific IgG (immunoglobulin G) or IgG_4 in food, but it does not show much diagnostic value for food allergens [69]. Filep et al. [26] detected principal food allergens by applying multiplex fluorescent array with lower limit of 0.06 ng/ml in samples of egg, soy, shrimp, peanut, cashew, hazelnut, and cow's milk.

11.6 REDUCTION OF FOOD ALLERGY

Nowadays, many techniques with which the amount of allergen in food can be reduced. Non-thermal processes, e.g., pulsed light is effective for reduction in gluten immune reactivity by structural rearrangement, formation of a wide range of molecular weight of proteins, unfolding of monomeric along with depolymerization of oligomeric [53]. Irradiation process decreases the allergenicity of α-lactalbumin by denaturation of protein and aggregation using doses of 1–10 kGy [46].

The reduction of allergenic compounds in paste of treated cashew and pistachio nuts have been observed, when autoclaving and hydrolysis (enzymatic) processes were carried out under sonication. Moreover, thermal treatment was added in enzymatic hydrolysis under sonication process and the effectiveness was higher in cashew nuts. The reduction in IgE binding capability of processed cashew was greater than unprocessed cashew and it was confirmed by ELISA [17]. It is reported that for protection against peanut allergy, feed peanuts to infants in the first year of life [18, 29].

Yang et al. [76] found that pulsed UV light treatment is an efficient method for reduction of peanut allergenicity. They found that treated peanuts have low amounts of Ara h 1, Ara h 2 and Ara h 3. The IgE binding capacity was reduced from 12.9 to 6.7% [76]. However, this method has food safety concerns similar to other irradiation methods. Lastly, it is advisable from many studies that infants provided with allergic food in the beginning of life may reduce the risk of allergy [52].

11.7 SUMMARY

Symptoms of food allergy include nausea, vomiting, pain, the skin irritation and irritation in the respiratory tract. Eight major foods (egg, soy, shrimp, peanut, cashew, hazelnut, and cow's milk) have been identified as common allergic foods, which contribute around 90% in the allergic reactions. Different methods are available for detection of food allergens; however, the ELISA method is the most validated method for the determination of allergens in different foods.

KEYWORDS

- **analytical method**
- **cardiovascular problems**
- **ELIZA**
- **food allergens**
- **immunoglobulin E**
- **skin prick test**

REFERENCES

1. Abreu, C. S. P., & Alves, M. J. L., (2020). Contribution of recombinant parvalbumin gad c 1 in the diagnosis and prognosis of fish allergy. *Journal of Investigational Allergology and Clinical Immunology, 30*(5), 1–19.

2. Arshad, S. H., (2001). Food allergen avoidance in primary prevention of food allergy. *Allergy, 56*, 113–116.
3. Bastiaan-Net, S., & Reitsma, M., (2019). IgE cross-reactivity of cashew nut allergens. *International Archives of Allergy and Immunology, 178*(1), 19–32.
4. Becker, W., Brasseur, D., & Bresson, J. L., (2004). Opinion of scientific panel on dietetic products, nutrition and allergies on a request from the commission relating to the evaluation of allergenic foods for labelling purposes. *European Food Safety Authority Journal, 32*, 180–197.
5. Bellioni-Businco, B., & Paganelli, R., (1999). Allergenicity of goat's milk in children with cow's milk allergy. *Journal of Allergy and Clinical Immunology, 103*(6), 1191–1194.
6. Bernhisel-Broadbent, J., Dintzis, H. M., Dintzis, R. Z., & Sampson, H. A., (1994). Allergenicity and antigenicity of chicken egg ovomucoid (Gal d III) compared with ovalbumin (Gal d I) in children with egg allergy and in mice. *Journal of Allergy and Clinical Immunology, 93*(6), 1047–1059.
7. Beyer, K., (2003). Characterization of allergenic food proteins for improved diagnostic methods. *Current Opinion in Allergy and Clinical Immunology, 3*(3), 189–197.
8. Beyer, K., Morrowa, E., & Li, X. M., (2001). Effects of cooking methods on peanut allergenicity. *Journal of Allergy and Clinical Immunology, 107*(6), 1077–1081.
9. Breiteneder, H., & Ebner, C., (2000). Molecular and biochemical classification of plant-derived food allergens. *Journal of Allergy and Clinical Immunology, 106*(1), 27–36.
10. Burman, J., Palosuo, K., & Kukkonen, A. K., (2019). Case report on importance of molecular allergy diagnostics in hen's egg allergy. *Biomedical Journal of Scientific & Technical Research, 14*(5), 10923, 10924.
11. Burks, A. W., & Williams, L. W., (1992). Identification and characterization of a second major peanut allergen (Arah-II), with use of the sera of patients with atopic dermatitis and positive peanut challenge. *Journal of Allergy and Clinical Immunology, 90*(6), 962–969.
12. Businco, L., Dreborg, S., & Einarsson, R., (1993). Hydrolyzed cow's milk formulae: Allergenicity and use in treatment and prevention. An ESPACI position paper. *Pediatric Allergy and Immunology, 4*(3), 101–111.
13. Businco, L., & Giampietro, P. G., (2000). Allergenicity of mare's milk in children with cow's milk allergy. *Journal of Allergy and Clinical Immunology, 105*(5), 1031–1034.
14. Cetinkaya, P. G., & Karaguzel, D., (2019). Pistachio and cashew nut allergy in childhood: Predictive factors towards development of a decision tree. *Asian Pacific Journal of Allergy and Immunology, 37*, 1–12.
15. Chokshi, N. Y., & Sicherer, S. H., (2016). Interpreting IgE sensitization tests in food allergy. *Expert Review of Clinical Immunology, 12*(4), 389–403.
16. Cianferoni, A., (2016). Wheat allergy: Diagnosis and management. *Journal of Asthma and Allergy, 9*, 13–25.
17. Cuadrado, C., Cheng, H., & Sanchez, A., (2018). Influence of enzymatic hydrolysis on the allergenic reactivity of processed cashew and pistachio. *Food Chemistry, 241*, 372–379.
18. Du Toit, G., Sayre, P. H., & Roberts, G., (2016). Effect of avoidance on peanut allergy after early peanut consumption. *New England Journal of Medicine, 374*(15), 1435–1443.

19. Eggesbo, M., Botten, G., Halvorsen, R., & Magnus, P., (2001). The prevalence of CMA/CMPI in young children: Validity of parentally perceived reactions in a population-based study. *Allergy, 56*(5), 393–402.
20. Eiwegger, T., Hung, L., San, D. K. E., O'Mahony, L., & Upton, J., (2019). Recent developments and highlights in food allergy. *Allergy, 74*(12), 2355–2367.
21. Ekezie, F. G. C., Cheng, J. H., & Sun, D. W., (2018). Effects of non-thermal food processing technologies on food allergens: Review of recent research advances. *Trends in Food Science & Technology, 74*, 12–25.
22. FAO (Food and Agriculture Organization of the United Nations). (2018). *The State of World Fisheries and Aquaculture at 2018 Meeting the Sustainable Development Goals*. Rome. http://www.fao.org/3/i9540en/I9540EN.pdf (accessed on 22 September 2021).
23. FAO (Food and Agriculture Organization of the United Nations). (2016). *The State of World Fisheries and Aquaculture Contributing to Food Security and Nutrition for All* (p. 204). Rome. http://www.fao.org/3/a-i5555e.pdf (accessed on 22 September 2021).
24. FAO (Food and Agriculture Organization of the United Nations), (2017). *FAOSTAT*. https://www.fao.org/faostat/en/#data/QC (accessed on 22 September 2021).
25. FAO (Food and Agriculture Organization of the United Nations), (2017). *FAOSTAT*. http://www.fao.org/poultry-production-products/production/en/ (accessed on 22 September 2021).
26. Filep, S., Smith, B., Black, K. R., & Chapman, M. D., (2018). Simultaneous detection of four major food allergens using a multiplex array. *Journal of Allergy and Clinical Immunology, 141*(2). Article ID: AB255.
27. Fiocchi, A., Bouygue, G. R., Sarratud, T., & Terracciano, L., (2004). Clinical tolerance of processed foods. *Annals of Allergy, Asthma & Immunology, 93*(5), S38–S46.
28. Fisher, H. R., Keet, C. A., Lack, G., & Du Toit, G., (2019). Preventing peanut allergy: Where are we now? *The Journal of Allergy and Clinical Immunology: In Practice, 7*(2), 367–373.
29. Fox, A. T., Sasieni, P., Du Toit, G., Syed, H., & Lack, G., (2009). Household peanut consumption as a risk factor for the development of peanut allergy. *Journal of Allergy and Clinical Immunology, 123*(2), 417–423.
30. Fu, Q., Wu, Z., Li, X., Yao, C., Yu, S., Xiao, W., & Tang, Y., (2016). Novel versatile smart phone-based microplate readers for on-site diagnoses. *Biosensors and Bioelectronics, 81*, 524–531.
31. Guillen, J., Natale, F., & Carvalho, N., (2019). Global seafood consumption footprint. *AMBIO, 48*(2), 111–122.
32. Heinzerling, L., Mari, A., & Bergmann, K. C., (2013). The skin prick test-European standards. *Clinical and Translational Allergy, 3*(1), 1–10.
33. Host, A., (2002). Frequency of cow's milk allergy in childhood. *Annals of Allergy, Asthma & Immunology, 89*(6), 33–37.
34. Host, A., & Halken, S., (1990). Prospective study of cow milk allergy in Danish infants during the first 3 years of life: Clinical course in relation to clinical and immunological type of hypersensitivity reaction. *Allergy, 45*(8), 587–596.
35. Hosu, O., Selvolini, G., & Marrazza, G., (2018). Recent advances of immunosensors for detecting food allergens. *Current Opinion in Electrochemistry, 10*, 149–156.

36. Hourihane, J. O. B., & Aiken, R., (2007). The impact of government advice to pregnant mothers regarding peanut avoidance on the prevalence of peanut allergy in United Kingdom children at school entry. *Journal of Allergy and Clinical Immunology, 119*(5), 1197–1202.
37. Jiang, D., Ge, P., Wang, L., Jiang, H., & Yang, M., (2019). novel electrochemical mast cell-based paper biosensor for the rapid detection of milk allergen casein. *Biosensors and Bioelectronics, 130*, 299–306.
38. Kuehn, A., & Arumugam, K., (2017). Fish allergy. Chapter 5. In: Lopata, A. L., (ed.), *Food Allergy: Molecular and Clinical Practice* (pp. 95–121). Boca Raton - FL: CRC Press.
39. Lee, A. J., Gerez, I., Shek, L. P. C., & Lee, B. W., (2012). Shellfish allergy: Asia-Pacific Perspective. *Asian Pacific Journal of Allergy and Immunology, 30*(1), 3–10.
40. Lemon-Mule, H., & Sampson, H. A., (2008). Immunologic changes in children with egg allergy ingesting extensively heated egg. *Journal of Allergy and Clinical Immunology, 122*(5), 977–983.
41. Leung, P. S., Chen, Y. C., & Chu, K. H., (1999). Seafood allergy: Tropomyosins and beyond. *Journal of Microbiology, Immunology and Infection, 32*(3), 143–154.
42. Majamaa, H., Moisio, P., Majamaa, H., Turjanmaa, K., & Holm, K., (1999). Wheat allergy: Diagnostic accuracy of skin prick and patch tests and specific IgE. *Allergy, 54*(8), 851–856.
43. Maleki, S. J., & Viquez, O., (2003). The major peanut allergen, Ara h 2, functions as a trypsin inhibitor and roasting enhances this function. *Journal of Allergy and Clinical Immunology, 112*(1), 190–195.
44. Matricardi, P. M., & Bockelbrink, A., (2008). Primary versus secondary immunoglobulin E sensitization to soy and wheat in the multi-center allergy study cohort. *Clinical & Experimental Allergy, 38*(3), 493–500.
45. Mendes, C., Costa, J., & Vicente, A. A., (2019). Cashew nut allergy: Clinical relevance and allergen characterization. *Clinical Reviews in Allergy & Immunology, 57*(1), 1–22.
46. Meng, X., Li, X., Wang, X., Gao, J., Yang, H., & Chen, H., (2018). Potential allergenicity response to structural modification of irradiated bovine α-lactalbumin. *Food & Function, 7*(7), 3102–3110.
47. Monaci, L., De Angelis, E., Montemurro, N., & Pilolli, R., (2018). Comprehensive overview and recent advances in proteomics MS based methods for food allergens analysis. *Trends in Analytical Chemistry, 106*, 21–36.
48. Moneret-Vautrin, D. A., (1998). Modifications of allergenicity linked to food technologies. *Allergic et Immunologic, 30*(1), 9–13.
49. Moonesinghe, H., & Mackenzie, H., (2016). Prevalence of fish and shellfish allergy: A systematic review. *Annals of Allergy, Asthma & Immunology, 117*(3), 264–272.
50. National Dairy Development Board (NDDB)-New Delhi. (2019). p. 1. https://www.nddb.coop/information/stats/across (accessed on 21 September 2021).
51. Neethirajan, S., Weng, X., Tah, A., Cordero, J. O., & Ragavan, K. V., (2018). Nano-biosensor platforms for detecting food allergens: New trends. *Sensing and Bio-sensing Research, 18*, 13–30.
52. Palmer, D. J., Sullivan, T. R., & Gold, M. S., (2017). Randomized controlled trial of early regular egg intake to prevent egg allergy. *Journal of Allergy and Clinical Immunology, 139*(5), 1600–1607.

53. Panozzo, A., Manzocco, L., Lippe, G., & Nicoli, M. C., (2016). Effect of pulsed light on structure and immunoreactivity of gluten. *Food Chemistry, 194*, 366–372.
54. Ramesh, S., (2008). Food allergy in children. *Clinical Reviews in Allergy & Immunology, 34*(2), 217–230.
55. Ranhotra, G. S., (1994). Wheat: Contribution to world food supply and human nutrition. Chapter 2. In: Bushuk, W., & Rasper, V. F., (eds.), *Wheat* (pp. 12–24).Boston: Springer.
56. Rozenfeld, P., & Docena, G. H., (2002). Detection and identification of a soy protein component that cross-reacts with caseins from cow's milk. *Clinical & Experimental Immunology, 130*(1), 49–58.
57. Ruethers, T., Raith, M., & Sharp, M. F., (2018). Characterization of ras-k1: Novel major allergen in Indian mackerel and identification of parvalbumin as the major fish allergen in 33 Asia-Pacific fish species. *Clinical & Experimental Allergy, 48*(4), 452–463.
58. Sampson, H. A., (2004). Update on food allergy. *Journal of Allergy and Clinical Immunology, 113*(5), 805–819.
59. Savage, J. H., Matsui, E. C., Skripak, J. M., & Wood, R. A., (2007). The natural history of egg allergy. *Journal of Allergy and Clinical Immunology, 120*(6), 1413–1417.
60. Shreffler, W. G., & Beyer, K., (2002). Use of peptide microarray immunoassay for the analysis of IgE-binding epitopes of major peanut allergens. *Journal of Allergy and Clinical Immunology, 109*(1), S286–S289.
61. Sicherer, S. H., (2001). Clinical implications of cross-reactive food allergens. *Journal of Allergy and Clinical Immunology, 108*(6), 881–890.
62. Swoboda, I., & Bugajska-Schretter, A., (2002). Recombinant carp parvalbumin, the major cross-reactive fish allergen: A tool for diagnosis and therapy of fish allergy. *The Journal of Immunology, 168*(9), 4576–4584.
63. Tan, J. W. L., Valerio, C., & Barnes, E. H., (2017). Randomized trial of egg introduction from 4 months of age in infants at risk for egg allergy. *Journal of Allergy and Clinical Immunology, 139*(5), 1621–1628.
64. Taylor, S. L., Hefle, S. L., & Bindslev-Jensen, C., (2002). Factors affecting the determination of threshold doses for allergenic foods: How much is too much? *Journal of Allergy and Clinical Immunology, 109*(1), 24–30.
65. Thalayasingam, M., Gerez, I. F. A., & Yap, G. C., (2015). Clinical and immunochemical profiles of food challenge proven or anaphylactic shrimp allergy in tropical Singapore. *Clinical & Experimental Allergy, 45*(3), 687–697.
66. Van, D. V. J. P. M. J., & Dubois, A. E., (2014). Systematic review on cashew nut allergy. *Allergy, 69*(6), 692–698.
67. Van, O. J., Kull, I., & Borres, M. P., (2003). Breastfeeding and allergic disease: A multidisciplinary review of the literature (1966–2001) on the mode of early feeding in infancy and its impact on later atopic manifestations. *Allergy, 58*(9), 833–843.
68. Varjonen, E., Vainio, E., & Kalimo, K., (2000). Antigliadin IgE-indicator of wheat allergy in atopic dermatitis. *Allergy, 55*(4), 386–391.
69. Verma, A. K., Kumar, S., Das, M., & Dwivedi, P. D., (2013). A comprehensive review of legume allergy. *Clinical Reviews in Allergy & Immunology, 45*(1), 30–46.
70. Wal, J. M., (2004). Bovine milk allergenicity. *Annals of Allergy, Asthma & Immunology, 93*(5), S2–S11.
71. Wal, J. M., (2001). Structure and function of milk allergens. *Allergy, 56*, 35–38.

72. Weng, X., Gaur, G., & Neethirajan, S., (2016). Rapid detection of food allergens by microfluidics ELISA-based optical sensor. *Biosensors, 6*(2), 24.
73. Wild, L. G., & Lehrer, S. B., (2005). Fish and shellfish allergy. *Current Allergy and Asthma Reports, 5*(1), 74–79.
74. Wong, L., Huang, C. H., & Lee, B. W., (2016). Shellfish and house dust mite allergies: Is the link tropomyosin? *Allergy, Asthma & Immunology Research, 8*(2), 101–106.
75. Yamada, K., Urisu, A., & Koyama, H., (2000). IgE-binding activity to enzyme-digested ovomucoid distinguishes between patients with contact urticaria to egg with and without overt symptoms on ingestion. *Allergy, 55*(6), 565–569.
76. Yang, W. N., & Mwakatage, R., (2012). Mitigation of major peanut allergens by pulsed ultraviolet light. *Food and Bioprocess Technology, 5*(7), 2728–2738.
77. Zhou, J., Qi, Q., Wang, C., Qian, Y., Liu, G., Wang, Y., & Fu, L., (2019). Surface plasmon resonance (SPR) biosensors for food allergen detection in food matrices. *Biosensors and Bioelectronics, 142*. Article ID: 111449

CHAPTER 12

ANTI-NUTRITIVE AND THERAPEUTIC PROPERTIES OF LECTINS

CH. V. K. SUDHEENDRA, TANMAY HAZRA, AKASH SOLANKI, and VIMAL M. RAMANI

ABSTRACT

The digestibility of the leguminous protein is low due to the presence of toxins, which are produced by plants as secondary metabolites. The role of lectins in self-defense against predators and during stress is significant, with anti-nutritive properties, red blood cell agglutination, interact with exact blood groups, induce the division of cells, and agglutinating tumor cells. Higher intake of lectins can lead to toxic effects, such as severe intestinal damage, provoking immunoglobulin production leading to allergies and other immune-related responses and hemagglutination and anemia. Lectins are being studied for their cancer and tumor detecting abilities and even in the treatment of cancer. They are exhibiting preclinical and clinical significance, which encourages their application as a therapeutic factor in cancer treatment.

12.1 INTRODUCTION

Apart from carbohydrates, proteins, fats, macronutrients, vitamins, and minerals, and micronutrients, plants also contain contaminants that at times may lead from minor to major health hazards for consumers. These hazardous compounds can enter into foods either as naturally occurring toxins or during handling and processing.

Naturally occurring toxins are endogenous toxins of plant origin or natural contaminants like toxic substances from animals or microbial toxins. These toxins can be termed as anti-nutritive factors or anti-nutrients, which are synthesized in plants as secondary metabolites and defense proteins against predators or during stress. Some major group of chemicals secreted as secondary metabolites are: tannins, flavonoids, alkaloids, phytoecdysteroids, terpenoids, cardenolides, iridoid glycosides, coumarins, glucosinolates, cyanogenic glycosides, phenolics, non-protein amino acids [40, 41, 71, 95]. Proteinase inhibitors, amylase inhibitors, proteinase, polyphenol oxidases and peroxidase, lipoxygenase, lectins, jasmonic acid-inducible proteins are some important defense proteins [127].

These metabolites act on sensory aspects or affect specific systems of herbivores, such as: nervous, digestive, and endocrine organs [97]. Chemical compounds that act on the specific systems are termed as anti-nutritive compounds. Ingestion of these secondary metabolites can cause some health implications, while a few are known to have beneficial effects and are widely used in nutrition and in pharmacy [105].

It is reported that these natural toxins in food may cause nausea and headache, gastrointestinal (GI) problems like bloating, skin-related issues, difficulty in absorption of minerals causing deficiencies, etc. [27] in both humans and animals. Sometimes, they may even cause death.

The anti-nutritional factors are mainly found in raw as well as food products prepared from pulses and legumes [31]. Legumes are rich in carbohydrates, proteins, minerals, and vitamins and hence occupy an important place in the human diet. However, the presence of these anti-nutritive compounds may hamper the digestibility of leguminous protein. Few anti-nutritional factors other than lectins present in legumes are: alkaloids, tannins, isoflavones, saponins, cyanogenic glycosides, phytates, non-protein amino acids, pyrimidine glycosides, and oligosaccharides [14].

In the present chapter, authors have discussed plant lectins, their classification, their Anti-nutritive and therapeutic properties, and their benefits.

12.2 LECTINS

Lectins (known as phytohemagglutinin (PHA) or plant agglutinins) are a group of glycoproteins with agglutination properties. This group of proteins can be found abundantly in nature. Lectins are isolated from plant-based products, animals, and humans, even from bacteria, fungi, algae, yeasts, and viruses. Presently more than 1,000 lectins have been discovered and studied. This group of proteins can attach to specific sugars, especially sugar moieties of glycoproteins at single or multiple binding sites present on animal cell surfaces and erythrocytes or lymphocytes [54].

Approximately 2–10% of the total protein constitutes lectins and of which 4 to 10% is carbohydrate. They are omnipresent in plant species, but are profusely found in legumes and are most studied lectins. Over 800 plant species and more than 600 species of Leguminosae family are known to contain these lectins [103]. Nearly 300 plant lectins have been identified, purified, and characterized. Their molecular structures, biochemical properties, and carbohydrate specificity have been studied [112]. Highest concentrations of lectins can be found in whole grains (wheat, rice), legumes (beans, peanuts, soybeans) and nuts (Almonds, cashews), leaves (*Allium sativum*, *Euphorbia tithymalo*), roots, and tuber (Garlic, potatoes) and fruits (Tomato) and flowers [17, 82, 84].

Plant lectins that have been widely studied are from families of Leguminosae (Legumes: red kidney bean, soya bean, Jack bean, etc.), Graminea (Wheat, rice, etc.), Solanaceae (Tomato, Potato, etc.) [91]. Several factors are found to be responsible for inducing the expression of lectins, such as pathogen infection [90], insect herbivory [108, 116], salt stress [131] and jasmonic acid treatment [56]. These induced lectins are found in cytoplasm and nucleus of plant cells and are different from the naturally formed lectins [65].

The lectins are also produced as a defense mechanism against microorganisms, insects, and plant-eating animals. Hence, these lectins have toxic nature and are known to have severe *in vivo* effects on animals and humans, causing damage to intestinal cells and disrupting the nutrient absorption, especially minerals like calcium, iron, phosphorus, and zinc. Lectins can cause allergy by stimulating the production of IgG and IgM, and can cause hemagglutination by binding to erythrocytes.

12.2.1 HISTORY

Peter Hermann Stillmark in 1888 first discovered a toxic hemagglutinin named "Ricin" from castor seed, which has a capability to agglutinate the red blood cells. Later, H. Hellin isolated a similar toxic hemagglutinin, named abrin, from the jequirity bean. Concanavalin A was the first hemagglutinin isolated to the purest crystalline form by James B. Sumner.

Lectin is derived from the Latin word "Legere" meaning "to select." Boyd and Shapleigh first used this term, due to the specificity of agglutinins' choice of binding to selective carbohydrate structures [15]. Later on, this definition has been revised with new discoveries of lectins and advances in science.

Finally, Peumans, and Van Damme defined lectin as "All plant proteins possessing at least one non-catalytic domain, which binds reversibly to a specific mono- or oligosaccharide" [83]. Nowadays, proteins having the capability to agglutinate red blood cells of known sugar specificity are generalized and referred to as "lectins" [29]. When there is no known specificity of sugars, then they are called "hemagglutinins." Hemagglutinins and lectins are those proteins/glycoproteins, which consist of at least one domain to exhibit reversible binding capability towards exact carbohydrate moiety.

12.2.2 CLASSIFICATION OF LECTINS

Basic classification of lectins is based on the source from which lectins are isolated and according to this, there are six groups: algal lectins, animal lectins, fungal lectins, bacterial lectins, plant lectins, and viral lectins. Based on the common structural features, lectins can be classified into three classes: Simple lectin, Mosaic lectin and Macromolecular assemblies.

Most of the plant lectins fall under the class simple lectins. Plant lectins based on their three-dimensional structures are classified into 48 families [32]. Plant lectins, which can only bind to carbohydrate moiety and based on their overall structure, can be divided into four groups, namely: super lectins, merolectins, hololectins, and chimero lectins [12, 113]. Initially, lectins were divided into seven groups based on different carbohydrate-binding domains and structures [111, 117]. These seven groups can be placed into two categories, namely, membrane-bound and soluble proteins [55].

With advances in science and new lectins discovered, new families of lectins have emerged, and presently there are 12 different families of lectins based on different carbohydrate-binding domains [12, 111]. Molecular and three-dimensional structures of various plant lectins were studied and elucidated [33, 62, 96, 113, 115, 123]. Perez et al. [81] compiled a database of more than 1,000 lectin structures.

12.3 ANTI-NUTRITIVE PROPERTIES OF LECTINS

As mentioned in earlier literature, lectins toxicity can be traced back to Stillmark's work on ricin. Due to lack of public knowledge about the toxic or anti-nutritional properties of lectins on the human health, many foods poisoning outbreaks have occurred [11, 30, 35, 76].

12.3.1 ORAL TOXICITY

Experimental animals when fed with plant lectins had lost their appetite and reduction in the body weight was observed, ultimately leading to death. Oral acute toxicity in humans may cause nausea, vomiting, bloating, and diarrhea. The mechanism behind this deleterious effect is still not understood [118]. It is also predicted that lectin consumption may lead to dental caries [67, 94].

12.3.2 RESISTANCE OF LECTINS TO PROTEOLYSIS

Lectins are known to evade the proteolysis by proteases both *in vitro* and *in vivo* [18, 89] studies. HCl of stomach or proteolytic enzymes secreted during digestion cannot degrade them practically making them resistant to digestion [5, 10, 44]. It was observed that when lectin was fed to rats, the excretions were found to contain nearly 90% of the fed lectin. In humans, WGA (wheat germ agglutinin) was found in the ileum and fecal samples when wheat germ was consumed as part of daily diet [16]. As a result of this effect, ingested lectins remain active in the intestinal tract and can attach themselves to glycosyl groups that are present in the cells of the GI tract.

12.3.2.1 EFFECT ON INTESTINES

The lectins bind to microvilli, which are finger-like structures on epithelial cells [2] and cause lesions through which they can enter the blood and lymph system [22]. This is called endocytosis. From here, they may reach the pancreas, liver, urinary tract and into the systemic circulation. WGA can obstruct with the protein breakdown and increases gut penetrability causing IBS (irritable bowel syndrome), Crohn's and Celiac diseases

[106], due to the inhibition of plasma membrane repair leading to death of gut epithelial cells [70]. Abnormal mucus growth due to protozoa and bacteria or removal of mucous from the small intestine could be observed, when rats were fed with lectins [8]. Lectin ingestion can also lead to peptic ulcers due to overproduction of acid, reduced mucus or no mucous layer and bacterial colonization.

12.3.2.2 EFFECT OF ABSORPTION OF NUTRIENTS

When lectins bind to intestinal epithelial cells, they cause change in the morphology of the bound cells and alter metabolism occurring in the stomach and small intestine. This altered metabolism has its effect on digestion, absorption, functions of the whole digestive system. Carbohydrate absorption and transportation of nutrients is interfered during digestion [98] as explained by Jaffe [45]. It was found that lectins are present in soybean and it was responsible for the reduction of growth response in rats [66]. Reduction in the nutrient absorption is due to the reduction in the surface area of the intestinal absorptive area, and reduction on microvilli size, which will lead to cellular hyperplasia [78, 99]. Lack of absorption of nutrients leads to the loss of weight and poor growth and may ultimately lead to death. Irregular hormonal secretions are also observed [49].

12.3.2.3 COLONIZATION OF BACTERIA

Another toxic effect of lectins in the intestine is the colonization of bacteria [7], due to the damage of intestinal cells [36]. Coliforms are found to proliferate in excess [10, 88, 125]. *Helicobacter pylori* are also found to grow in the intestines due to loss of defense mechanism [38]. Due to colonization of bacteria and attachment of lectins to bacterial receptors on the bacterial surface induces the production of enterotoxin/endotoxin that may be detrimental to animals or humans [46, 121].

12.3.3 SYSTEMIC EFFECTS OF LECTINS

Lectins can provoke immunochemical effects in the blood of rats and pigs [37, 124]. Partially digested or intact lectins may enter the blood circulation and can affect the spleen, pancreas, and kidney, which affects the lipid and protein catabolism. These lectins also have an effect on muscle tissues and are known to cause an elevation in insulin content [10, 89, 130]. Entry into the circulation system may also cause rheumatoid arthritis [1, 20]. IgG and IgM are provoked by lectins causing allergy and other immune-related problems [13]. It may also have a long-lasting effect on nervous systems, immunity, reproductive, and cancer. Ingestion of lectin may also cause obesity and can increase blood pressure [48], and infertility [120].

Raw grains, fruits, and vegetables can lead to the above-said problems. The 70,000 HAU (hemagglutinating units) are found in raw red kidney beans; and lectins can be inactivated during the food processing [26, 102]. Different lectins have different heat stability but most of the lectins can be inactivated by heating properly, i.e., using moist heat not less than 100°C/5 min [42, 43]. Soaking, sprouting, and fermentation also have a promising effect on the inactivation of lectins [13, 21].

12.4 THERAPEUTIC PROPERTIES OF LECTINS

Lectins were demonized due to their anti-nutritive properties. However, lectins have promising bioactive compounds with their beneficial applications [54]. Examples of legumes with lectin-therapeutic properties are: mistletoe *(Viscum album),* lentils (*Lens culinaris*), soybean (*Glycine max*), fava beans (*Vicia faba*), chickpeas (*Cicer arietinum*), and peas (*Pisum sativum*). These lectins are known for their anti-carcinogenic, antimicrobial, antifungal, and antiviral protein activities.

12.4.1 CLINICAL MICROBIOLOGICAL APPLICATIONS OF LECTINS

12.4.1.1 EFFECT ON INTESTINES

Lectins are known as imitators of growth hormones and cytokines of different cells. As these lectins are hormone mimetics, therefore there are reports that orally given lectins may normalize the GI tract functioning by release of pancreatic juices and proliferation of injured epithelial cells [36]. They have an impact on human blood pressure as they reduce the production of angiotensin [39].

12.4.1.2 EFFECT ON BACTERIA AND VIRUS

Lectin's capability to bind to specific glycosides on the epithelial cells and bacteria's surface is used in the medical world to treat patients against pathogens and detect pathogens. The *salmonella* [73], *streptococci, coliforms*, especially *E. coli.* can be detected using lectins. Fungi, protozoa, and bacteriophage binding sites on the bacterial surface can also be checked [28, 59, 60, 75]. Flow cytometry methods based on lectins and lectin-arrays can help in identifying different carbohydrates and live-cells [119]. Plant lectins have bactericidal and bacteriostatic activities.

Antimicrobial activity [72, 80] of lectins was observed against Gram-positive bacteria (*E. faecalis, S. aureus, B. subtilis*), and Gram negative bacteria (*P. aeruginosa, K. pneumoniae, E. coli*). Antifungal activity on *Candida albicans, Aspergillus flavus* and *F. oxysporum* was observed [3, 50, 51, 85, 101]. *Helicobacter pylori* can also be inhibited by a lectin from mung beans sprouts. [92]. Lectins have shown an antiviral effect on HIV and can inactivate the viral enzymes [24, 87]. There are various mechanisms for the antiviral activity [64, 69].

12.4.1.3 EFFECT OF LECTINS ON TUMOR AND CANCER

The lectins are prolifically used in the medical field. Lectin's ability to attach to a particular binding site on a cell makes it the best choice in tumor and cancer studies. This unique characteristic of lectin helps in identifying normal cells from tumor or cancer cells, such as, liver cirrhosis [12, 52] and hepatocellular carcinoma (HCC) [104]. Lectin treatment in combination with chemotherapy is the effective impact to treat tumors [74]. The plant lectins are being tested for their activity against cancer cells. *In vivo* study of lectins demonstrates their anti-cancerous properties [47].

Cell apoptosis and autophagy reduces the synthesis of enzymes (such as: telomerase), which cause the development of cancer cells, hamper the development of new blood vessels that supply blood to cancer cells and may cause their death. Another mechanism is that these lectins bind on to surfaces of immune cells, leading to the production of cytokines that have an anti-tumoral effect on cancer cells [107]. Lectins have shown to reduce melanoma [110, 128], leukemia [19, 132], pancreatic cancer [100], colon cancer [23, 132], lung cancer [57, 126], and breast cancer [23, 53, 58, 79].

12.4.1.4 APPLICATION IN BONE TREATMENT

Animal study suggests that lectins can promote the development and growth of bones [129]. Their role in the treatment of non-union bones has also been established [114]. Immobilized lectins can proliferate osteoblast-like cells [77].

12.4.1.5 LECTINS AS DRUG DELIVERY AGENTS

Lectin's ability to bind on to specific targets has made them into an ideal compound, which may be employed as a targeted drug transport system in cancer and tumor therapies. Ricin is known to target leukemia [25], WGA targets urinary bladder cancer cells [6, 86]. To enhance the effectiveness of the activity of lectins, an encapsulation process was developed to resist harsh acidic environment in the stomach [63], and it was successful in treating tumors of the liver and kidney [4]. Efficacy of lectins can be further improved by using them in combination with different agents. WGA in combination with anticancer drugs (such as: doxorubicin and daunorubicin) ameliorated the cancer cells [6]. Lectins, along with radio-isotopes and photo-activated antioxidants are another type of combinations, which have yielded the best results in treating cancer.

12.4.1.6 LECTINS AND NANOTECHNOLOGY

Nanomaterials based on lectins are being developed for biomedical applications, such as biosensors for carbohydrates, cancer, and pathogens, drug delivery systems, and gene therapy. Lectins coated onto nanoparticles can be orally given, which could be a potential tool for the controlled release of drugs at targeted sites [61, 93] and could reduce the transit time of the formulations [68]. Nanoparticles coated with lectins are being developed to deliver vaccine targets on breast cancer cells; and ocular drugs [34] to treat tuberculosis in lungs [109] and periodontal diseases [122]. Nano-structured ricin had a positive effect on acute myeloid leukemia and had potent anticancer activity [25].

12.5 SUMMARY

The isolation and characterization of the new lectins from wild plants, the study of their structures and different isoforms, is currently an active field of research. The mechanism behind antitumor activities of plant lectins still needs to be characterized. Although lectin-arrays may be a biomedical tool, yet the variation in the isoform may give unpredictable results. Hence novel strategies are needed to study the effect of lectins during clinical trials, in combination with other therapies. Novel recombinant lectins are being isolated, which can be used as biomarkers and can also induce apoptosis. Lectins have anti-nutritive and therapeutic values.

KEYWORDS

- **anti-nutritive properties**
- **cancer**
- **hemagglutinins**
- **lectins**
- **macronutrients**
- **therapeutic agents**

REFERENCES

1. Albani, S., & Carson, D. A., (1996). Multistep molecular mimicry hypothesis for the pathogenesis of rheumatoid arthritis. *Immunology Today, 17*(10), 466–470.
2. Almeida, N. G., Calderon, D. A. M., & Valencia, M. E., (1991). Effect of different heat treatments on antinutritional activity of *Phaseolus vulgaris* lectin. *Journal of Agricultural and Food Chemistry, 39*, 1627–1630.
3. Al-Snafi, A. E., (2016). Antimicrobial effects of medicinal plants (Part 3): Plant based review. *IOSR Journal of Pharmacy, 6*(10), 67–92.
4. Andrade, C. A., & Correia, M. T., (2004). Antitumor activity of *Cratylia mollis* lectin encapsulated into liposomes. *International Journal of Pharmaceutics, 278*, 435–445.
5. Anil-Kumar, H. V., & Muralidhar, T. S., (2019). Beneficial and harmful properties of lectins. *International Journal of Herbal Medicine, 7*(3), 29–32.
6. Apfelthaler, C., Skoll, K., Ciola, R., Gabor, F., & Wirth, M., (2018). A doxorubicin loaded colloidal delivery system for the intravesical therapy of non-muscle invasive

bladder cancer using wheat germ agglutinin as targeter. *European Journal of Pharmaceutics and Biopharmaceutics, 130*, 177–184.

7. Banwell, J. G., Boldt, D. H., Meyers, J., & Weber, F. L., (1983). Phytohemagglutinin derived from red kidney bean (*Phaseolus vulgaris*): Cause for intestinal malabsorption associated with bacterial overgrowth in the rat. *Gastroenterology, 84*, 506–515.
8. Banwell, J. G., Howard, R., Kabir, I., & Costerton, J. W., (1988). Bacterial overgrowth by indigenous microflora in the pha-fed rat. *Canadian Journal of Microbiology, 34*, 1009–1013.
9. Bardocz, S., Ewen, S. W. B., Grant, G., & Pusztai, A., (2005). Lectins as growth factors of the small intestine and the gut. Chapter 6. In: Pusztai, A., & Bardocz, S., (eds.), *Lectins Biomedical Perspectives* (pp. 86–97). London: Taylor and Francis.
10. Bardocz, S., Grant, G., & Pusztai, A., (1996). The effect of phytohemagglutinin at different dietary concentrations on growth, body composition and plasma insulin of the rat. *British Journal of Nutrition, 76*, 613–626.
11. Bender, A. E., & Reaidi, G. B., (1982). Toxicity of kidney beans (*Phaseulus vulgaris*) with particular reference to lectins. *Journal of Plant Foods, 4*, 15–22.
12. Bhutia, S. K., Panda, P. K., & Sinha, N., (2019). Plant lectins in cancer therapeutics: Targeting apoptosis and autophagy-dependent cell death. *Pharmacological Research, 144*, 8–18.
13. Boehm, S., & Huck, S., (1998). Presynaptic inhibition by concanavalin: Are alpha-latrotoxin receptors involved in action potential-dependent transmitter release? *Journal of Neurochemistry, 71*, 2421–2430.
14. Bora, P., (2014). Anti-nutritional factors in foods and their effects. *Journal of Academia and Industrial Research, 3*(6), 285–290.
15. Boyd, W. C., & Shapleigh, E., (1954). Specific Precipitating activity of plant agglutinins (lectins). *Science, 119*(3091), 419.
16. Brady, P. G., Vannier, A. M., & Banwell, J. G., (1978). Identification of the dietary lectin, wheat germ agglutinin, in human intestinal contents. *Gastroenterology, 75*, 236–239.
17. Broekaert, W. F., Allen, A. K., & Peumans, W. J., (1987). Separation and partial characterization of isolectins with different subunit compositions from datura stramonium seeds. *FEBS Letters, 220*(1), 116–120.
18. Carbonaro, M., Cappelloni, M., Nicoli, S., Lucarini, M., & Carnovale, E., (1997). Solubility- digestibility relationship of legume proteins. *Journal of Agricultural and Food Chemistry, 45*(9), 3387–3394.
19. Carvalho, F. C., Soares, S. G., & Tamarozzi, M. B., (2011). The recognition of n-glycans by the lectin ArtinM mediates cell death of a human myeloid leukemia cell line. *PLoS One, 6*(11). Article ID: e27892.
20. Cordain, L., Toohey, L., Smith, M. J., & Hickey, M. S., (2000). Modulation of immune function by dietary lectins in rheumatoid arthritis. *British Journal of Nutrition, 83*(3), 207–217.
21. Cuadrado, C., Hajos, G., & Burbano, C., (2002). Effect of natural fermentation on the lectin of lentils measured by immunological methods. *Food and Agricultural Immunology, 14*(1), 41–49.
22. De Oliveira, A. C., De Campos, V. B., & Sgarbieri, V. C., (1989). Lesions of intestinal epithelium by ingestion of bean lectins in rats. *Journal of Nutritional Science and Vitaminology, 35*(4), 315–322.

23. Deepa, M., Sureshkumar, T., Satheeshkumar, P. K., & Priya, S., (2012). Purified mulberry leaf lectin (MLL) induces apoptosis and cell cycle arrest in human breast cancer and colon cancer cells. *Chemico-Biological Interactions, 200*(1), 38–44.
24. Dent, M., & Matoba, N., (2020). Cancer biologics made in plants. *Current Opinion in Biotechnology, 61*, 82–88.
25. Díaz, R., Pallarès, V., & Cano-Garrido, O., (2018). Selective $CXCR_4^+$ cancer cell targeting and potent antineoplastic effect by a nanostructured version of recombinant ricin. *Small, 14*(26). Article ID: 1800665.
26. Dominguez, H., Nunez, M. J., & Lema, J. M., (1993). Chlorogenic acid removal during aqueous processing of sunflower kernels. *Grasasy Aceites, 44*(4, 5), 235–242.
27. Essack, H., Odhav, B., & Mellem, J. J., (2017). Screening of traditional South African leafy vegetables for specific anti-nutritional factors before and after processing. *Food Science and Technology, 37*(3), 462–471.
28. Etzler, M. E., (1985). Plant lectins: Molecular and biological aspects. *Annual Review of Plant Physiology, 36*(1), 209–234.
29. Fereidoon, S., (2014). Beneficial health effects and drawbacks of antinutrients and phytochemicals in foods. *Applied Microbiology and Biotechnology, 97*, 45–55.
30. Freed, D., (1991). Lectins in food: Their importance in health and disease. *Journal of Nutritional Medicine, 2*(1), 45–64.
31. Friedman, M., & Brandon, D. L., (2001). Nutritional and health benefits of soy proteins. *Journal of Agricultural and Food Chemistry, 49*(3), 1069–1086.
32. Fujimoto, Z., Tateno, H., & Hirabayashi, J., (2014). Lectin structures: Classification based on the 3-D structures. Chapter 46. In: Hirabayashi, J., (ed.), *Lectins: Methods in Molecular Biology (Methods and Protocols)* (Vol. 1200, pp. 579–606). New York: Humana Press.
33. Gabius, H. J., André, S., Jiménez, B. J., Romero, A., & Solís, D., (2011). From lectin structure to functional glycomics: Principles of the sugar code. *Trends in Biochemical Sciences, 36*(6), 298–313.
34. Gavrovic, J. M., & Prodanovic, R., (2011). Drug delivery: Plant lectins as bio-adhesive drug delivery systems. *Journal of Biomaterials and Nanobiotechnology, 2*(5), 614–621.
35. Gilbert, R., (1988). Healthy eating day. *Communicable Disease Report, 33*, 3, 4.
36. Grant, G., Edwards, J. E., & Ewan, E. C., (1999). Secretion of pancreatic digestive enzymes induced in rats by first-time oral exposure to kidney bean E2L2 lectin is mediated only in part by cholecystokinin (CCK). *Pancreas, 19*, 382–389.
37. Grant, G., Greer, F., McKenzie, N. H., & Pusztai, A., (1985). The nutritional response of mature rats to kidney bean (*Phaseolus vulgaris*) lectins. *Journal of the Science of Food and Agriculture, 36*, 409–414.
38. Hamid, R., & Masood, A., (2009). Dietary lectins as disease-causing toxicants. *Pakistan Journal of Nutrition, 8*(3), 293–303.
39. Hamid, R., Masood, A., Wani, I. H., & Rafiq, S., (2013). Lectins: Proteins with diverse applications. *Journal of Applied Pharmaceutical Science, 3*(4), S95–S102.
40. Harborne, J., (1993). *Introduction to Ecological Biochemistry* (4th edn., p. 317). London: Academic Press.
41. Harborne, J., (1999). Plant chemical ecology. In: Mori, K., (ed.), *Comprehensive Natural Products Chemistry, Miscellaneous Natural Products Including Marine*

Natural Products, Pheromones, Plant Hormones, and Aspects of Ecology (Vol. 8, pp. 263–376). Oxford: Elsevier.

42. He, S., Simpson, B. K., Ngadi, M. O., & Ma, Y., (2015). *In vitro* studies of the digestibility of lectin from black turtle bean (*Phaseolus vulgaris*). *Food Chemistry, 173*, 397–404.
43. He, S., Simpson, B. K., Sun, H., Ngadi, M. O., Ma, Y., & Huang, T., (2018). *Phaseolus vulgaris* lectins: A systematic review of characteristics and health implications. *Critical Reviews in Food Science and Nutrition, 58*(1), 70–83.
44. Helleland, S. B. L., & Pryme, I. F., (2012). The importance of dietary plant lectins: The blood type diet revisited. Chapter 5. In: Gupta, V. K., Singh, G. D., & Singh, S., (eds.), *Medicinal Plants: Phytochemistry, Pharmacology and Therapeutics* (Vol. 2, pp. 69–79). New Delhi: Daya Publishing House.
45. Jaffe, W. G., Planchart, A., Ραεζ Pumar, J. I., Torrealba, R., & Franceschi, D. N., (1955). Nuevos estudios sobre un factor tóxico de las caraotas crudas (*Phaseolus vulgaris*) (study on toxic factor of black beans). *Archives of Venezuela Nutrition, 6*, 195–205.
46. Jayne-Williams, D., (1973). Influence of dietary jack beans (*Canavalia ensifortnis*) and of concanavalin a on the growth of conventional and gnotobiotic Japanese quail (*Coturnix coturnix japonica*). *Nature New Biology, 243*(126), 150, 151.
47. Jiang, Q. L., Zhang, S., & Tian, M., (2015). Plant lectins, from ancient sugar-binding proteins to emerging anti-cancer drugs in apoptosis and autophagy. *Cell Proliferation, 48*(1), 17–28.
48. Jönsson, T., Ahrén, B., & Pacini, G., (2006). Paleolithic diet confers higher insulin sensitivity, lower C - reactive protein and lower blood pressure than a cereal-based diet in domestic pigs. *Nutrition & Metabolism, 3*(1), 39–45.
49. Jordinson, M., Goodlad, R. A., & Brynes, A., (1999). Gastrointestinal responses to a panel of lectins in rats maintained on total parenteral nutrition. *American Journal of Physiology-Gastrointestinal and Liver Physiology, 276*(5), G1235–G1242.
50. Kan, A., Özçeli, B., Kartal, M., Özdemir, Z. A., & Özgen, S., (2010). *In vitro* antimicrobial activities of *Cicer arietinum* L (Chickpea). *Tropical Journal of Pharmaceutical Research, 9*(5), 475–481.
51. Karkouch, I., Tabbene, O., & Gharbi, D., (2017). Antioxidant, antityrosinase and antibiofilm activities of synthesized peptides derived from *Vicia faba* protein hydrolysate: A powerful agents in cosmetic application. *Industrial Crops and Products, 109*, 310–319.
52. Kuno, A., Ikehara, Y., & Tanaka, Y., (2011). Multi lectin assay for detecting fibrosis-specific glyco-alteration by means of lectin microarray. *Clinical Chemistry, 57*(1), 48–56.
53. Lam, S. K., & Ng, T. B., (2010). First report of a haemagglutinin-induced apoptotic pathway in breast cancer cells. *Bioscience Reports, 30*(5), 307–317.
54. Lam, S. K., & Ng, T. B., (2013). Purification and characterization of an antifungal peptide with potent antifungal activity but devoid of antiproliferative and HIV reverse transcriptase activities from *Legumi secchi* beans. *Applied Biochemistry and Biotechnology, 169*(7), 2165–2174.
55. Lannoo, N., & Van, D. E. J., (2014). Lectin domains at the frontiers of plant defense. *Frontiers in Plant Science, 5*, 397–402.

56. Lannoo, N., & Vandenborre, G., (2007). The jasmonate-induced expression of the *Nicotiana tabacum* leaf lectin. *Plant and Cell Physiology, 48*(8), 1207–1218.
57. Li, C., Chen, J., Lu, B., & Shi, Z., (2014). Molecular switch role of Akt in *Polygonatum odoratum* lectin-induced apoptosis and autophagy in human non-small cell lung cancer A549 cells. *PloS One, 9*(7). Article ID: e101526.
58. Lin, P., Ye, X., & Ng, T. B., (2008). Purification of melibiose-binding lectins from two cultivars of Chinese black soybeans. *Acta Biochimica et Biophysica Sinica, 40*(12), 1029–1038.
59. Lis, H., & Sharon, N., (1986). Application of lectins. Chapter 5. In: Liener, I. E., Sharon, N., & Goldstein, I. J., (eds.), *The Lectins: Properties, Functions and Applications in Biology and Medicine* (pp. 293–370). New York - USA: Academic Press Inc.
60. Lis, H., & Sharon, N., (1986). Lectins as molecules and as tools. *Annual Review of Biochemistry, 55*(1), 35–67.
61. Lizotte, P. H., Wen, A. M., & Sheen, M. R., (2016). *In situ* vaccination with cowpea mosaic virus nanoparticles suppresses metastatic cancer. *Nature Nanotechnology, 11*(3), 295–303.
62. Loris, R., (2002). Principles of structures of animal and plant lectins. *Biochimica et Biophysica Acta (BBA)-General Subjects, 1572*(2, 3), 198–208.
63. Lyu, S. Y., Kwon, Y. J., Joo, H. J., & Park, W. B., (2004). Preparation of alginate/chitosan microcapsules and enteric-coated granules of mistletoe lectin. *Archives of Pharmacal Research, 27*(1), 118–122.
64. Mazalovska, M., & Kouokam, J., (2018). Lectins as promising therapeutics for the prevention and treatment of HIV and other potential coinfections. *BioMed Research International, 2018*. Article ID: 3750646.
65. Michiels, K., Van, D. E. J., & Smagghe, G., (2010). Plant-insect interactions: What can we learn from plant lectins? *Archives of Insect Biochemistry and Physiology, 73*(4), 193–212.
66. Miller, J. B., Hsu, R., Heinrikson, R., & Yachnin, S., (1975). Extensive homology between the subunits of the phytohemagglutinin mitogenic proteins derived from *Phaseolus vulgaris. Proceedings of the National Academy of Sciences, 72*(4), 1388–1391.
67. Mirth, D. B., Miller, C. J., Kingman, A., & Bowen, W. H., (1979). Inhibition of saliva-induced aggregation of *Streptococcus mutans* by wheat germ agglutinin. *Caries Research, 13*, 121–131.
68. Mishra, A., Behura, A., & Mawatwal, S., (2019). Structure-function and application of plant lectins in disease biology and immunity. *Food and Chemical Toxicology, 134*, 110827.
69. Mitchell, C. A., Ramessar, K., & O'Keefe, B. R., (2017). Antiviral lectins: Selective inhibitors of viral entry. *Antiviral Research, 142*, 37–54.
70. Miyake, K., Tanaka, T., & McNeil, P. L., (2007). Lectin-based food poisoning: A new mechanism of protein toxicity. *PloS One, 2*(8). Article ID: e687.
71. Morgan, E. D., & Wilson, I. D., (1999). Insect hormones and insect chemical ecology. Chapter 5. In: Mori, K., (ed.), *Comprehensive Natural Products Chemistry, Miscellaneous Natural Products Including Marine Natural Products, Pheromones, Plant Hormones, and Aspects of Ecology* (Vol. 8, pp. 263–376). Oxford: Elsevier.

72. Nair, S. S., Madembil, N. C., Nair, P., Raman, S., & Veerabadrappa, S. B., (2013). Comparative analysis of the antibacterial activity of some phytolectins. *International Current Pharmaceutical Journal, 2*(2), 18–22.
73. Naughton, P. J., Grant, G., Bardocz, S., & Pusztai, A., (2000). Modulation of salmonella infection by the lectins of *Canavalia ensiformis* (Con A) and *Galanthus nivalis* (GNA) in a rat model *in vivo. Journal of Applied Microbiology, 88*(4), 720–727.
74. Neutsch, L., Wirth, E. M., Spijker, S., & Pichl, C., (2013). Synergistic targeting/prodrug strategies for intravesical drug delivery: Lectin-modified PLGA microparticles enhance cytotoxicity of stearoyl gemcitabine by contact-dependent transfer. *Journal of Controlled Release, 169*(1, 2), 62–72.
75. Nicolson, G. L., (1974). The interactions of lectins with animal cell surfaces. *International Review of Cytology, 39*, 89–100.
76. Noah, D. N., Bender, A. E., Reaidi, G. B., & Gilbert, R. J., (1980). food poisoning from raw kidney beans. *British Medical Journal, 281*, 236, 237.
77. Oliveira, W. F., Silva, G. M., & Cabral, F. P. E., (2018). Titanium dioxide nanotubes functionalized with *Cratylia mollis* seed lectin, cramoll, enhanced osteoblast-like cells adhesion and proliferation. *Materials Science and Engineering - C, 90*, 664–672.
78. Otte, J. M., Chen, C., & Brunke, G., (2001). Mechanisms of lectin (Phyto hemagglutinin)-induced growth in small intestinal epithelial cells. *Digestion, 64*, 169–178.
79. Ouyang, L., Chen, Y., & Wang, X. Y., (2014). *Polygonatum odoratum* Lectin induces apoptosis and autophagy via targeting EGFR - mediated Ras - Raf - MEK - ERK pathway in human MCF-7 breast cancer cells. *Phytomedicine, 21*(12), 1658–1665.
80. Paiva, P. M. G., & Gomes, F. S., (2010). Antimicrobial activity of secondary metabolites and lectins from plants. *Current Research, Technology and Education Topics in Applied Microbiology and Microbial Biotechnology, 1*, 396–406.
81. Pérez, S., Sarkar, A., Rivet, A., Breton, C., & Imberty, A., (2015). Glyco3D: Portal for structural glycosciences. Chapter 18. In: Lütteke, T., & Frank, M., (eds.), *Glyco Informatics* (Vol. 1273, pp. 241–258). New York: Humana Press.
82. Peumans, W. J., Smeets, K., & Van, N. K., (1997). Lectin and alliinase are the predominant proteins in nectar from leek (*Allium porrum* L.) flowers. *Planta, 201*(3), 298–302.
83. Peumans, W. J., & Vandamme, E. J. M., (1995). Lectins as plant defense proteins. *Plant Physiology, 109*, 347–352.
84. Peumans, W. J., Winter, H. C., & Bemer, V., (1997). Isolation of a novel plant lectin with an unusual specificity from *Calystegia sepium. Glycoconjugate Journal, 14*(2), 259–265.
85. Pina-Pérez, M. C., & Pérez, M., (2018). Antimicrobial potential of legume extracts against foodborne pathogens: Review. *Trends in Food Science & Technology, 72*, 114–124.
86. Plattner, V. E., Wagner, M., Ratzinger, G., Gabor, F., & Wirth, M., (2008). Targeted drug delivery: Binding and uptake of plant lectins using human 5637 bladder cancer cells. *European Journal of Pharmaceutics and Biopharmaceutics, 70*(2), 572–576.
87. Pusztai, A., & Bardocz, S., (1995). *Lectins: Biomedical Perspectives* (p. 277). London: Taylor and Francis.

88. Pusztai, A., (1993). Dietary lectins are metabolic signals for the gut and modulate immune and hormone functions. *European Journal of Clinical Nutrition, 47*, 691–699.
89. Pusztai, A., (1991). *Plant Lectins* (p. 369). New York: Cambridge University Press.
90. Qin, Q., Zhang, Q., Zhao, W., Wang, Y., & Peng, Y., (2003). Identification of a lectin gene induced in rice in response to *Magnaporthe grisea* infection. *Acta Botanica Sinica, 45*(1), 76–81.
91. Rahaie, M., & Kazemi, S. S., (2010). Lectin-based biosensors: As powerful tools in bioanalytical applications. *Biotechnology, 9*(4), 428–443.
92. Randhir, R., Lin, Y. T., & Shetty, K., (2004). Stimulation of phenolics, antioxidant and antimicrobial activities in dark germinated mung bean sprouts in response to peptide and phytochemical elicitors. *Process Biochemistry, 39*(5), 637–646.
93. Rodrigues, J. S., & Santos-Magalhaes, N. S., (2003). Novel core (polyester)-shell (polysaccharide) nanoparticles: Protein loading and surface modification with lectins. *Journal of Controlled Release, 92*(1, 2), 103–112.
94. Rogers, A. H., (1979). Biological methods of dental caries prevention-review. *Australian Dental Journal, 24,* 153–158.
95. Rosenthal, G. A., & Berenbaum, M. R., (1991). *Herbivores* (2nd edn., p. 452) San Diego: Academic Press.
96. Rüdiger, H., (1998). Plant lectins-more than just tools for glyco scientists: Occurrence, structure, and possible functions of plant lectins. *Cells Tissues Organs,* 161(1–4), 130–152.
97. Sánchez, S. H., & Morquecho, C. A., (2017). Chemical plant defense against herbivores. Chapter 1. In: Shields, V. D. C., (ed.), *Herbivores* (pp. 3–28) IntechOpen, online.
98. Santiago, J., Levy-Benshimol, A., & Carmona, A., (1993). Effect of *Phaseolus vulgaris* lectins on glucose absorption, transport, and metabolism in rat everted intestinal sacs. *The Journal of Nutritional Biochemistry, 4*(7), 426–430.
99. Sasaki, M., Fitzgerald, A. J., & Grant, G., (2002). Lectins can reverse the distal intestinal atrophy associated with elemental diets in mice. *Alimentary Pharmacology & Therapeutics, 16*(3), 633–642.
100. Schwarz, R. E., & Wojciechowicz, D. C., (1999). Wheat germ agglutinin-mediated toxicity in pancreatic cancer cells. *British Journal of Cancer, 80*(11), 1754–1762.
101. Sharma, A., Ng, T. B., Wong, J. H., & Lin, P., (2009). Purification and characterization of a lectin from *Phaseolus vulgaris* (cv. Anasazi beans). *Journal of Biomedicine and Biotechnology, 2009*. Article ID: 929568.
102. Shi, L., Arntfield, S. D., & Nickerson, M., (2018). Changes in levels of phytic acid, lectins and oxalates during soaking and cooking of Canadian pulses. *Food Research International, 107*, 660–668.
103. Shibamoto, T., & Bjeldanes, L. F., (2009). *Introduction to Food Toxicology* (2nd edn., p. 320). New York - USA: Academic Press.
104. Shiraki, K., Takase, K., Tameda, Y., Hamada, M., & Kosaka, Y., (1995). Clinical study of lectin-reactive alpha-fetoprotein as an early indicator of hepatocellular carcinoma in the follow-up of cirrhotic patients. *Hepatology, 22*(3), 802–807.
105. Soetan, K. O., (2008). Pharmacological and other beneficial effects of anti-nutritional factors in plants: Review. *African Journal of Biotechnology, 7*(25), 4713–4721.

106. Sollid, L. M., Kolberg, J., Scott, H., Ek, J., Fausa, O., & Brandtzaeg, P., (1986). Antibodies to wheat germ agglutinin in coeliac disease. *Clinical and Experimental Immunology, 63*(1), 95–99.
107. Souza, M. A., Carvalho, F. C., Ruas, L. P., Ricci, A. R., & Roque, B. M. C., (2013). The Immunomodulatory effect of plant lectins: Review with emphasis on ArtinM properties. *Glycoconjugate Journal, 30*(7), 641–657.
108. Subramanyam, S., Smith, D. F., & Clemens, J. C., (2008). Functional characterization of HFR1, a high-mannose N-glycan-specific wheat lectin induced by hessian fly larvae. *Plant Physiology, 147*(3), 1412–1426.
109. Sung, J. C., Pulliam, B. L., & Edwards, D., (2007). Nanoparticles for drug delivery to the lungs. *Trends in Biotechnology, 25*, 563–570.
110. Une, S., Nonaka, K., & Akiyama, J., (2018). Lectin isolated from Japanese red sword beans (*Canavalia gladiata*) as a potential cancer chemopreventive agent. *Journal of Food Science, 83*(3), 837–843.
111. VanDamme, E. J. M., Lannoo, N., & Peumans, W. J., (2008). Plant lectins. *Advances in Botanical Research, 48*, 107–209.
112. VanDamme, E. J. M., Peumans, W. J., Barre, A., & Rouge, P., (1998). Plant lectins: A composite of several distinct families of structurally and evolutionary related proteins with diverse biological roles. *Critical Review in Plant Sciences, 17*, 575–692.
113. VanDamme, E. J. M., Rougé, P., & Peumans, W. J., (2007). Carbohydrate-protein interactions: Plant lectins. In: Kamerling, J. P., Boons, G. J., Lee, Y. C., Suzuki, A., Taniguchi, N., & Voragen, A. G. J., (eds.), *Comprehensive Glycoscience - from Chemistry to Systems Biology* (Vol. 3, pp. 563–599). Oxford, UK: Elsevier.
114. Vander, E. J., VanBaardewijk, L. J., Sier, C. F. M., & Schipper, I. B., (2013). Bone healing and mannose-binding lectin. *International Journal of Surgery, 11*(4), 296–300.
115. VanDriessche, E., (1988). Structure and function of Leguminosae lectins. In: Franz, H., (ed.), *Advances in Lectin Research* (Vol. 1, pp. 73–134). Berlin: Springer-Verlag.
116. Van, D. G., Miersch, O., & Hause, B., (2009). *Spodoptera littoralis* induced leptin expression in tobacco. *Plant and Cell Physiology, 50*, 1142–1155.
117. Van, D. G., Smagghe, G., & Van, D. E. J., (2011). Plant lectins as defense proteins against phytophagous insects. *Phytochemistry, 72*(13), 1538–1550.
118. Vasconcelos, I. M., & Oliveira, J. T. A., (2004). Antinutritional properties of plant lectins. *Toxicon, 44*(4), 385–403.
119. Verhelst, X., & Dias, A. M., (2019). Protein glycosylation as a diagnostic and prognostic marker of chronic inflammatory gastrointestinal and liver diseases. *Gastroenterology, 158*(1), 95–110.
120. Vojdani, A., (2015). Lectins, agglutinins, and their roles in autoimmune reactivities. *Alternative Therapies, 21*(1), 46–51.
121. Vojdani, A., Gushgari, L. R., & Vojdani, E., (2020). Interaction between food antigens and the immune system: Association with autoimmune disorders. *Autoimmunity Reviews, 19*(3). Article ID: 102459.
122. Vyas, S. P., Sihorkar, V., & Dubey, P. K., (2001). Preparation, characterization and *in vitro* antimicrobial activity of metronidazole-bearing lectinized liposomes for intraperidontal Pocket delivery. *Pharmazie, 56*, 554–560.

123. Weis, W. I., & Drickamer, K., (1996). Structural basis of lectin-carbohydrate recognition. *Annual Review of Biochemistry, 65*(1), 441–473.
124. Williams, P. E., Pusztai, A. J., MacDearmid, A., & Innes, G. M., (1984). The use of kidney beans (*Phaseolus vulgaris*) as protein supplements in diets for young rapidly growing beef steers. *Animal Feed Science and Technology, 12*(1), 1–10.
125. Wilson, A. B., King, T. P., Clarke, E. M. W., & Pusztai, A., (1980). Kidney bean (*Phaseolus vulgaris*) lectin-induced lesions in rat small intestine: 2. Microbiological studies. *Journal of Comparative Pathology, 90*(4), 597–602.
126. Wu, L., Liu, T., Xiao, Y., Li, X., Zhu, Y., Zhao, Y., Bao, J., & Wu, C., (2016). *Polygonatum odoratum* lectin induces apoptosis and autophagy by regulation of microRNA-1290 and microRNA-15a-3p in human lung adenocarcinoma A549 cells. *International Journal of Biological Macromolecules, 85*, 217–226.
127. Yamane, H., Konno, K., Sabelis, M., Takabayashi, J., Sassa, T., & Oikawa, H., (2010). Chemical defense and toxins of plants. Chapter 4. In: Mander, L., & Lui, H. W., (eds.), *Comprehensive Natural Products II Chemistry and Biology* (Vol. 4, pp. 339–385). Amsterdam: Elsevier.
128. Yang, Y., Xu, H. L., Zhang, Z. T., & Liu, J. J., (2011). Characterization, molecular cloning, and in silico analysis of a novel mannose-binding lectin from *Polygonatum odoratum* (Mill.) with anti-HSV-II and apoptosis-inducing activities. *Phytomedicine, 18*(8, 9), 748–755.
129. Yue, R., Shen, B., & Morrison, S., (2016). Clec-11a/ osteonectin is an osteogenic growth factor that promotes the maintenance of the adult skeleton. *Elife, 13*(5), e18782.
130. Zang, J., Li, D., Piao, X., & Tang, X., (2006). Effects of soybean agglutinin on body composition and organ weights in rats. *Archives of Animal Nutrition, 60*, 245–253.
131. Zhang, W., Peumans, W. J., & Barre, A., (2000). Isolation and characterization of a jacalin-related mannose-binding lectin from salt-stressed rice (*Oryza sativa*) plants. *Planta, 210*, 970–978.
132. Zou, L. B., & Zhan, J. B., (2005). Purification and anti-cancer activity of ricin. *Medical Sciences, 34*(3), 217–219.

CHAPTER 13

RECALL OF FOOD PRODUCTS: PRACTICES AND LESSONS

SARANG D. POPHALY, MANORAMA CHAUHAN, AWADHESH K. TRIPATHI, RAGHAV PANDEY, TEJINDER P. SINGH, and KUSHAL K. SANDEY

ABSTRACT

Greater food safety compliance and consumer awareness have resulted in more and more food recall around the globe. A food recall is an unavoidable situation if a hazard has entered the food chain and has the ability to cause undesirable health effects to consumers. Many companies have faced bankruptcy due to a recall event and ensuing costs. It is, therefore, in the best interest of the manufacturers to implement modern food safety systems, which may even incur a considerable cost rather than facing a looming food recall crisis. The recall requires beforehand preparation, provisions, and plans. A strong and robust traceability system will simplify the logistical complications of a recall event. With increased global food trade mandating harmony between countries, emerging economies should bring about the structural and cultural changes required for effective planning, execution, and reporting of food recalls.

13.1 INTRODUCTION

Modern food safety practices and systems have changed the scenario of food-related outbreaks around the world. Many countries have revamped their food laws and regulatory agencies in the last two decades. With heightened sensitivity of consumers, stricter laws in place and increased international trade of food and related items, the manufacturers are pushed towards greater compliance to food safety laws. This has resulted in a number of food-related incidents coming to light vis-a-vis a previous state, where food outbreaks and non-compliance were hushed away mostly in developing countries.

One of the worst nightmares of a food business operator is to face a situation, wherein the prepared product is in movement or at consumer's end and a potential hazard is detected at this stage. The manufacturer, in such a case, has to undergo the process of food recall. In its simplest form, a food recall is a self-initiated action on the part of a manufacturer to prevent the consumer from having a food item, which may contain a potential hazard. The food product needs to be removed from commerce and this results in legal, social, economic, technological, and logistical challenges for the organization.

A recall plan is usually part of Pre-requisite Programs (PRPs), which should be implemented by an organization to ensure a basic working system with respect to hygiene, sanitation, and contingency situations, in a food facility. Many countries have laid down detailed framework of a recall plan and have a systematic approach to recall events. On the other hand, most countries lack a dedicated framework and procedures to undertake a recall.

Microbial pathogens and undeclared allergens are two major reasons for food recalls in the US and Canada [9], others being spoilage, packaging, and labeling issues. *Listeria monocytogenes*, *E. coli* (especially Enterohemorrhagic *E. coli*) and Salmonella are major causes of recalls of processed foods. Detection and control of microbial pathogens, prevention of cross-contact for allergen control and detection of physicochemical hazards will go a long way in the prevention of recalls. However, complacency in mindset and working in safety-related issues sets in quite easily, when critical operations become routine work [25]. A food recall is often not a result of split-second decisions but rather a long-term behavioral and cultural reflection of the inner working and understanding of crucial food safety matters. This book chapter deals with different incidences, recall procedures and other aspects of food recall.

13.2 REASONS FOR FOOD RECALL

The leading cause of food recalls are undeclared allergens, microbial contamination, foreign material in food, quality-related issues (Table 13.1).

Food allergy is a peculiar immune response triggered in hosts due to one or more food constituents. Even a miniscule amount of allergens present in food may elicit an immune misfire and cause symptoms, such as itchiness, hives, swelling, vomiting, and difficulty in breathing or lowered blood pressure. The *Codex Alimentarius Commission* (CAC) and United States-Food Allergen Labeling and Consumer Protection Act (FALCPA) recognizes eight major types of food allergens: eggs, crustacean shellfish, tree nuts, milk, fish, peanuts, soybean, and wheat [6]. These allergens are broad classification and the actual compound or ingredient causing the reaction may belong to any of these categories.

TABLE 13.1 Major Reasons of Food Recall

Reason	Description
Faulty packaging	Defective or damaged package, adding toys/ other non-food items inside the package.
Food spoilage	Chemically or microbiologically induced spoilage of food contents.
Foreign material	Physical objects: metal shavings, plastic, wood, insects, glass pieces, etc. Chemical contaminants: Antibiotic or pesticide residues, heavy metal residues, veterinary drugs.
Microbial contamination	Microbial spoilage or Bacterial contamination (*E. coli*, *Salmonella typhi*, *Clostridium botulinum*, *Listeria monocytogenes*)
Misleading label information	Undeclared allergens, incorrect ingredient information, missing essential logo (e.g., Vegetarian logo), misprinting.

A common consumer often lacks a broad scientific knowledge of technical terms and nomenclature, e.g., a consumer may or may not know that casein is derived from milk and lecithin from soy, and may end up buying a product, which have these molecules as ingredients. Hence, appropriate labeling regulations are in place in most countries to deal with such ambiguities for consumer protection. Individuals susceptible to specific allergic reactions should avoid altogether consuming the food containing the allergen. However, cross-contact during manufacturing can inadvertently lace the food with an unwanted allergen [11].

Cross contact refers to the phenomenon, wherein a product containing allergen comes in contact with a product that was not supposed to have that allergen, naturally or otherwise. Cross contact occurs due to using a common facility, equipment, surfaces for manufacturing and storage of two different types or variants of food, e.g., using the same facility to manufacture gluten-containing; and gluten-free products might result in cross-contact and transfer of trace amounts of gluten in the former. Even trace amounts of allergens transferred as a result of cross-contact can be a public health safety concern [28]. Rigorous cleaning and sanitation protocols and rational scheduling of operations are key control mechanisms to prevent inadvertent cross contact of allergens [11].

Microbial contamination by spoilage or pathogenic bacteria is another important cause of food recall. The diversity and number of microorganisms in the food system vary with type of food, biochemical composition, processing, manufacturing, and storage conditions, etc. Pathogens in foods can cause foodborne intoxication or infections. *Listeria monocytogenes*, *E. coli* (especially Enterohemorrhagic *E. coli*) and Salmonella are major cause of recalls of processed foods and spices [3, 9, 21, 31]. A wide range of systems and practices in form of pre-requisite programs, hazard analysis and critical control point (HACCP), good manufacturing practices (GMPs), good hygienic practices (GHPs), food safety management system, etc., are applied and followed in manufacturing facilities to contain entry and ensure destruction of pathogens in food.

Foreign material contamination is more apparent and visible cause of food recall as compared to other factors. Foreign material can be defined as an object or item that is alien to the food [4]. This broad definition of foreign material virtually makes the list of items limitless. Several food recalls in the past have been related to the presence of undesirable physical objects, such as plastic pieces, insects, metal, glass, plant material, stones, animal tissues and bones, rubber, wood, fibers, hairs, etc. (Table 13.2). Presence of such objects either can be a result of an inadvertent mistake or can be intentional adulteration. The source of entry of a foreign material can be at any stage of cultivation, harvesting, processing, packaging, and distribution.

Control of foreign materials entry is achieved by a wide variety of practices and technological interventions. Common operations and systems used are: metal detectors, X-ray machines, thermal imaging, magnets, electronic sorters, air blowers, flumes, sieves, and filters [8, 22].

13.3 CLASSES OF FOOD RECALL

1. **Firm Initiated Recall:** The company itself initiates the recall.
2. **Food Authority Initiated Recall:** The national Food safety authority notifies and initiates the recall.

Table 13.2. Some Common Foreign Material, which can be Potential Physical Hazards

Foreign Material	Sources/Entry	Particular Concern
Bone pieces	Common problem associated with meat products, particularly with fish.	Can cause choking
Glass	Pieces of glass broken from storage bottles and jars, glass thermometers and pressure gauges, glass spectacles and goggles.	Glass is fragile and easily shattered generating very small pieces, which are difficult to detect. Potential to cause intestinal damage to the consumer is very high.
Insects	Insects associated with the food and incorporated during harvesting or processing. Insects such as flies, mosquitoes, cockroaches, cricket, grasshoppers can be pestered in storage and processing environments.	Difficulty in detection especially in fresh produce.
Metals	Machinery parts, nuts, bolts, nails, and other fixtures, cutting, and dicing operations, splinters, and shavings, employee personal utility items such as coins, jewelry, stapler pins, and other stationary objects. Leaching of metal from storage containers to food can also be a source.	Wide type of metals can get entry in food such as iron, gold, silver, steel, aluminum.
Plant material	Bits of leaf, stalk, woody material, roots of carrot, stem of cabbage, etc.	Blends easily with food due to the soft nature of the tissues.
Plastics	Machinery parts, packaging, fragments of kitchen items such as pepper grinders, casing of food mixers and blenders, electric wires and cables, Personnel utility items such as pen caps, key chains, etc.	Difficulty in detection
Rubber pieces	Gaskets, rubber tubing in machines	Commonly associated with wear and tear and high temperature processing which damages the rubber
Stones, husk, and soil	Harvesting, intentional adulteration	Difficulty in detection if the color and size of stones matches that of grains

US-FDA have divided food recall on the basis of seriousness and severity of the hazard to cause public health effect:

- **Class I:** If the product in the commerce is an imminent threat and can cause death if consumed, then the recall is classified as Class I.
- **Class II:** This class of recall is done for products that pose moderate severity and can cause serious health issues or irreversible consequences.
- **Class III:** This class of recall is done for products which have mild severity and it is unlikely to cause any adverse health threat.

Food Standards Australia and New Zealand (FSANZ) have classified food recall on the basis of causative agent:

- Microbial contamination;
- Labeling;
- Foreign matter;
- Chemical/other contaminant;
- Undeclared allergen;
- Biotoxin;
- Other.

13.4 COMPARATIVE STATUS OF RECALLS IN DIFFERENT COUNTRIES

Table 13.3 enlists some major food recalls in the past two decades. In the US, meat/beef, poultry related recalls are monitored by USDA, whereas other products fall in the domain of US-FDA. In the last decade (2010 to 2019), the United States of America had reported 1076 recalls (for meat/beef, poultry), compared to 707 in Australia. Canadian Food Inspection Agency (CFIA) have reported 2094 recall events in Canada from 2000 to 2017, with a total of 10,432 products being recalled due to microbial contamination [9].

European food market is governed by EFSA and the recall alerts are issued by the rapid alert system for food and feed (RASFF). Countries like Japan, China, Brazil, and India with an expanding food market needs to focus on structural reforms to create single-point agencies for recall coordination. Unfortunately, a study conducted over a period of 11 months just before the Beijing declaration reported that the vast majority of food reports on outbreaks and recalls are made by just 10 countries [18]. It also highlighted faulty foodstuff originating from selected countries and lack of reporting in these countries is a cause of concern.

TABLE 13.3 Major Food Recalls in History

Company	Food Product	Recall Reason	Year	Affected	References
Cargill Meat Solutions Corporation, USA	Ground Turkey	Salmonella outbreak	2011	Around 75 confirmed illness: 1 Death	[15]
Fonterra, New Zealand	Whey powder	Clostridium spp.	2013	None	[20]
Hallmark/Westland Meat Packaging, USA	Raw and Frozen beef	Unhealthy animals. Slaughter practices	2007	None	[23]
Menu Foods, USA	Pet Foods	Melamine	2007	Several deaths and illness	[5]
Peanut Corporation of America, USA	Peanuts	Salmonella outbreak	2007	Around 714 confirmed illness: 9 Deaths	[13]
Pilgrim's Pride, USA	Frozen Turkey and Chicken	Listeria contamination	2002	46 cases, 7 deaths, 3 stillbirths	[29]
Sanlu Group, China	Infant formula	Melamine	2008	2,90,000 cases; 102 deaths.	[33]
Sara Lee Corporation, USA	Deli products and Hot dogs	Listeria contamination	1998	100 cases; 21 deaths	[32]
Wright County/Hillandale Farm, USA	Eggs	Salmonella outbreak	2010	Around 1,500 confirmed illness: No death	[14]
XL foods, Canada	Ground beef	*E. coli* contamination	2012	18 cases, no death	[2]

13.5 RECALL PROCEDURE

A recall event is an undesirable but inevitable occurrence and needs to be handled professionally. It may require liaising with local food safety authorities, media, civil courts, supply chain, consumers, etc. It should be done as a teamwork with a real time cognition of developing and ongoing situation(s). Thus, every organization should draw a properly drafted roadmap or a standard operating procedure to deal with such a situation. Many national food regulatory agencies categorically define the framework of steps to be taken in case of a recall event. The general protocol can be summarized as follows:

1. **Assembling a Recall Management Team:** The management should formulate an interdisciplinary recall team. The team should comprise experts from different spheres so that every aspect of the recall can be handled professionally. This includes personnel from higher management (decision-makers), quality assurance (QA) and food safety team, media communication team, complaint investigation (root cause analysis (RCA)), company representative who deals with local food authority, legal counsel, representative from accounts. Additional members from concerned manufacturing sections, marketing team, supply chain and HACCP team can also be made members if required. The team members, besides having technical competence, should also be motivated and dedicated towards the cause of the organization and also sensitive towards the health and well-being of their customers. The role and responsibilities of the team members should be clearly delineated and demarcated.
2. **Reporting the Problem to the Concerned Food Safety Authority:** The initiation of market recall is usually a result of either a detection of a problem in the product by the organization itself or a market complaint received from a customer. In the latter case, the customer may file a complaint on the contact details given on the package to the company website or email directly. But occasional customers can report the matter directly to the food authority as well. Either way, the concerned local food safety authority should be notified at the earliest about the problem and ensuing situation. The information provided to the authorities should have details on the product under recall process, hazard associated with the batch, brand, size,

and lot code affected, details of any complaint received, extent of distribution of the product, total quantity of product manufactured, firm's contact details and any other relevant information available or requested by the authorities.

3. **Identification of Product(s) to be Recalled:** The manufacturer should ensure that all the range of products and brands which are affected should be properly identified. Use of common ingredients, same packaging line, etc., can cause similar defects or hazards in other products manufactured by the organization. Hence a thorough check on all the product range and batches should be done.
4. **Segregation of All Products to be Recalled which are in Firm's Control:** It is imperative that the recalled product should be properly segregated and identified. Control of the recalled product with respect to its storage, accidental remixing with fresh lot should be done in an impeccable way.
5. **Prepare and Distribute the Information of Recall, Including Press Release:** Recall information needs to be communicated to all the stakeholders such as customers, distribution channel, media outlets. The customer should be informed about the specific product being recalled, batch size, manufacturing date of the product, reasons for the recall, health advisory with respect to the hazard, action required on the part of consumer, refund or replacement options, contact details for customer to contact the organization.
6. **Prepare the Distribution List:** Supply chains usually include zonal or state distributors, area distributors, retailers, and supermarkets and finally customers. The more the product has moved down the distribution channel, the bigger the logistical challenge to get back each and every unit of the recalled product. Managing an accurate distribution record with a possibility of a real time updated system is a key to retrieve the market penetrated product.
7. **Verify the Effectiveness of the Recall:** The effectiveness of an ongoing recall should be monitored to check if all the intended target of receiving back the product and stopping the accidental sale of the tainted item is achieved.
8. **Control the Recalled Product:** The recalled product should be pooled to a central location or in case of a wide distribution channel, to earmarked recovery sites. The recovered product should be

stored separately and should not be mingled with forward rolling products. Accurate record-keeping of recovered products is very important.

9. **Decide What to do with the Recalled Product:** Depending upon the problem for which it was recalled, the product may either be reworked, reprocessed, repacked or it may have to be discarded or destroyed. The technological challenges in reworking and reprocessing may limit its ability to repack and resale and can be done in certain circumstances only. If the product has a hazard which cannot be mitigated or controlled, it needs to be destroyed by incineration, landfill or any other suitable means. The action taken with respect to the recalled product must be properly documented.
10. **Fix the Cause of the Recall if the Problem Occurred at the Facility:** An RCA should be done to ascertain the reasons and conditions under which the hazard or defect occurred in the product. The problem should be fixed in a comprehensive manner.

13.6 TRACEABILITY

Traceability enables the tracking of any food item through different stages of production, processing, and distribution including retail points and possibly at consumer doorstep. It can ensure that movement of an article can be traced one step upwards or downwards in the distribution chain with absolute certainty at any given time [1].

Traceability is an important tool, which enables for an effective food recall when a food safety hazard is detected. When a potential food safety problem is detected, by a food business operator or a regulatory agency, an effective traceability system can help separate out and prevent contaminated products from reaching consumers. Technical instruments and interventions commonly used for traceability are alphanumeric codes, bar codes, radio frequency identification (RFID), wireless sensor network [1] and more recently introduced blockchain technology [16].

13.7 FINANCIAL IMPACT OF FOOD RECALL

A food recall event brings a financial burden on the organization and also has a long-term impact on the economy of a country [26]. A food borne outbreak adds to the healthcare bill of a country and increases in disability adjusted life years (DALYs) [30]. Assessment of damage to the company's brand value and reputation in the long run is difficult and not readily calculable as the ensuing factors are overwhelming and complex in nature, but the short-term impact of recall with tangible cost can be assessed [34].

The direct costs incurred as a result of a food recall include: cost towards assembling a recall team, logistical expenses for removing the product from the market, RCA, managing the media, etc. On the other hand, indirect costs include litigation, decline in stock value, penalties, drop in sales, impact on the sector in general, etc. Companies, such as Peanut Corporation of America, Sunland, Hudson Foods, Thorn Apple Valley, etc., have gone bankrupt due to recall events, as they could not overcome the costs associated with recalls. A firm's size, its previous experience of recall handling, media information and recall size are important drivers of the financial implication of the recall [24]. Given the high risk posed by food recalls certain insurance companies offer recall insurance policies to cover the direct and indirect cost of a recall event [17, 27].

Avoiding recall by following strict food safety norms and foresight in decision making is important. Implementation of technological solutions for food safety are business decisions and will definitely come up with a price tag. In hindsight, the manufacturer may think that a specific cost cutting step towards a food safety technology might have triggered a recall and a timely wise decision of implementing it, could have prevented the recall.

It is difficult and complex to assess the financial worth associated with food safety risk control step [7], thus predicting financial loss associated with a potential recall is a more realistic way of determining the impact of not choosing a food safety technology. Gomez and Marks [7], demonstrated that if the intervention by a food safety technological solution could decrease the risk of a recall by fivefold, the mean annual economic benefit would be greater than 2 million USD in reduced risk, for companies which are involved in manufacture of low moisture foods.

13.8 CONSUMER BEHAVIOR

The consumer is unforgiving and the return of the consumer to repurchase of a product after a recall event will depend on how the recall was perceived by the general masses. Public perception of companies undergoing a recall varies from product to product. For example, recalls in the food industry have a more negative reaction than the automobile industry [34]. Proactive recall and forthcoming behavior in refunds and compensation generates confidence among the consumers which is lacking in a passive recall [10, 34]. The willful delay and tactics of a company to avoid a recall and non-seriousness during the recall can widely cause public dissent as was in the case of 2007 pet food recall by Menu Foods in Canada and United States (US) [5].

Different corporate social responsibility (CSR) activities during and after recall events also can play an important role in a company's image building. Chipotle Mexican Grill (Chipotle) faced a food safety crisis in 2015, when several 100 people were reported ill due to contaminated Chipotle food. The company, in turn, responded to the crisis situation by introducing a food safety program and constituted the Chipotle Local Grower Support Initiative, which provided farmers with technical and financial assistance to develop best practices to meet enhanced food safety standards [19]. Thus, CSR helps to abate the negative effect of food recalls on the wealth of shareholders [12].

13.9 SUMMARY

Ensuring food safety is often described as a thankless job, with a pitfall of complacency. The modern food safety practices and systems have generated a robust environment of compliance and resulted in more cases of foodborne incidents being reported to authorities and the general public. In an unlikely situation of a hazard entering the food system or the food developing an unwanted spoilage condition, a food recall has to be taken voluntarily or can be mandated by the authorities. Recall is achieved by an interdisciplinary recall team with members from higher management (decision-makers), QA and food safety team, media communication team, complaint investigation (RCA), company representative who deals with local food authority, legal counsel, representative from accounts, etc. An effective recall plan considers provisions for reporting to authorities, media communications, control of the recalled products, re-processing or destruction of the product, proper documentation of action taken, and fixing of the problem.

KEYWORDS

- **allergen**
- **food recall**
- **foreign matter**
- **pathogen**
- **recall procedure**
- **traceability**

REFERENCES

1. Aung, M. M., & Chang, Y. S., (2014). Traceability in a food supply chain: Safety and quality perspectives. *Food Control, 39*, 172–184.
2. Charlebois, S., Von-Massow, M., & Pinto, W., (2015). Food recalls and risk perception: An exploratory case of the XL Foods and the biggest food recall in Canadian history. *Journal of Food Products Marketing, 21*(1), 27–43.
3. Dey, M., Mayo, J. A., Saville, D., Wolyniak, C., & Klontz, K. C., (2013). Recalls of foods due to microbiological contamination classified by the US food and drug administration, fiscal years 2003 through 2011. *Journal of Food Protection, 76*(6), 932–938.
4. Edwards, M. C., & Stringer, M. F., (2007). Observations on patterns in foreign material investigations. *Food Control, 18*(7), 773–782.
5. Fitzgerald, A. J., (2009). It's a horrible coincidence: Corporate responsibility and the 2007 pet food recall. *Critical Criminology, 17*(3), 195–215.
6. Gendel, S. M., (2012). Comparison of international food allergen labeling regulations. *Regulatory Toxicology and Pharmacology, 63*(2), 279–285.
7. Gomez, C. B., & Marks, B., (2020). Monetizing the impact of food safety recalls on the low-moisture food industry. *Journal of Food Protection, 83*(5), 829–835.
8. Graves, M., Smith, A., & Batchelor, B., (1998). Approaches to foreign body detection in foods. *Trends in Food Science & Technology, 9*(1), 21–27.
9. Herod, A., Goodridge, L., & Rohde, J., (2019). Recalls of foods due to microbial contamination classified by the Canadian food inspection agency: 2000 to 2017. *Journal of Food Protection, 82*(11), 1901–1908.
10. Hu, H., Djebarni, R., Zhao, X., Xiao, L., & Flynn, B., (2017). Effect of different food recall strategies on consumers' reaction to different recall norms. *Industrial Management & Data Systems, 117*(9), 2045–2063.
11. Jackson, L. S., & Al-Taher, F. M., (2008). Cleaning and other control and validation strategies to prevent allergen cross-contact in food-processing operations. *Journal of Food Protection, 71*(2), 445–458.

12. Kong, D., Shi, L., & Yang, Z., (2019). Product recalls, corporate social responsibility, and firm value: Evidence from the Chinese food industry. *Food Policy, 83*, 60–69.
13. Leighton, P., (2016). Mass salmonella poisoning by the peanut corporation of America: State-corporate crime involving food safety. *Critical Criminology, 24*(1), 75–91.
14. Li, T., Bernard, J. C., Johnston, Z. A., Messer, K. D., & Kaiser, H. M., (2017). Consumer preferences before and after a food safety scare: Experimental analysis of the 2010 egg recall. *Food Policy, 66*, 25–34.
15. Lorcheim, P., (2013). Reducing food recalls with microbial fumigation of food processing facilities. *Food Protection Trends,* 334–336.
16. Lu, Q., & Xu, X., (2017). Adaptable blockchain-based systems: Case study for product traceability. *IEEE Software, 34*(6), 21–27.
17. Meuwissen, M. P., Valeeva, N. I., Velthuis, A. G. J., & Huirne, R. B., (2006). The insurability of product recall in food supply chains. Chapter 11. In: Ondersteijn, C. J. M., Wijnands, J. H. M., Huirne, R. B. M., & Van, K. O., (eds.), *Quantifying the Agrifood Supply Chain* (pp. 145–157). Netherlands: Springer.
18. Nepusz, T., Petróczi, A., & Naughton, D. P., (2008). Worldwide food recall patterns over an eleven-month period: A country perspective. *BMC Public Health, 8*(1), 308.
19. Noack, D., Miller, D. R., & Smith, D., (2019). Understanding the mitigative ability of corporate social responsibility following product recalls. *Journal of Business Ethics, 157*(2), 431–446.
20. Pang, A., (2017). Product safety failure and restoring reputation across markets: Fonterra's management of the 2013 bacterial contamination crisis. *Journal of Marketing Channels, 24*(3, 4), 136–152.
21. Paramithiotis, S., Drosinos, E. H., & Skandamis, P. N., (2017). Food recalls and warnings due to the presence of foodborne pathogens: Focus on fresh fruits and vegetables, dairy and eggs. *Current Opinion in Food Science, 18*, 71–75.
22. Peariso, D., (2008). *Preventing Foreign Material Contamination of Foods* (p. 305). USA: Blackwell Publishing.
23. Perry, N., & Brandt, P., (2008). Case study on cruelty to farm animals: Lessons learned from the hallmark meat packing case. *Michigan Law Review First Impressions, 106*, 117–122.
24. Pozo, V. F., & Schroeder, T. C., (2016). Evaluating the costs of meat and poultry recalls to food firms using stock returns. *Food Policy, 59*, 66–77.
25. Regan, Á., McConnon, Á., & Holah, J., (2016). Food hygiene and food workers: From complacency to compliance. Chapter 14. In: Lelieveld, H., Holah, J., & Gabric, D., (eds.), *Handbook of Hygiene Control in the Food Industry* (2nd edn., pp. 197–203). Duxford, UK: Elsevier.
26. Salin, V., & Hooker, N. H., (2001). Stock market reaction to food recalls. *Review of Agricultural Economics, 23*(1), 33–46.
27. Skees, J. R., Botts, A., & Zeuli, K. A., (2001). The potential for recall insurance to improve food safety. *The International Food and Agribusiness Management Review, 4*(1), 99–111.
28. Spanjersberg, M. Q. I., & Knulst, A. C., (2010). Concentrations of undeclared allergens in food products can reach levels that are relevant for public health. *Food Additives and Contaminants, 27*(2), 169–174.

29. Teratanavat, R., & Hooker, N. H., (2004). Understanding the characteristics of US meat and poultry recalls: 1994–2002. *Food Control, 15*(5), 359–367.
30. Thomas, M. K., Vriezen, R., Farber, J. M., Currie, A., Schlech, W., & Fazil, A., (2015). Economic cost of a *Listeria monocytogenes* outbreak in Canada, 2008. *Foodborne Pathogens and Disease, 12*(12), 966–971.
31. Vij, V., Ailes, E., Wolyniak, C., Angulo, F. J., & Klontz, K. C., (2006). Recalls of spices due to bacterial contamination monitored by the US food and drug administration: The predominance of salmonellae. *Journal of Food Protection, 69*(1), 233–237.
32. Wing, E. J., & Gregory, S. H., (2002). *Listeria monocytogenes*: Clinical and experimental update. *The Journal of Infectious Diseases, 185*(1), S18–S24.
33. Xiu, C., & Klein, K. K., (2010). Melamine in milk products in China: Examining the factors that led to deliberate use of the contaminant. *Food Policy, 35*(5), 463–470.
34. Zhao, X., Li, Y., & Flynn, B. B., (2013). The financial impact of product recall announcements in China. *International Journal of Production Economics, 142*(1), 115–123.

INDEX

D

I

J

K

L

M

P